W9-BNL-150

Population and Community Biology

DYNAMICS OF CORAL COMMUNITIES

Population and Community Biology Series

VOLUME 23

The study of both populations and communities is central to the science of ecology. This series of books explores many facets of population biology and the processes that determine the structure and dynamics of communities. Although individual authors are given freedom to develop their subjects in their own way, these books are scientifically rigorous and a quantitative approach to analysing population and community phenomena is often used.

The titles published in this series are listed at the end of this volume.

DYNAMICS OF CORAL COMMUNITIES

by

Ronald H. Karlson

Department of Biological Sciences,
University of Delaware, Newark, Delaware, U.S.A.

KLUWER ACADEMIC PUBLISHERS
DORDRECHT / BOSTON / LONDON

A C.I.P. Catalogue record for this book is available from the Library of Congress.

ISBN 0-412-79550-7

Published by Kluwer Academic Publishers,
P.O. Box 17, 3300 AA Dordrecht, The Netherlands.

Sold and distributed in North, Central and South America
by Kluwer Academic Publishers,
101 Philip Drive, Norwell, MA 02061, U.S.A.

In all other countries, sold and distributed
by Kluwer Academic Publishers,
P.O. Box 322, 3300 AH Dordrecht, The Netherlands.

Printed on acid-free paper

01-0902-50 ts

Graphics by Kenobi.Productions@pandora.be

Printed in the Netherlands.

CONTENTS

PREFACE

Coral communities are among the most fascinating of all biotic assemblages on earth. It is their rich diversity and the strong biological interactions which characterize these communities that provides the focus for this book. Here I describe patterns of diversity, species interactions, and community organization as well as the processes which influence these structural attributes. Although this treatment of the subject will to some degree blend evolutionary and ecological phenomena, I am primarily interested in the dynamic properties of living coral communities. Hence, such processes as succession, competition, predation, herbivory, and disturbances will be emphasized in ecological terms, but not to the exclusion of evolutionary considerations. The former influence the maintenance of diversity in coral communities and local distribution and abundance patterns. The latter deal primarily with the origins of diversity, adaptations to the local environment, biogeographic distributions, and longevity in the fossil record. With the recent resurgence of interest in historical and large-scale geographical effects on the local diversity of ecological communities, ecological and evolutionary perspectives are beginning to be integrated into our understanding of community organization and dynamics. Hence, a synthesis of these perspectives is attempted in the final chapter of this book.

This effort emerges as a consequence of academic experiences, research interests, and the strong influence of several individuals. My first exposure to ecology occurred at Pomona College where three faculty members guided my early explorations into this subject. Dick MacMillen exposed my cohort of undergraduates to the vertebrate assemblages of the semi-arid deserts of southern California, the special adaptive responses of these organisms to this extreme environment, and to the convergent evolution of the desert fauna of Australia and North America. Mike Hadfield introduced me to invertebrate zoology and encouraged me through two ecological research projects at the shore. He also facilitated my admission to the Biological Oceanography Program at Stanford University during the summer of 1968 (a training cruise to the eastern tropical Pacific and the Galapagos Islands on board the research vessel *Te Vega*). Following graduation, Mike hosted me for a summer research experience on metamorphic induction in *Phestilla sibogae* at the University of Hawaii. Don Bentley was the first instructor to introduce me to the use of mathematical models as tools for blending theory with empirical research. His influence continues to affect how I approach biological questions to this day.

The mix of mathematical theory and field experimentation drew me to community ecology as an academic discipline in graduate school. John Sutherland advised my dissertation research at Duke University on the influence of grazing by sea urchins on the structure of subtidal, epibenthic communities in North Carolina. Virtually all aspects of coral community ecology presented in the pages which follow can be traced to this graduate school experience. In spite of his passing, John's influence on my thinking seems to grow with time. From 1973-1975, a collaboration between John and Jeremy Jackson provided an opportunity for me to join them in studies on epibenthic communities at Discovery Bay in Jamaica. This stimulated my interest in coral communities and eventually lead to a postdoctoral experience with Jeremy at Johns Hopkins University and a fifteen-year preoccupation with the dynamics of the *Zoanthus* zone on Jamaican coral reefs. In recent years, collaborations with Don Levitan, Larry Hurd and Buck Cornell at the University of Delaware and Terry Hughes at James Cook University of North Queensland have sustained my interest in coral communities. Without these interactions, this book would not have been written. In fact, my latest interaction with Buck has dealt specifically with the integration of historical and geographical phenomena with local factors controlling the number of species occurring within coral assemblages. The influence of this collaboration will be evident in the final two chapters of this book.

I thank Michael Usher of the Scottish Natural Heritage, who has served as the principal editor for the Population and Community Biology Series, and Bob Carling, senior editor at Chapman and Hall, for their encouragement and help throughout the writing of this book. I thank Rene Mijs at Kluwer Academic Publishing for his helpful assistance during the transition from Chapman and Hall. I also thank Margie Barrett and my wife Susan for their efforts in drawing, modifying, and redrawing figures. My family was extremely supportive and patient throughout the time I worked on this project. To all I am grateful. In the pages which follow, I hope to convey to readers the real beauty of coral communities and the elegance of theory in helping us understand the natural world around us. Lastly, I would like to acknowledge the scientific community which is responsible for what we know about coral communities today. In writing this book, I have ventured into several areas of study which are largely unfamiliar territory. I thank all of you who helped me through this process by providing reprints, material in press and preparation, insights into your work, and constructive criticism of what I've written. It is my hope that I have represented this material adequately and that most would agree that we still have much to learn. Any errors of interpretation are solely my own.

1 INTRODUCTION

1.1 Ecological communities, guilds, assemblages and webs

The study of the dynamics of coral communities focuses on processes which control community structure. Although there would appear to be little confusion as to what is meant by the terms structure, coral, and community, these seemingly simple terms take on different meanings depending on the specific context in which they are used. Thus in order to understand the dynamics of coral communities, one needs to begin by identifying the relevant properties of community structure which are of interest, the spatial and temporal scale at which these properties should be considered, and the number and kinds of species to be included in the analysis. These sorts of considerations are fundamental to any scientific inquiry regardless of discipline. One needs to define the system and the question of interest.

Communities can be viewed from multiple perspectives. For example, Pimm (1982) distinguished between "horizontal" and "vertical" patterns of community structure. The former focuses on the influence of species interactions within a single trophic level, while the latter emphasizes the interactions among multiple trophic levels. This distinction has a long history in the field of community ecology (*e.g.*, see Elton 1948). If one were to focus on the role of resource competition on species coexistence, one would restrict the number of trophic levels to be considered. On the other hand, a focus on the influence of predators on the diversity of prey species would expand this number. Since my intent in what follows is to be as inclusive as possible, this treatment of the dynamics of coral communities will not be restricted to any single set of trophic constraints. Instead, I will attempt to specify the species and trophic levels of interest as each topic is developed.

Community ecology deals with patterns and processes involving multiple species. In its broadest sense, communities "are simply defined as a collection of species occurring in the same place at the same time" (Fauth *et al.* 1996). Here I use this geographic designation of coral communities to include the collection of corals and other species which occur together. These communities may occur in reefal environments where corals contribute to the physical structure and flow of energy through the community. They may also occur in non-reefal environments where corals may play a lesser role in community processes. Although much of what we know about coral communities is from work in the tropics on coral reefs, temperate coral communities will not be excluded here nor will they be heavily emphasized.

Besides geographic distinctions, one might restrict consideration of the dynamics of collections of species to only those which are related by descent or to only those which utilize a common resource. The former distinction specifies a taxonomic unit of interest and, when applied to a specific geographic location, the collection of species may be called an assemblage (Fauth *et al.* 1996). The latter distinction specifies a guild (Root 1967) which may include taxonomically unrelated species. When one specifies a guild of related species occurring together at the same place and time, Fauth *et al.* (1996) recommended we refer to the collection of species as an ensemble (Figure 1.1). Although these terms are often used inconsistently by community ecologists, I will attempt to use the conventions suggested by Fauth *et al.* (1996) below.

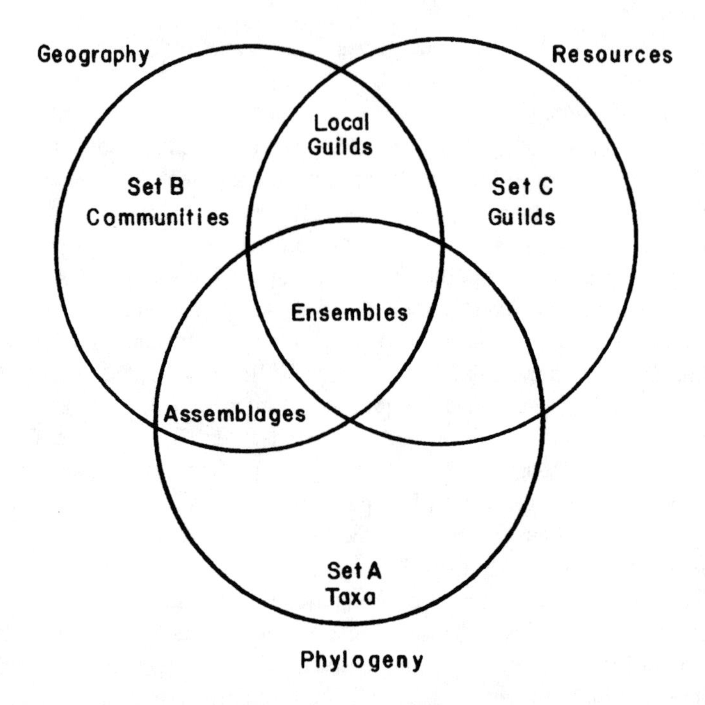

Figure 1.1 Venn diagram illustrating how collections of species can be defined on the basis of phylogeny (Set A), geography (Set B), and resources (Set C). Although no term denotes the intersection of sets based on phylogeny and resources, "such phylogenetically restricted groups using a common resource in different communities are generally referred to by a compound descriptor defining resource and taxon" (*e.g.*, herbivorous fishes) (redrawn after Fauth *et al.* 1996). © 1996 by The University of Chicago.

Much of community ecology deals with interactions among a collection of co-occurring species. At the simplest level, these interactions may involve competitors using a single food resource, parasites exploiting a single host species, and so on. More complex collections of species may include increased numbers of species using multiple resources on a single trophic level, increased numbers of species on different trophic levels, or both. Thus the complexity of ecological communities can increase horizontally and vertically as the collection of trophic links between species form food chains, cycles, and webs (Elton 1927, Pimm 1982, Cohen 1989a).

Although the complexity of food webs in natural communities is well known, there are often so many species across a large range of sizes, trophic levels, and microhabitats that one cannot consider all species in a community. Hence, investigators usually limit their consideration of species based on a variety of constraints (*e.g.*, question of interest, technological or logistical limitations, professional experience, etc.). Below, I specify the limits for this evaluation of coral communities based on taxonomic and trophic level considerations as well as the scale at which community dynamics will be evaluated. In most cases, the structure of these communities will be defined by the number of species, the simplest measure of species diversity. Although the number of species provides only a limited basis for evaluating community structure, it has been widely used in the ecological literature. Other important aspects of community structure include species composition, relative abundance patterns, patch structure, habitat associations, resource utilization patterns, food-web structure, and interaction webs. These aspects of community structure are patchily distributed throughout the pages which follow.

1.2 Taxonomic and trophic constraints

It has been estimated that 4-5% of all described species on earth occur on tropical coral reefs (Reaka-Kudla 1995). Most of these approximately 91,000 species are macroscopic forms (73%) representing a large number of higher taxa. Given this apparent prevalence of macrofauna, I will deal primarily with these larger taxa below. Most diversity studies on coral reefs have focused on the more conspicuous of these macrofaunal organisms (Reaka-Kudla 1995), so there is a bias in the literature which largely ignores cryptic invertebrate taxa. These forms comprise a rich and fascinating collection of inconspicuous species on coral reefs (Reaka-Kudla 1995), yet their role in coral communities has not been well studied. Some of these forms will be considered here especially in terms of resource competition in coral communities.

The trophic base for coral reef communities largely derives from autotrophy by symbiotic zooxanthellae (dinoflagellates which live within the tissues of many cnidarians and molluscs), benthic algae, and nearby seagrass communities (Meyer *et al.* 1983). The trophic structure is complicated by the fact that polytrophic mechanisms for acquiring resources are common among many species in these communities. Thus specification of trophic levels and web relationships is not a simple matter. For example, reef-building (hermatypic) corals feed on zooplankton, yet also acquire much of their food from photosynthesis by symbiotic zooxanthellae (Porter 1976). Other polytrophs in coral communities like sponges use particulate organic carbon (Reiswig 1971) as well as photosynthetically fixed carbon from their symbionts (Wilkinson 1987). Recently, even herbivory (grazing on phytoplankton) has been found to be an important feeding mode among some reef cnidarians, a taxon long known for zooplanktivory (Fabricius *et al.* 1995). The focus here will be on shallow-water coral communities. I will not consider the coral communities of deep waters where azooxanthellate corals are common and heterotrophic consumption of nutritional resources predominate. Although these deep water communities are quite fascinating, we are only just beginning to explore their dynamic properties.

1.3 Scale-dependent dynamics

As with any scientific endeavor, the scale at which one makes observations can influence the patterns one can detect and the interpretation as to how these patterns are generated and maintained. In physics, for example, the relative importance of inertial, frictional, and electromagnetic forces on particle motion clearly depends on particle mass and velocity as well as the observer's frame of reference. Likewise, inferences regarding the relative importance of viscous and turbulent flow to the feeding activities or locomotion of aquatic organisms depend on the scale of measurement, flow velocity, and the sizes of the organisms and structures of interest. Ecological patterns and processes are also strongly scale-dependent thus necessitating the use of cross-scale methodologies in the study of ecological phenomena (Levin 1992). In fact, scale-dependence has reached global dimensions as ecologists consider issues of climatic change and habitat fragmentation (Karieva *et al.* 1993) as well as the influence of large-scale historical and geographical phenomena on ecological communities (Ricklefs and Schluter 1993).

So how wide a range of scales should one employ for studying a particular community? To answer this, I turn to a symposium volume which directly addressed the issue of scale on ecological patterns and processes in aquatic systems (Giller *et al.* 1994). Although coral communities were not considered specifically in this collection of papers, an international group of twenty nine contributors covered the vast majority of freshwater and marine habitats. These contributors represented a range of academic subdisciplines in such subjects as fluid mechanics, thermodynamics, population ecology, community ecology, evolutionary biology, and paleontology. This range of perspectives inevitably led to different perceptions as to the relative importance of specific spatial and temporal scales on ecological processes. For example, Statzner and Borchardt (1994) evaluated the impact of water flow on four ecological variables (trout spawning, animal drift, habitat suitability for insects, and predator-prey interactions) in four near-bottom stream communities using a hydrological modelling approach. They concluded that >50% of the variation in ecological variables was explained by shear stress variation within streams. These effects of shear stress were evaluated at 1 km intervals over a 20 km reach in two streams, and over a range of 2-4 orders of magnitude in discharge (l sec^{-1}) in two other streams. Likewise, Reynolds (1994) concluded that variation in the structure of phytoplankton communities in lakes and rivers can be best understood as responses to the abruptly fluctuating temporal scales of fluid motion over the range of 10^2 - 10^8 sec. In contrast, the structure of microfossil assemblages of oceanic plankton (foraminifera, coccolithophorida, and radiolaria) examined over more than a 200,000-year record fluctuates in response to components of orbital forcing acting at very long time scales (Molfino 1994). Low latitude assemblages fluctuate in response to orbital precession with a period of 23 ky (see McIntyre and Molfino 1996 for evidence of even shorter period oscillations in addition to dominant forcing at 19 and 22 ky). High latitude assemblages are strongly influenced by all three major components of orbital forcing at periods of 100 (eccentricity), 41 (obliquity), and 23 (precession) ky (Figure 1.2). Polar planktonic assemblages vary primarily in response to variation in orbital obliquity (Ruddiman and McIntyre 1984). Hence, global geography and cyclical astronomical phenomena appear to have the greatest impact on variation in the structure of planktonic assemblages when evaluated using microfossils as evidence.

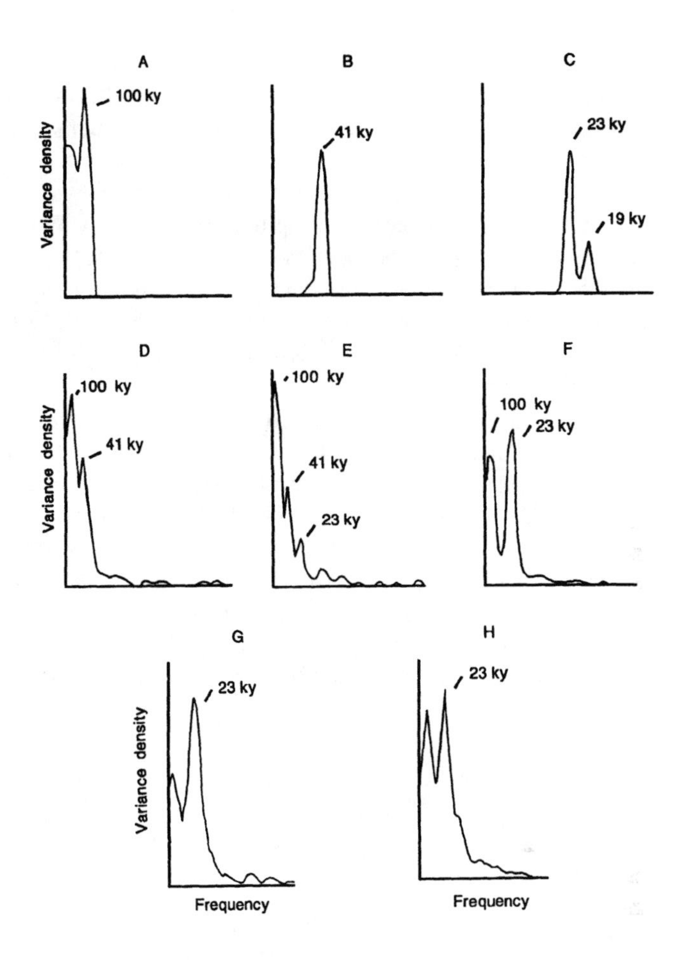

Figure 1.2 Variance spectra from time-series analyses of (A-C) orbital eccentricity, obliquity, and precession (after Imbrie *et al.* 1984, with kind permission from Kluwer Academic Publishers), (D-F) estimated palaeotemperatures inferred from foraminiferan counts in high latitude cores from the North Atlantic (after Ruddiman and McIntyre 1984, with permission of the publisher, the Geological Society of America, Boulder, Colorado USA. Copyright © 1984 Geological Society of America), and (G) percentage of the coccolithophore *Florisphaera profunda* and (H) a foraminiferal assemblage in low latitude cores from the equatorial Atlantic (all figures redrawn after Molfino 1994).

The contrast in the perception as to which temporal and spatial scales are important in a particular community was highlighted by the symposium reviews by Fisher (1994) and Platt and Sathyendranath (1994) for freshwater and marine systems, respectively. The latter authors concentrated on the scale dependence of ecosystem responses to physical forcing in the ocean. This emphasis was justified in terms of the perceived need to scale up ecological phenomena hierarchically from local to larger regional scales in order to understand, for example, biogeochemical fluxes in the sea (Platt and Sathyendranath 1994). However, not all ecological phenomena can be scaled up in this manner. Fisher (1994) cautioned against making overly simplistic generalizations about complex, scale-dependent relationships in freshwater systems; "simple unqualified statements, such as that competition controls lake communities or that predation is not important in streams, are likely to be incorrect". Given the prevalence of spatial and temporal heterogeneity in ecological systems, Fisher (1994) argued that our challenge "is not to determine which of several competing processes control patterns of interest, but to resolve the spatial and temporal boundaries of each and the causes of transitions among them". He noted that "several processes may operate together or sequentially to shape the spatially and temporally explicit patterns which are observed". Such patterns may then result from additive, compensatory, or even synergistic effects of different processes operating on different scales.

In the final chapter, the organizers of the symposium reiterated their message that physical processes operating on an enormous range of temporal and spatial scales emerge as the "architects of heterogeneity in virtually all aquatic systems" (Raffaelli *et al.* 1994). They acknowledged that this perspective may be biased for two reasons. First, the background and choice of scale used by several of the contributors to the symposium could bias one's perspective. Second, the organizers had an expectation that the physics of water movement "would be a central, and perhaps unifying" theme for the symposium. Hence, they noted the scales on which a number of physical processes operate in aquatic systems (Figure 1.3). These scales span at least 8 and 12 orders of magnitude for temporal and spatial effects, respectively.

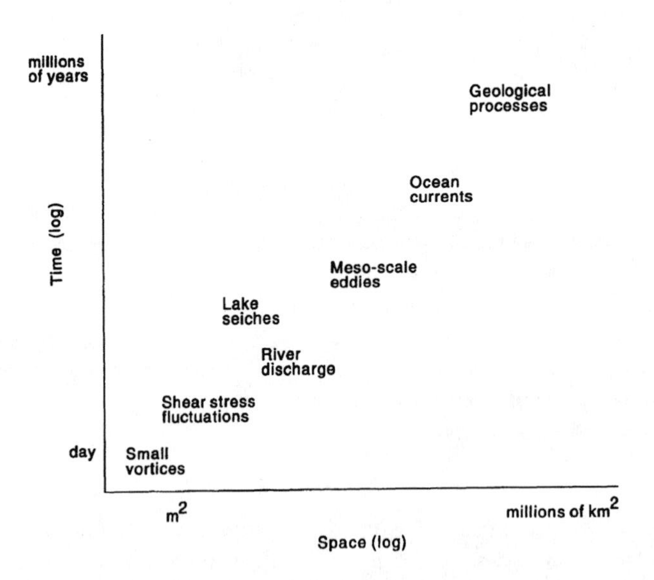

Figure 1.3 Temporal and spatial scales at which physical processes operate in aquatic systems (redrawn after Raffaelli *et al.* 1994).

Raffaelli *et al.* (1994) began the integration of biological processes into this scale-dependent perspective by noting the major role disturbances play in generating patchiness and regenerating resources in biological systems. They noted the large range in the size of aquatic organisms and the fact that life spans and the space used over the life-time of an organism are highly dependent on an organism's size (Figure 1.4). These scales span 6 and 8 orders of magnitude for temporal and spatial variation, respectively. Any particular disturbance will differentially influence the species within a biological community over these scales. Thus, small-scale disturbances may have dramatic effects on the dynamics of smaller species but none at all on larger species. Using relative temporal (based on the generation time of an organism) and spatial (based on the life-time range of an organism) scales, Raffaelli *et al.* (1994) designated the scales at which several biological processes operate (Figure 1.5). For any given type of organism, community processes were depicted to operate at scales larger than those influencing individuals and populations, but smaller than those influencing biogeographic processes and evolutionary events. Since communities are composed of species of many different sizes, these biological processes all operate simultaneously at any particular place and time. For example, factors influencing evolutionary events among planktonic bacteria may also influence the "distributional dynamics of zooplankton and the individual behavior of large nekton" (Raffaelli *et al.* 1994). Thus, unlike physical processes which clearly operate at specific spatiotemporal scales, biological processes operate at relative scales.

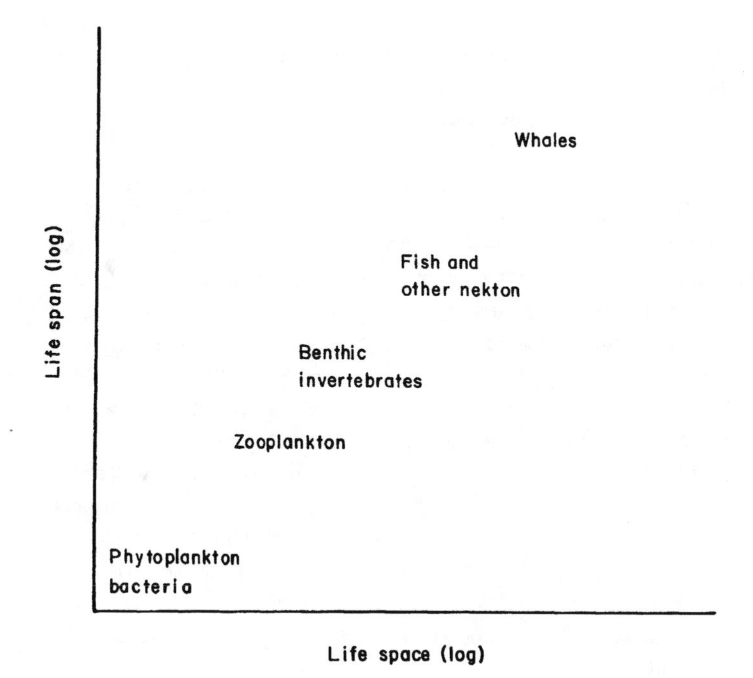

Figure 1.4 Life span and spatial scales which characterize a size range of aquatic organisms. No units are given as in the original (redrawn after Raffaelli *et al.* 1994).

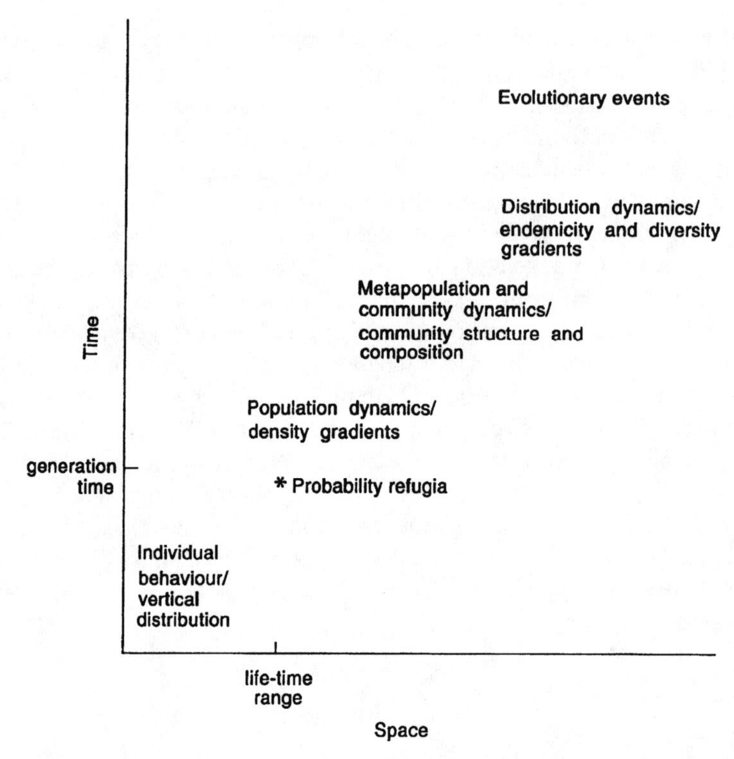

Figure 1.5 Temporal and spatial scales at which biological processes operate in aquatic systems relative to generation time and the life-time spatial range of an organism. No units are given as in the original (redrawn after Raffaelli *et al.* 1994).

Furthermore, the essence of the problem of scale when dealing with communities of organisms is that the interaction between the environment and each species within the community is species-specific and subject to change over evolutionary time. It is this dynamic feature which distinguishes biological phenomena from those of the physical world. In aquatic communities, organisms "may have significant capacity to ameliorate, redirect or override the effects of physical processes occurring over a wide range of scales" (Raffaelli *et al.* 1994). Thus, an evolved innovation can radically alter how an organism interacts with either biological or physical aspects of its environment. The unique life history strategy of each species is manifested by age- and size-dependent fecundity and mortality patterns (Begon and Mortimer 1986) as populations respond to selective regimes (*i.e.*, to disturbances, predators, competitors, the presence of refugia, other aspects of environmental heterogeneity, etc.). These evolving life history attributes have important consequences on the temporal and spatial scales over which organisms are distributed. This is especially true of many aquatic organisms which can disperse over enormous oceanic scales.

Another way to view the interaction between physical and biological processes is to consider the impact of physical processes on the community across a range of scales simultaneously. For example, an environmental disturbance of some given magnitude and frequency, will have its greatest effects at local spatial and relatively short temporal scales (Figure 1.6).

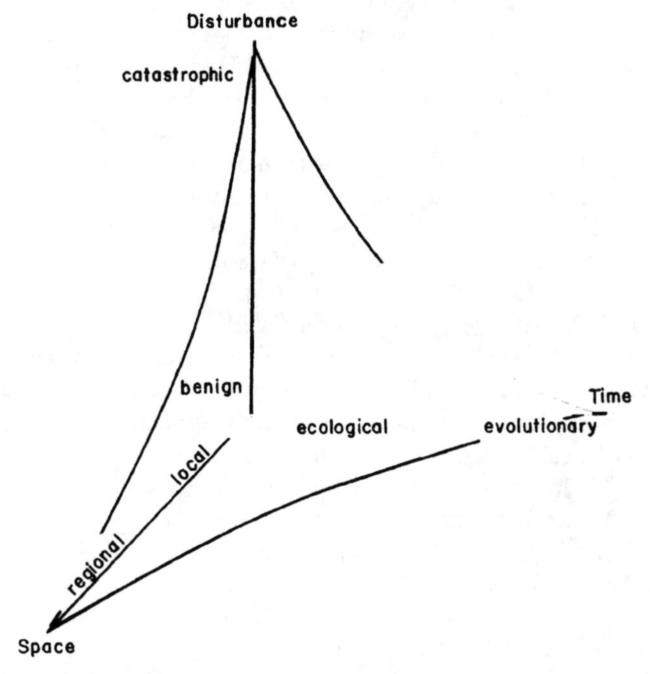

Figure 1.6 The impact of disturbance on a community attribute (*e.g.*, species richness, dominance patterns, taxonomic membership) across temporal and spatial scales (redrawn after Karlson and Hurd 1993). © Springer-Verlag 1993.

At regional spatial scales, variation among local communities will buffer the disruptive effects of disturbance on a community. At evolutionary time scales, adaptation to the disturbance regime also reduces the detrimental impact of disturbance (Karlson and Hurd 1993). This perspective considers community attributes separately without confounding the effects of multiple biological processes with variation in body size, generation time, and life-time spatial range across multiple spatiotemporal scales (as depicted in Figure 1.5). Given the capacity of organisms to adapt to the physical and biological attributes of the environment and the fact that this environment is in a perpetual state of change, the following consideration of coral communities will attempt to integrate the available information across a wide range of scales.

1.4 Marine epibenthic communities

Coral communities are but one of several types of marine epibenthic communities which occur on hard bottom substrates throughout the world. Such communities have a number of common features which have intrigued ecologists for many years (*e.g.*, see Jackson 1977a). So before delving specifically into the dynamics of coral communities, it is appropriate to introduce this topic using three well-studied examples of these types of communities, namely rocky intertidal, kelp, and encrusting communities. As part of the symposium noted above, Dayton (1994) considered the role of large scale phenomena on the dynamics of these communities. He noted that experimental work has clearly documented the importance of physical stress, biological interactions, physical disturbances, and recruitment on rocky intertidal communities. However, there remains a need to integrate oceanographic processes more fully into our understanding of community dynamics. Patch dynamics in kelp beds are dominated by episodic disturbances and the predictable recovery of stable canopy guilds. Some species complexes represent alternative stable states thus contributing to patchiness in kelp communities. The outcome of competition among plant taxa is strongly affected by depth gradients in light, grazing, and physical disturbances. Encrusting communities are even more patchy and are often dominated by long-lived clones which are resilient and/or resistant to perturbations. For future work, Dayton (1994) suggested that the study of dispersal processes can act as a unifying theme in helping us integrate the influences of local and larger scale phenomena on ecological communities. Below I attempt to highlight our emerging appreciation of how multiple scale-dependent phenomena influence these three epibenthic communities. For more depth, I refer the reader not only to Dayton (1994), but also to Paine (1994) and the earlier reviews by Roughgarden (1986, 1989) and Buss (1986).

1.4.1 ROCKY INTERTIDAL COMMUNITIES

Perhaps rocky intertidal communities are best known for their well developed zonation patterns and disturbance-mediated patchiness within zones. Prior to the now classic contributions of Connell (1961a,b) and Paine (1966), many investigators believed that the observed zonation patterns were primarily caused by the differential tolerance of species to the sharp physical gradients characterizing intertidal habitats (Connell 1972). After three decades of careful experimentation by numerous investigators at multiple sites around the world, it is clear that strong biological interactions can limit intertidal distributions within broader physiological tolerance limits particularly at low and mid-intertidal heights [but see Underwood and Denley (1984) for an alternative perspective]. The role of disturbance in intertidal communities may be understood in terms of how it offsets competitive processes and promotes the coexistence of competitively inferior species with competitive dominants in a patchy landscape. The empirical work of Dayton (1971) together with the theoretical contributions by Levin and Paine (1974) [and Paine and Levin (1981)] were early developments to this perspective.

Paine and Levin (1981) considered the dynamics of the mid-intertidal zone typical on the wave-swept shores of Tatoosh Island off the coast of Washington. Just above the vertical limit set by the predatory sea star *Pisaster ochraceus*, mussel beds formed by *Mytilus californianus* dominate the landscape. In the absence of destabilizing disturbances caused by "a variety of exogenous factors" (*e.g.*, logs, shear stress from large waves, mortality due to freezing during unusually cold weather, etc.), *M. californianus* monopolizes the space in this zone indefinitely. Disturbance generates a spatial mosaic of patches composed of various other members of this community including three species of barnacles, another mussel species, and a variety of algal species. Although Paine and Levin (1981) acknowledged that the composition of this community at any point in time is highly stochastic, they were able to reduce this variability to the essential elements of a temporal replacement process (Figure 1.7). Disturbance generates bare patches within the mussel beds which are rapidly colonized by ephemeral algae. After a few months, the algae begin to be displaced by the mussel *M. edulis.* At this intertidal height, *M. edulis* persists as a dominant species for 2-4 years, but is then replaced by *M. californianus.* After four years of undisturbed conditions, patches become dominated by this superior competitor.

The generalized pattern of the interplay between disturbance and competitive displacement processes depicted in Figure 1.7 greatly simplifies both the role of disturbance and biological interactions in this natural community. Disturbances can have adaptive consequences which can clearly favor some species such as the intertidal alga *Postelsia palmaeformis* (Paine 1979, Paine and Levin 1981). This alga can replace barnacles and mussels on disturbed intertidal shores where it requires bare substrate for colonization and persistent disturbance to remain abundant. It does not occur at undisturbed locations nor does it completely resist eventual patch domination by *M. californianus* at disturbed sites. Thus patches dominated by *P. palmaeformis* represent an alternative state of the community not depicted in the generalized Paine-Levin model (Figure 1.7). Although these algal patches do not indefinitely persist in the absence of disturbance, such disturbance-free conditions are not common on these unsheltered, wave-swept shores.

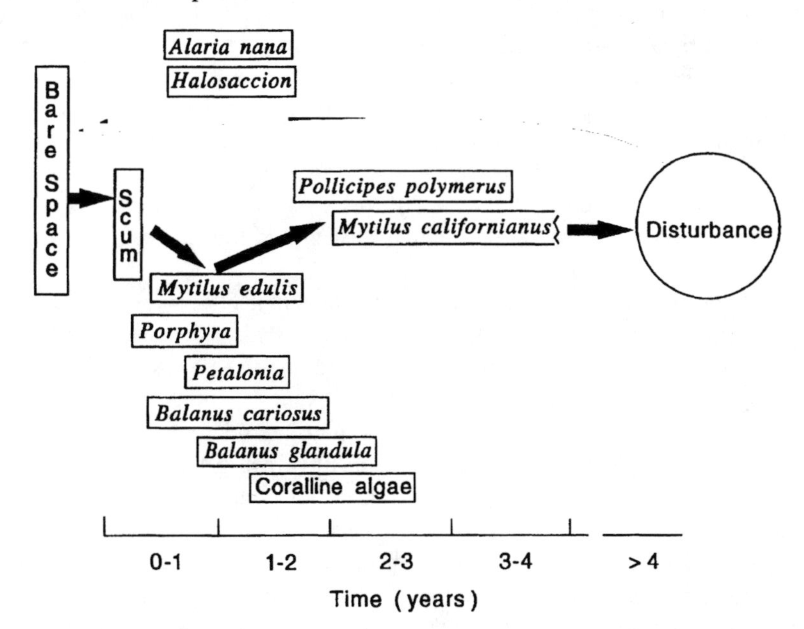

Figure 1.7 The general replacement sequence typical of the patch dynamics within mussel beds at Tatoosh Island (redrawn after Paine and Levin 1981).

Furthermore, the Paine-Levin model includes only colonization, competition, and exogenous disturbance processes as determinants of intertidal community structure. Over the last several years, increasingly broader views of biological processes controlling intertidal community structure have incorporated the role of predator-prey interactions across a wider intertidal scale (*e.g.*, Menge and Sutherland 1987), the role of both strong predator-prey and competitive interactions between pairs of species (*e.g.*, Menge *et al.* 1986, Menge and Sutherland 1987), and even the role of indirect effects whereby an interaction with one species can have positive or negative consequences on other species (*e.g.*, Wootton 1993, 1994, Menge 1995).

Wootton (1993) identified two basic ways in which indirect effects can occur in natural communities. Firstly, a species can directly alter the density of a second species (*e.g.*, a competitor or resource species) which in turn directly affects a third species thus forming an "interaction chain". Secondly, a species can alter a direct interaction between two other species in an "interaction modification" without directly influencing the density of either species. For example, a predator might cause some prey species to alter its behavior (*e.g.*, use of temporal and spatial refuges) thus making it more or less susceptible to predation by a third species. Alternatively, a predator's foraging behavior might be influenced by the presence of non-prey species in a patch of habitat thus altering the risk encountered by actual prey species in that patch.

Several types of indirect effects involving interaction chains have been demonstrated in natural communities including exploitative competition, trophic cascades, apparent competition, and indirect mutualisms (Wootton 1994, Menge 1995). Exploitative competition is widely known to occur when two species indirectly interact by altering the availability of a common resource species (Figure 1.8). Species on three or more trophic levels can indirectly interact through trophic cascades and multiple consumer-resource interactions. Two prey species with a common predator can indirectly interact in apparent competition when an increase in the abundance of one prey species results in increased numbers of predators and a decline in the abundance of the other prey. Indirect mutualisms typically involve consumer-resource interactions in which two different resource species are linked through interference or exploitative competition. Increases in one consumer species result in declines in its resource species, which causes an increase in the competing resource species, and thus increases in the second consumer species.

Although interaction modifications may be common in natural communities (Wootton 1994), they can be more difficult to detect and evaluate. This is because their effects can be obscured by alternative interaction pathways and one must use complicated experimental designs to separate the alternatives. As an example of this complexity, we can examine an interaction web constructed for the mid-intertidal zone of Tatoosh Island (Wootton 1992, 1994, Figure 1.9). Algae are competitively displaced by mussels or goose barnacles and are eaten by three species of limpets. Gulls eat goose barnacles and thus competitively release mussels. Dark-colored limpets (*Lottia pelta*) preferentially select mussels as substrate, while light-colored limpets (*L. digitalis*) select goose barnacles. These two limpet species thereby become cryptic and less susceptible to predation by birds. However, bird predation still reduces limpet biomass thus increasing algal biomass and competitively releasing a third limpet species (*L. strigatella*). Thus indirect effects involving both interaction chains as well as interaction modifications have important consequences in this community. In fact, indirect effects can significantly modify community structure and "counteract (or reinforce) direct interactions between species pairs" (Wootton 1992).

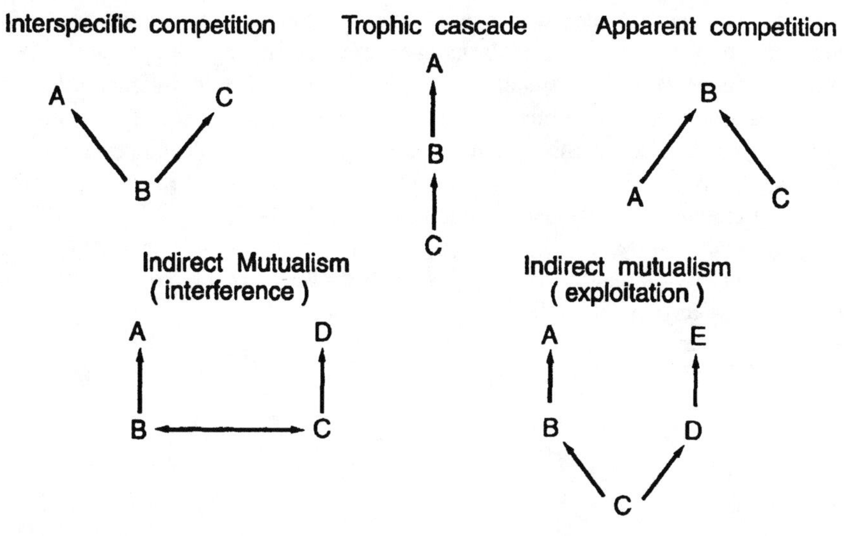

Figure 1.8 Five common indirect effects found in natural communities. The horizontal arrow represents interference competition. All other arrows indicate the direction of energy flow in consumer-resource interactions (redrawn after Wootton 1994 with permission, from the Annual Review of Ecology and Systematics, Volume 25, © 1994, by Annual Reviews Inc. and from the author).

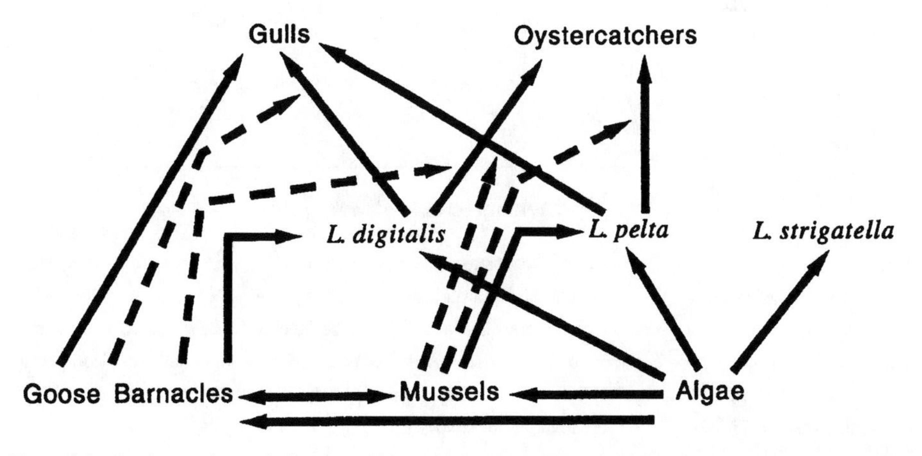

Figure 1.9 An interaction web for the mid-intertidal zone at Tatoosh Island. Solid arrows indicate direct interactions between species involving consumption, interference competition, and habitat preferences. Dashed arrows indicate interaction modifications of predator-prey interactions due to crypsis. Interaction chains are indicated by solid arrows in series (redrawn after Wootton 1994 with permission, from the Annual Review of Ecology and Systematics, Volume 25, © 1994, by Annual Reviews Inc. and from the author).

The emphasis above on the role of biological interactions on the dynamics of rocky intertidal communities represents a local perspective of how interacting species can influence community structure at a specific location. In recognition of the importance of gradients in environmental stress, wave action, currents, and supply of larvae in rocky intertidal communities (see Connell 1972, 1985, Roughgarden 1986, 1989, Dayton 1994), there has been renewed interest in how biological interactions vary over larger spatial scales. In particular, variation in environmental stress (including both physical stress due to external mechanical forces and physiological stress due to internal biochemical reactions) and recruitment regimes have been integrated into the conceptual framework for studying natural communities. Menge and Sutherland (1987) developed a community model which assumed the increasing importance of complex interaction webs on community structure with decreasing environmental stress (Figure 1.10). In addition to specifying how disturbance, competition, and consumer-resource interactions should vary along an environmental stress gradient, they also incorporated even larger scale effects along a recruitment gradient.

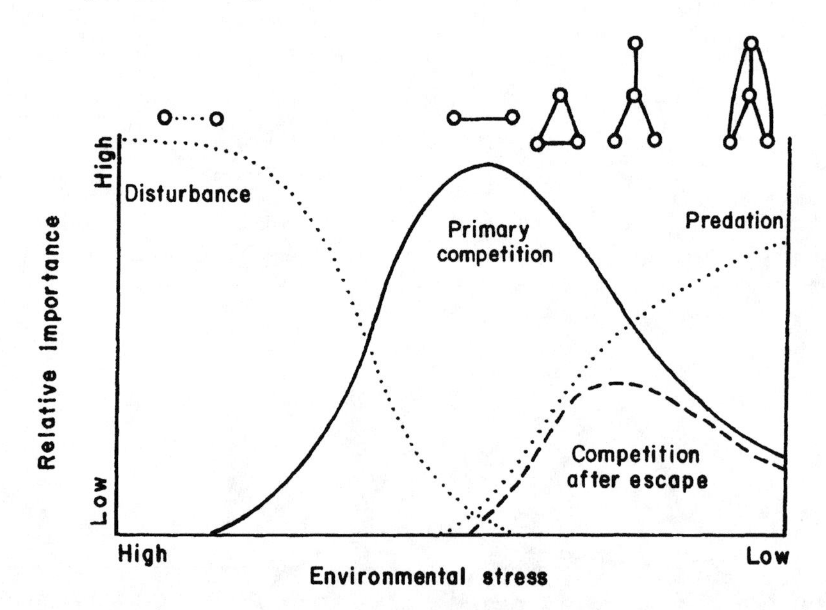

Figure 1.10 The relative importance of disturbance, competition, predation, and omnivory (indicated in the diagram in the upper right) over a gradient in environmental stress. As disturbance becomes less important, the complexity of interaction webs become more important. Primary competition at low trophic levels is most important at intermediate levels of environmental stress. At lower levels of environmental stress, competition among well-defended prey at intermediate and low trophic levels is more important (redrawn after Menge and Sutherland 1987). © 1987 by The University of Chicago.

At the population level, recruitment is the addition of juveniles to an adult population. Several quite different processes can influence recruitment (*e.g.*, larval transport, mortality, and settlement as well as the post-settlement mortality of juveniles).

At the community level, Menge and Sutherland (1987) defined recruitment density as a composite variable which is estimated from the recruitment densities of a few dominant species in a community and is assumed to be positively correlated among trophic levels. Thus communities are considered to occur along a recruitment gradient. High recruitment characterizes communities in which high densities are achieved and competition is strong. Low recruitment characterizes communities in which densities are not sufficient for competition to be important.

Menge and Sutherland (1987) generated a set of predictions for how species diversity should vary in communities along these gradients in recruitment density and environmental stress. High levels of environmental stress are predicted to result in low species diversity (due to physical disturbance) and the prevalence of only a few opportunistic or resistant species in a community regardless of recruitment density (Figure 1.11). At low recruitment levels, decreasing environmental stress results in increasing species diversity due to colonization processes and the reduction in mortality from physical disturbance. At intermediate environmental stress levels, a peak in species diversity results from moderation in the mortality losses due to both predation and physical disturbance. At low levels of environmental stress, severe predation reduces prey populations to levels where local extinctions occur and only prey with adequate refuges can persist.

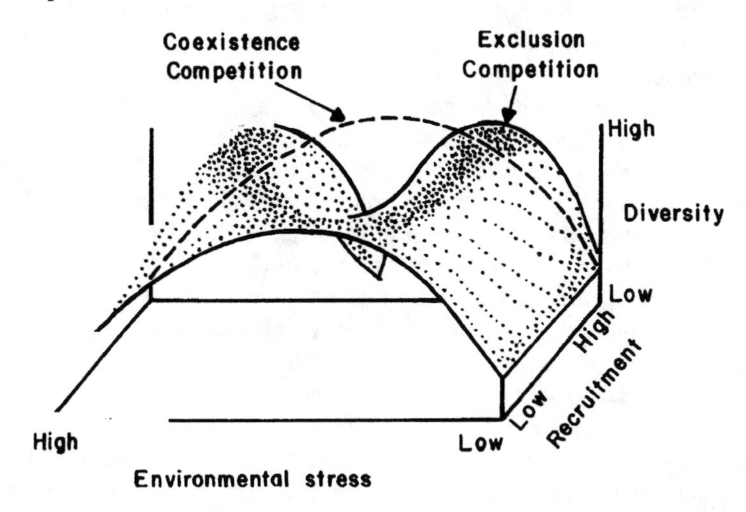

Figure 1.11 The Menge-Sutherland model of species diversity along gradients in environmental stress and recruitment. At high recruitment levels and intermediate levels of environmental stress, species diversity may be low due to "exclusion competition" or high due to "coexistence competition". Competition is not important at low recruitment levels. The diversifying effects of disturbance and predation occur at different levels of environmental stress (redrawn after Menge and Sutherland 1987). © 1987 by The University of Chicago.

According to the Menge-Sutherland model, competition only becomes important at high recruitment levels. Competitive exclusion results in low species diversity at intermediate environmental stress levels (Figure 1.11). From this point, higher stress levels result in elevated species diversity due to the effects of disturbance (Connell 1978). Lower stress levels result in increased species diversity as predation becomes more important (Paine 1966). This bimodal pattern in species diversity is contrasted with a unimodal pattern predicted for communities in which intense competition results in the coexistence of competitors. From the peak in diversity indicated at intermediate stress levels under "coexistence competition" (Figure 1.11), diversity decreases with increasing environmental stress due to greater mortality from disturbance and with decreasing stress due to increased predation.

The significance of the Menge-Sutherland model is that it provides a conceptual framework which incorporates spatial gradients in stress and recruitment as well as the diversifying effects of disturbance, predation, and "coexistence competition" on community dynamics. Thus, elements of the physical environment, biological interactions, and spatial heterogeneity are integrated into a single model which transcends the limitations of earlier models. In recognizing disturbance and predation as separate processes in communities, their effects become decoupled and the predicted patterns of community structure become more complex. Furthermore, the explicit inclusion of recruitment gradients encapsulates the growing body of evidence that community dynamics can be contingent on the physical and biological processes which influence the supply of larvae to a site (*e.g.*, Underwood and Denley 1984, Roughgarden *et al.* 1988).

Although recruitment into rocky intertidal communities can be an extremely variable process (*e.g.*, Underwood *et al.* 1983), predictable recruitment gradients and their influence on community structure have been detected at several spatial scales. For example, the rocky intertidal community in the Bay of Panama, on the central Pacific coast of Costa Rica, and in the Gulf of California appear to occur along a latitudinal gradient in recruitment (Sutherland 1990a, Dayton 1994). In the Bay of Panama, recruitment limitation is strongest. This is the condition in which adult densities are limited by extrinsic factors controlling recruitment rather than by intrinsic factors. Recruitment limitation is intermediate in Costa Rica and rare in the Gulf of California. The high densities achieved in the Gulf of California result in strong biological interactions (competition and predation) which can have evolutionary consequences. For example, a shell dimorphism in the barnacle *Chthamalus anisopoma* is maintained by the feeding activities of the predatory gastropod *Acanthina angelica* (Lively 1986a,b). A bent morph is better protected from predation, but grows more slowly and produces fewer offspring than the more typical conical morph. Due to the patchy distribution of this predator which tends to remain in crevices and under boulders when inundated, there is sufficient spatial heterogeneity to promote coexistence of the two morphs.

At the other extreme of this apparent latitudinal gradient in the Bay of Panama, recruitment rates are quite low, populations in the intertidal zone occur at very low densities, and competition appears to be unimportant. In addition, these organisms are exposed to high levels of predation by fish at high tide and to extreme heat and desiccation at low tide. In Costa Rica, intermediate rates of recruitment [(orders of magnitude higher than in Panama (Sutherland 1990a)] result in higher densities of sessile animals than observed in Panama, yet recruitment limitation is still prevalent (Sutherland 1987, 1990a). Although biological interactions occur here (*e.g.*, predation by *Acanthina brevidentata* on *Chthamalus fissus*), most barnacle mortality is density independent, survivorship is generally low, and recruitment largely reflects highly variable, episodic settlement events. Hence, recruitment patterns and the hydrodynamic processes controlling the supply of larvae to the intertidal zone at these lower latitudes strongly influence the structure of the community.

Variation in the recruitment rate of intertidal organisms over smaller regional scales has also been linked to larval transport mechanisms, predation on larvae during the dispersal process, and to settlement patterns. Along the central coast of California in Monterey Bay, for example, the competitive dominance of *Balanus glandula* over *Chthamalus fissus* and *C. dalli* occurs only when the abundance of the balanoid barnacle exceeds 75% of the surface area (Gaines and Roughgarden 1985). This occurs at the most seaward margin of the high intertidal zone where the settlement rates of *B. glandula* are high and clear zonation patterns are evident. At other sites, settlement rates are lower and these barnacle species "have completely overlapping distributions" (Gaines and Roughgarden 1985). In Monterey Bay, low settlement rates result from predation on barnacle larvae by rockfish in a nearby kelp forest, to a settlement shadow primarily affecting inshore sites, and to yearly variation in wind-driven currents, upwelling along this coast, and the transport of barnacle larvae offshore (Roughgarden 1986, Gaines and Roughgarden 1987, Roughgarden *et al.* 1987, Roughgarden *et al.* 1988).

At the conceptual level, intertidal community structure emerges as a consequence of processes operating within this community of adult organisms (*i.e.*, physical stress, biological interactions, and disturbances) as well as those operating within the water column (*i.e.*, currents, larval mortality, etc.) (Roughgarden *et al.* 1988, Roughgarden 1989). Thus in the intertidal zone of central California, predation by the sea star *Pisaster ochraceus* on barnacles and mussels and competitive interactions among these prey species dominate the structure in areas exposed to high settlement rates. Physical transport processes and predation on larvae dominate areas exposed to lower settlement rates (Figure 1.12). The relative importance of these processes is then a direct consequence of the supply of larvae reaching the adult habitat. Furthermore, the influence of transport processes on benthic community structure can vary significantly among species; note how these processes are depicted separately for *Pisaster*, *Balanus*, and *Chthamalus* in Figure 1.12. Complex distribution and abundance patterns within the rocky intertidal community can be generated by temporal variability among species in the timing of larval release and duration of the larval stage as well as in the strength of local upwelling and current systems.

The inclusion of transport processes into the conceptual framework for studying rocky intertidal communities mandates the consideration of the large spatial scales over which larvae can be dispersed (Dayton 1994, Paine 1994). Larval release, transport, and settlement link multiple sites by influencing local population growth, source-sink relationships among sites (Pulliam 1988), and the probabilities of local colonization and extinction events (Harrison 1991, Hastings and Harrison 1994). This linkage expands the scale of interest from a single site (*e.g.*, Figure 1.12) to encompass multiple sites in what is called a metacommunity (Hanski and Gilpin 1991). Recently, Paine (1994) depicted such an array of linked patches in order to illustrate the added level of complexity faced by ecologists attempting to understand rocky intertidal communities (Figure 1.13). The complexity of trophic structure within patches is shown to vary with patch size. A single consumer-resource interaction occurs in a small isolated patch, while the largest patch has the most trophic levels and the highest incidence of omnivory. A top predator feeds not only in this large patch, but also within a small patch representing a net sink for larvae from the large patch (Figure 1.13). Thus when viewed at a large spatial scale, the dynamics of rocky intertidal communities are dictated by the dispersal of larvae and predators as well as by local biological interactions, life history events, and disturbance-mediated patch dynamics (Paine 1994).

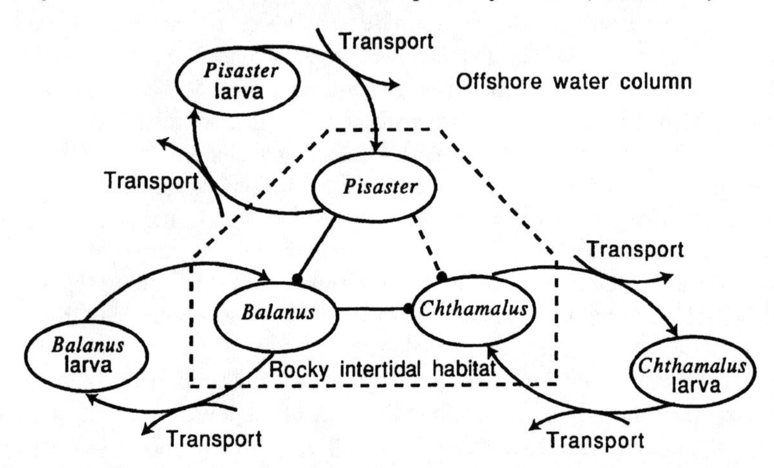

Figure 1.12 Biological interactions among three primary components of the rocky intertidal community of the central California coast and linkage with the larval stages in the water column by offshore transport mechanisms. Adult interactions include predation by *Pisaster* on barnacles and competition between *Balanus* and *Chthamalus*. All adults release larvae from the rocky intertidal habitat. Following transport, larvae may return to settle and engage in biological interactions (redrawn with permission from Roughgarden *et al.* 1988. Recruitment dynamics in complex life cycles. *Science* **241**, 1460-1466). Copyright 1988 American Association for the Advancement of Science.

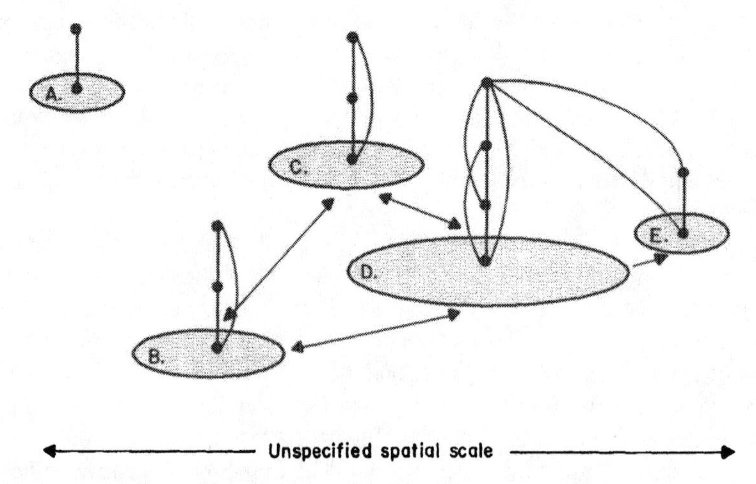

Figure 1.13 Trophic structure and the linkage of habitat patches by larval exchange. Arrows indicate direction of larval transport. Trophic structure is indicated by interconnected nodes. The complexity of this structure varies with patch size. A single consumer-resource interaction occurs in the small isolated patch (A), while the largest patch (D) has the most trophic levels and the highest incidence of omnivory. Another small patch (E) is a net sink for larvae from the largest patch (redrawn after Paine 1994).

1.4.2 KELP COMMUNITIES

Most epibenthic communities located on hard substrate in shallow subtidal, temperate waters are dominated by marine algae (Dayton 1994). Among the most conspicuous of these are the kelps (Phaeophyta, Laminariales) which can form extensive underwater forests. Like rocky intertidal communities, kelp communities are influenced by relatively unpredictable recruitment events, but they also exhibit predictable patterns of temporal succession and interspecific competition (Dayton 1994). Furthermore, there is increasing evidence that these communities are influenced by not only local phenomena, but also by atmospheric and oceanic processes operating on quite large scales. Hence, we are becoming more aware of the need to expand the scale of ecological studies beyond the limits set by physiological constraints, demographic attributes, and species interactions.

Kelps are thought to have radiated rather recently with the shift towards cooler water temperatures in the North Pacific during the late Cenozoic (Estes and Steinberg 1988). While 27 of the 28 recognized genera in this order occur in the North Pacific, only 5 and 4 genera occur in the North Atlantic and the southern oceans, respectively. Similarly, kelp communities in the North Pacific tend to be much more species rich [with 64-83 kelp species (Estes and Steinberg 1988)] than those found elsewhere. Dayton (1985a) speculated that the depauperate kelp communities of southern South America are "composed of species which have invaded during the last few thousand years of the Holocene". Hence, the observed low species richness may result from the young age of this fauna and the effective isolation of the region (Dayton 1985a).

Perhaps the best studied kelp community is off the coast of southern California at Point Loma (Dayton *et al.* 1984, 1992). The kelp forest was first surveyed in 1857 and first harvested for commercial purposes in 1916. Kelps and other algae occur here from the intertidal to a depth of approximately 30 m (Dayton *et al.* 1992). As with most kelp communities, the algae at this site occur in five, recognized canopy guilds (Dayton and Tegner 1984b, Dayton 1985b, 1994). These include 1) a floating canopy (*Macrocystis pyrifera*), 2) a stipitate, erect understorey canopy (*Pteryogophora californica* and *Eisenia arborea*), 3) a short prostrate canopy over the substrate (*Laminaria farlowii*, *Cystoseira osmundacea*, and *Dictyoneurum californicum*), 4) algal turf composed mostly of red algae, and 5) encrusting coralline algae.

These canopy guilds are composed of species which have become differentially adapted to different aspects of the selective regime encountered in kelp forests (Dayton 1985b). At Point Loma, the most important sources of algal mortality are associated with the dislodgment of large kelp plants during winter storms, the entanglement of other plants with these drifting individuals, and grazing by sea urchins (Dayton *et al.* 1984, 1992). *M. pyrifera* in the floating canopy guild is primarily adapted to compete for light (and/or nutrients), but tends to be susceptible to both physical disturbance from wave stress and grazing by sea urchins (Dayton *et al.* 1984, Dayton 1985b, 1994). Understorey species in both the erect and prostrate canopies are adapted primarily to tolerate physical stress, are poor competitors for light, and are susceptible to grazing (yet less so than is *M. pyrifera*). Encrusting corallines (and chemically defended algae like *Agarum fimbriatum* in the prostrate canopy) are adapted to intense grazing by sea urchins. More opportunistic species like *Desmarestia ligulata* in the understorey canopy are favored by physical disturbance in that they are good at colonizing available substrate following storms and inhibiting colonization by superior competitors like *M. pyrifera* (Dayton *et al.* 1992).

In a 10-year study of the kelp community at Point Loma, Dayton *et al.* (1984) utilized a combination of observational and experimental techniques to document the dynamics of distinctive algal patches dominated by some of the canopy guilds noted above. In particular, this study focused on the stability of these patches. Dayton *et al.* (1984) concluded that algal patches at Point Loma have "sharply defined and apparently stable boundaries" and these patches tend to persist over several generations of the dominant species. Clearing experiments indicated that colonization by the competitively superior floating canopy species (*M. pyrifera*) into understorey canopies is often inhibited due to interference with spore settlement and competition for light, but this inhibition can sometimes be overcome by large numbers of spores. In addition, recruitment by *L. farlowii* is strongly inhibited within patches dominated by *P. californica* as is recruitment by *M. pyrifera*, *L. farlowii*, and *P. californica* into patches dominated by algal turf. Furthermore, recruitment by *M. pyrifera*, *P. californica*, and *E. arborea* also appears to be limited by dispersal ability thus contributing to variable recruitment success. Successful development of patches dominated by *M. pyrifera* depends on a combination of factors promoting substrate colonization and survival of this dominant species. These include the generation of cleared areas by disturbance, proximity of fertile adult plants, high spore densities, reductions in the strength of inhibitory interactions, and reduced grazing by sea urchins. These processes can maintain the kelp community as a relatively stable mosaic of canopy patches.

From 1971-1981 at Point Loma, grazing by the sea urchins *Strongylocentrotus purpuratus* and *S. franciscanus* appears to have been restricted. They primarily ate drifting algae and maintained barren zones dominated by encrusting coralline algae near underwater ledges and in boulder fields (Dayton *et al.* 1984). Although these sea urchins can eat through holdfasts and stipes to dislodge kelp from the substrate and "...almost every temperate kelp community has been found to be particularly sensitive to over-exploitation by grazing sea urchins" (Dayton 1994), this was not a major source of mortality or disturbance to the kelp community during this study. Dayton *et al.* (1984) documented how the removal of sea urchins in the barren zone resulted in colonization by *Dictyota flabellata*, *Desmarestia ligulata*, articulated coralline algae, and three of the more dominant understorey algal species at Point Loma. They also noted particularly intense urchin grazing on young holdfasts of *M. pyrifera* at a shallow (10-11 m) inshore site following two episodes of heavy kelp recruitment.

In 1981, another 10-yr study was initiated at Point Loma along a depth gradient in the center of the kelp bed as well as along an 18-m longshore transect (Dayton *et al.* 1992). In striking contrast with the previous study, the dynamics of the kelp bed during this decade was dominated by large disturbances which "obliterated much of the structure in the kelp forest". During the winter of 1982-1983, 11 powerful storms hit southern California reducing the floating canopy of *M. pyrifera* from 600 to only 40 ha (Dayton and Tegner 1984a). Understorey patches were generally resistant to this physical disturbance. These storms were followed by spring recruitment of *M. pyrifera*, *P. californica*, and *L. farlowii* and competitive interactions among the canopy guilds which would normally have favored *M. pyrifera* (Dayton and Tegner 1984a). However, the most severe California El Niño event of the century occurred in 1982-1984 (Dayton and Tegner 1984a, Dayton *et al.* 1992). El Niño is the name given to periodic and potentially catastrophic warming events in the eastern tropical Pacific. These events are strongly associated with cyclical oscillations in the relative atmospheric pressure over the Pacific and Indian Oceans (the Southern Oscillation) and a general weakening of the trade winds driving the westward circulation of the North and South Equatorial Currents of the Pacific Ocean (Philander 1990, Diaz and Markgraf 1992). At Point Loma, El Niño events result in a reduction in the strength of the southward California Current, reduced upwelling, warmer than normal water temperatures, and nutrient limited conditions which are severely stressful to *M. pyrifera.* During the 1982-1984 event, the spring recruitment of *M. pyrifera* following the winter storms "had virtually no survivors" (Dayton and Tegner 1984a) due to this physiological stress and the effects of interference competition with the more tolerant understorey algal species.

By 1985, *M. pyrifera* had recovered much of the canopy area it had occupied prior to the winter storms and the El Niño event, but was hit again in January of 1988 by perhaps the most severe storm at this site in 200 years (Dayton *et al.* 1992, Dayton 1994). This storm eliminated the floating canopy, caused the first observed large-scale mortality to some of the understorey patches, and removed virtually all the drift algae from the kelp forest. Consequently, grazing by sea urchins shifted from drift algae to attached kelp and even encrusting coralline algae thus devastating the patch structure of the kelp forest. Kelp recovery did not begin until after an unknown disease caused extensive mortality to sea urchins in 1991 (Dayton 1994).

The dynamics of the kelp community at Point Loma then is dictated by local phenomena (*i.e.*, recruitment, storms, and biological interactions) and by periodic atmospheric-oceanographic processes influencing large regions of the Pacific Ocean. Thus the interplay between disturbance and recovery of the dominant floating canopy involves a suite of cross-scale phenomena which transcend the local and regional processes known to influence rocky intertidal communities (*e.g.*, Figures 1.7 and 1.12). Disturbances at multiple scales, regional transport processes, and local biological processes all have consequences which must be integrated in order to understand the dynamics of this specific kelp community. And as noted previously by Dayton (1985a), regional differences among kelp communities are likely to result from important geographic and historical phenomena.

1.4.3 ENCRUSTING COMMUNITIES

Encrusting communities the world over are often dominated by a variety of clonal organisms which tend to be highly resistant to invasion and mortality caused by competitors, predators, and physical disturbances (Jackson 1977a, Dayton 1994). Consequently, these communities tend to be even more patchy than rocky intertidal and kelp communities. Patch structure reflects a history of disturbances, recruitment events, and biological interactions which may or may not be indicative of normal present day conditions. In some encrusting communities, patches of substrate can be dominated by one or a few species for long periods of time. For example, in the normally ephemeral collection of species comprising the "fouling" community of North Carolina, most species exhibit seasonal recruitment patterns and relatively short life spans of less than one year (Sutherland and Karlson 1977). However, intense grazing by sea urchins in this community can favor the recruitment, growth, and persistence of the longer-lived hydroid *Hydractinia echinata* and the sponge *Xestospongia halichondroides* for several years (Karlson 1978). Domination of the substrate by *H. echinata*, the ascidian *Styela plicata*, and the bryozoan *Schizoporella errata* were identified by Sutherland (1974) as "multiple stable points" in this community in recognition of their ability to resist invasion and the importance of history on community development. Other encrusting communities dominated by long-lived clonal organisms can result in similar historical effects lasting decades or even centuries (Dayton 1994).

As a consequence of the longevity and sessile nature of the clonal organisms dominating many encrusting communities, competition for the space they hold is generally intense. In fact, competition in this community has resulted in the evolution of a number of attributes which either favor competitive success or enhance the probability of avoiding competitive encounters. These include 1) aggressive growth and interference mechanisms promoting the acquisition of space, 2) defensive capabilities facilitating the retention of acquired space, and 3) directional growth and larval settlement behaviors promoting escape from spatial competition. Attributes increasing the likelihood of favorable competitive encounters have been predicted to occur (Buss and Jackson 1979, Karlson and Jackson 1981), but have been demonstrated in nature only rarely (*e.g.*, see Wahle 1980).

The acquisition of space by growing clonal organisms occurs as they increase colony size and the area of their attachment to the substrate. In fact, competitive success in some cases is purely dependent of the relative size of interacting competitors (Buss 1980, Sebens 1982a, Russ 1982). In the subtidal encrusting community off the coast of Nahant, Massachusetts at East Point, the outcome of encounters between the octocoral *Alcyonium siderium* and the colonial ascidian *Aplidium pallidum* favor the former only when its colonies exceed 15 mm in diameter. Smaller octocoral colonies are generally overgrown by the ascidian (Sebens 1982a). Such growth over the substrate and competitors may take on stereotypical shapes or be modified by the presence of competitors. Jackson (1979a) identified six morphological strategies for sessile competitors which could be ranked in terms of their relative competitive success. Sheets, mounds, plates, and trees represent morphologies with decreasing "commitments to survival within and around the areas of settlement and to maintenance of the integrity of their colony surfaces". In contrast, "runners and vines are entirely committed to a fugitive (escape or refuge-oriented) morphological strategy" (Jackson 1979a). Thus, very aggressive colonies growing as sheets across the substrate are much like plants exhibiting a "phalanx" strategy (Lovett Doust 1981). This military analogy is most appropriate here given the intensity of spatial competition and the possibility that encrusting organisms engage in chemical warfare (Jackson and Buss 1975).

Defensive capabilities among encrusting organisms enhance their longevity and the persistence of local patch structure by delaying the expression of competitive dominance. Several studies have indicated that many competitive encounters result in the cessation of growth in the region of contact between adjacent colonies and suggest the possibility that growth inhibitors may be responsible for the observed phenomenon. Such encounters are commonly referred to as standoffs (Connell 1976). These can be quite temporary situations, but they can also persist for very long periods of time (Connell & Keough 1985). Connell (1976) noted one such interaction between two coral species at Heron Island that persisted throughout a 9-year period of observation. In an analysis of interactions among many of the colonial invertebrates dominating the temperate fouling community of Port Phillip Bay, Victoria, Australia, Russ (1982) reported frequent standoffs and delays in the expression of competitive dominance. In another very thorough 2-year study of hundreds of competitive encounters occurring on subtidal rock walls off the East Point site mentioned above, Sebens (1986) found that standoffs represented 41% of all encounters observed in this encrusting community and 79% of those encounters involving the common octocoral *A. siderium* (Sebens 1986). Thus the prevalence of some species in these communities is favored in spite of their inability to overgrow adjacent competitors. Other examples, include *Hydractinia echinata* in the fouling community of North Carolina (Karlson 1978) and *Zoanthus solanderi* on Jamaican coral reefs (Karlson 1980).

Attributes which favor escape from spatial competition include directional growth (Buss 1979) and larval settlement behaviors (Grosberg 1981). The former involve runner and vine-like growth strategies (Jackson 1979a) which promote relatively rapid growth over the substrate and the probability of encountering spatial refuges (Buss 1979). As an example, Buss (1979) described how the spatial distribution of the bryozoan *Stomatopora* sp. shifted from predominantly primary substrate to cover mostly bivalve shells after 26 months. The substrates in this case were settlement panels submerged at a depth of 40 m on the reef at Discovery Bay, Jamaica (described in Jackson 1977a). Portions of *Stomatopora* runners not locating the elevated bivalve shells were overgrown by a variety of other, competitively superior, colonial species. Hence, enhanced survivorship on these spatial refuges was linked to the relative rates of directional growth and overgrowth in this encrusting community. Sometimes directional growth by runners also promotes the capture of primary substrate and the formation of high-density aggregations in which the risks of mortality are significantly reduced (Karlson *et al.* 1996). In such cases, the growth strategy appears to be more strongly correlated with risks associated with predation and/or physical disturbances rather than spatial competition.

Intense spatial competition can select for larval settlement behaviors which, like directional growth, can promote exploitation of spatial refuges and avoidance of potentially lethal competitive encounters. For example, the colonial ascidian *Botryllus schlosseri* is a successful competitor for space in the hard substrate community located in the Eel Pond at Woods Hole, Massachusetts. Grosberg (1981) manipulated the density of this dominant species on settlement plates in order to assess differences in larval behavior of 19 other species. Furthermore, he assessed the outcome of spatial competition between each of these species and *B. schlosseri* in over 2000 encounters occurring elsewhere in Eel Pond. Nine species selectively avoided settlement on plates with high ascidian densities, while 10 were nonselective. As adults, the selective species all have feeding structures near the substrate which are easily overgrown by

B. schlosseri. This ascidian did so in 96.2% of the observed encounters. In contrast, the nonselective species, which generally have elevated feeding structures, were overgrown by *B. schlosseri* in only 17.1% of 961 encounters. Thus, invertebrate larvae can detect and avoid settlement on substrates where post-settlement spatial competition is likely to kill them (Grosberg 1981).

The above mentioned attributes all contribute to local success at competing for space or avoiding spatial competition and thus to small-scale positional effects in the observed structure of encrusting communities. These effects can be quite strong and, because of the longevity of so many of these organisms, they can persist for long periods of time. For example, the designation of a community dominated by *Hydractinia echinata* as a stable equilibrium point by Sutherland (1974) was based on the success of a single colony which persisted for 3.5 years (Sutherland 1981). Longer term success of multiple colonies of this hydroid contributed to greater patch heterogeneity on larger substrates where *H. echinata* co-occurred with other members of this community (Karlson 1978). Because of the low recruitment rate of this hydroid, its abundance is heavily influenced by substrate size. In general, substrate size can affect colonization patterns, the rate of recruitment into the adult community, the subsequent suite of competitive interactions (Jackson 1977a,b), and the susceptibility of these encrusting communities to physical disturbances (Osman 1977). In spite of higher risks of disturbance, small substrates can act as refuges from spatial competition for organisms with high larval colonization rates and relatively poor adult competitive abilities. Although solitary (aclonal) organisms are more likely to exploit such refuges than are colonial organisms (Jackson 1977a), there are multiple examples of colonial organisms also doing so (*e.g.*, Sutherland 1974, Osman 1977, Buss 1979, 1986, Keough 1984).

As a consequence of the small-scale phenomena noted above, biological interactions involving contact between adjacent sessile organisms or relatively slow moving consumers have been emphasized in these encrusting communities. In some cases, there is a complex interplay between predation/herbivory and competitive processes. Buss (1980, 1986) described a simple assemblage of two bryozoans (*Onychocella alula* and *Antropora tincta*) and a coralline alga (*Neogoniolithum rugulosum*) found on cobbles in tidal channels at Punta Paitilla, Panama. The results of overgrowth observations indicate that *A. tincta* and *N. rugulosum* are nearly competitively equivalent, while *A. tincta* generally overgrows *O. alula* which in turn generally overgrows *N. rugulosum*. Therefore, there is no competitively dominant species in this assemblage. Although relative competitive abilities are determined primarily by colony size (Buss 1980), the joint occurrence of multiple species on individual cobbles appears to be promoted by "coexistence competition" (Menge and Sutherland 1987). In order to determine the impact of predation on the structure of this assemblage, Buss (1986) performed a 9-month long experiment in which the predatory isopod *Paraleptosphaeroma glynni* was removed from cobbles. Large isopods feed on both bryozoan species, while small individuals are more selective feeding only on *A. tincta*. In the absence of the isopod, the fast growing *A. tincta* increased in abundance and eventually dominated the assemblage (Figure 1.14). Predation by isopods modified the colony-size dependent basis for spatial competition, reduced the growth rate of *A. tincta*, and promoted the coexistence of competitors. If one supposes that this qualifies as an example of "coexistence competition", the results directly contradict the prediction that predation reduces diversity (Menge and Sutherland 1987). If, on the other hand, one supposes that competitive relationships should be evaluated in the absence of consumers (Paine 1984, 1994), *A. tincta* would be designated as the competitive dominant and predation would then promote coexistence and reduce the effects of competitive exclusion [see Paine (1984) and Buss (1986) for further discussion of these alternative perspectives as well as an analysis of a similar situation involving herbivory and spatial competition among coralline algae].

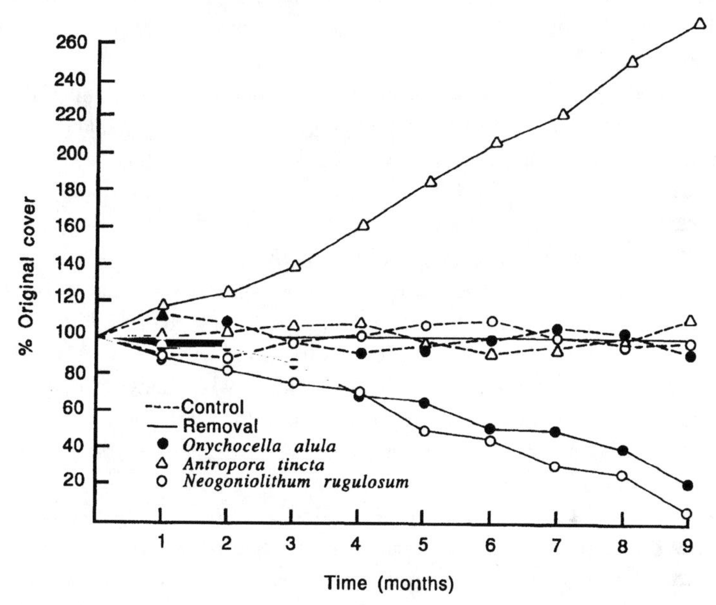

Figure 1.14 Effects of isopod removal on the structure of the Panamanian cobble assemblage. The mean percent of the original cover of *A. tincta*, *O. alula*, and *N. rugulosum* are indicated for control and removal treatment groups. All three species co-exist in the presence of the isopod. *A. tincta* becomes the dominant species as a consequence of isopod removal (redrawn after Buss 1986). From *Community Ecology* by Jared Diamond and Ted J. Case. Copyright (©) 1986 by Harper & Row Publishers, Inc. Reprinted by permission of Addison - Wesley Educational Publishers.

When viewed over evolutionary time, biological interactions in encrusting communities can take on an even more dynamic nature as the interacting organisms respond to strong selection pressures. Limpet-algal and bryozoan-hydroid associations provide two illustrations of how typical herbivore-plant and competitive interactions can evolve into mutualistic relationships. Example 1: The dominance of encrusting coralline algae in heavily grazed, hard-substrate communities is recognized as a global distributional pattern (Steneck 1986). Grazing by sea urchins, gastropods, and fish has selectively favored the convergent evolution of thick, herbivore-resistant morphologies as well as sunken reproductive structures and meristem. Thickness also improves competitive success among these algae. In the Gulf of Maine, the alga *Clathromorphum circumscriptum* and the herbivorous limpet *Acmaea testudinalis* are commonly found together (Steneck 1982). In fact, the limpet is more abundant on this algal species than on other substrates and it appears to feed preferentially on it. *C. circumscriptum* requires grazing by limpets to remove epiphytic algae. In the absence of grazing, these plants become overgrown by epiphytes and die. Although intense grazing (observed in laboratory aquaria) can completely remove the upper, multilayered epithallium, the rates of epithallial production in the field match the rate at which limpets remove these cells. The algal reproductive structures are buried deep below the surface and are largely unaffected by grazing limpets. Thus, the alga and limpet mutually benefit from their association. The normally detrimental effects associated with herbivory have been lost over evolutionary time.

Example 2: The bryozoan-hydroid associations described by Osman and Haugsness (1981) involve two taxa which typically constitute part of the guild of spatial competitors known to dominate encrusting communities. In Vineyard Sound, Massachusetts, the hydroid *Zanclea gemmosa* is found exclusively on the common bryozoan *Schizoporella errata.* Following selective settlement of hydroid larvae on the bryozoan (which is well defended from other potentially epizoic species), stolons of growing hydroid colonies extend over the surface of the bryozoan colony without occluding zooidal apertures from which the bryozoan protrudes its feeding organs. The bryozoan responds in a species-specific manner to *Z. gemmosa* by depositing a protective, calcareous tube around the hydroid stolons, while leaving openings for the hydroid polyps. Since hydroids are planktivorous and bryozoans feed on phytoplankton, this close association does not interfere with feeding by either organism. Although the precise benefit afforded to *S. errata* by this association remains unclear, Osman and Haugsness (1981) noted the considerable cost associated with tube deposition as an indication of the probable mutualistic nature of this interaction. Better evidence was provided for a mutualistic association between the bryozoan *Celloporaria brunnea* and the hydroid *Zanclea* sp. off the coast of southern California. Over a 19-month period of observation, 171 competitive encounters involving *C. brunnea* with 7 other epibenthic species were monitored. When *Zanclea* was present, competitive success of the bryozoan was significantly enhanced (successful overgrowth occurred in 24 of 25 cases). In the absence of *Zanclea*, competitive success occurred in only 17 of 146 cases and *C. brunnea* was quickly overgrown. Ancillary information indicates that in addition to enhancing competitive success, *Zanclea* also defends bryozoan colonies from predators (Osman and Haugsness 1981).

The strong biological interactions noted above represent local processes contributing to variability in the structure of encrusting communities. In fact, after evaluating the relative influence of local and regional factors on the number of species found encrusting 2374 settlement panels spread over multiple sites worldwide, Osman and Dean (1987) concluded that only local phenomena are important in such communities. In addition to noting that biological interactions contribute to local colonization and extinction rates on these substrates, they used stepwise regression procedures to find significant effects of immigration rates (the maximum number of colonizing species per month), the total number of local species available to colonize substrates (the local pool of species), and substrate size on the equilibrium number of species per substrate after at least 6 months of colonization. Thus they stressed processes affecting local dispersal and colonization and found no significant effects of latitude, temperature, salinity, depth, or the total number of potential colonizing species in the region (the regional pool of species). They rejected the notion that regional differences in the variables they explored could influence the distribution and abundance of species in these communities. Therefore, some encrusting communities may be so dominated by local processes which quickly equilibrate community structure that regional differences in geography and history may be unimportant. However, in a recent re-evaluation of this question again using multiple sites around the world, significant regional effects have actually been detected in these highly interactive communities (J. Witman personal communication).

I conclude this introductory chapter with one final example of an encrusting community which is strongly influenced by large-scale processes. As part of their research program at McMurdo Sound, Antarctica, Dayton *et al.* (1969) noted the destructive effects of uplifted anchor ice on epibenthic organisms to a depth of 33 m. Below this depth, the hard-substrate community is dominated by a zone of long-lived sponges and a variety of molluscan and asteroid predators (Dayton *et al.* 1969, Dayton 1972). Early research into the dynamics of this community highlighted the influence of biological interactions (the interplay of predation and competition) on community structure in this deep zone (*e.g.*, Dayton *et al.* 1974). At intermediate depths (15-30 m), long term shifts in the prevailing currents are thought to cause cyclical fluctuations on decadal scales [see Gu and Philander (1997) for linkage between high latitude and equatorial climate fluctuations] in the upwelling of extremely cold deep water, anchor ice formation, and abundances of the sponge *Homaxinella balfourensis* and its predators (Dayton 1989). Between 1967 and 1988, *H. balfourensis* exhibited massive recruitment capabilities, but also experienced catastrophic mortality events associated with these oceanographic climate changes. Explosive population growth of the sponge during relatively warm conditions in 1974-1977 resulted in as much as 80% of the substrate being covered by *H. balfourensis* at some sites. This was followed by its virtual elimination from this zone in 1984 and early tentative signs of recovery in 1988. The 1984 and 1988 censuses revealed persistent predator populations which had expanded with the earlier growth of the sponge population. Such climate-driven fluctuations clearly illustrate how large scale processes can contribute to local variation in community structure.

1.5 Overview

Much of community ecology deals with interactions among species, yet we are becoming increasingly aware that local communities of interacting species must be understood in a larger-scale context. Hence this treatment of the dynamics of coral communities specifically focuses on the constituent species, the interactions among them, the local environment, oceanographic transport processes, climatic fluctuations, and a range of other geographical and historical phenomena. In that coral communities are but one of several types of marine epibenthic communities occurring on hard substrate, I have introduced the dynamics of this general class of communities using three of our best-studied examples. Coral communities are likely to exhibit many comparable characteristics with these examples. 1) Rocky intertidal communities are characterized by strong local gradients in the physical environment (*e.g.*, intertidal exposure and wave energy) and relatively few species engaged in strong biological interactions. At larger spatial scales, recruitment gradients contribute to significant variation in community structure. These have been attributed to predation in nearby habitats and regional oceanographic transport processes. 2) Kelp communities have stable patch structure over multi-year time scales due to the influences of the local environment and strong biological interactions. However, variation in climate and oceanographic conditions over longer time scales can totally disrupt this structure. 3) Encrusting communities are unusual in that they can be dominated by long-lived clonal organisms with well developed defense mechanisms. These mechanisms contribute to the maintenance of patch structure and a reduction in the rate of species turnover as these organisms inhibit competitive overgrowth, discourage predators, and resist the negative effects of physical disturbances. However, the stability of these communities is also sensitive to variation in climate and oceanographic conditions.

2 DIVERSITY

2.1 Origins of diversity

The diversity of coral communities emerges as a consequence of evolutionary and ecological processes which vary in their importance depending on the scale of consideration. As new species are generated, they become part of local interacting communities as well as the regional biota. This process may occur gradually or as part of rapid adaptive radiations over evolutionary time. When viewed over increasing spatial scales, additions to the number of species exceeds the bounds of local ecological communities, species replacements occur over geographic distances, and the impact of specific biological interactions on diversity patterns weakens (Schluter and Ricklefs 1993a). Hence, the diversifying effects of predation or "coexistence competition" (Menge and Sutherland 1987) may be locally quite strong, but relatively insignificant when evaluated at much larger scales where regional transport processes, major disturbance events, and evolved niche differences can act as primary determinants of community structure (Jackson 1991).

Modern coral communities are but one of several types of reef communities which have flourished on earth over the last 3.2 billion years (Kauffman and Fagerstrom 1993). In their review of the fossil evidence from all such communities, Kauffman and Fagerstrom (1993) noted dynamic fluctuations in taxonomic diversity throughout this record (Figure 2.1). Their analysis revealed 1) abrupt declines in reef diversity associated with eleven global extinction events, 2) quite slow recovery immediately following these extinctions including long periods of time (1-8 my) without reef development, and 3) a gradual recovery in taxonomic diversity typically lasting several million years. The declines in diversity are correlated with both terrestrial forces (geological, climatic, and oceanographic) as well as extraterrestrial events which destabilized tropical ecosystems. Modern reef communities are the "product of only the last 45-50 my of evolution, and even during this interval have experienced major ecological crises" (Kauffman and Fagerstrom 1993).

One of the highest diversity peaks in the history of reef communities occurred in the late Triassic period when scleractinian corals displaced sponges as the dominant reef builders and the number of reef-associated taxa increased dramatically (Figure 2.1). In spite of several episodes of global extinctions since that time, scleractinian corals persist today as major reef builders providing the structural base for some of the richest ecological communities on earth. Recent molecular evidence suggests that living scleractinian corals may have evolved from two clades of soft-bodied ancestors which are thought to have diverged 240 my ago before they evolved hard skeletons (Romano and Palumbi 1996). The oldest known scleractinian fossils date back to 220-229 my (Stanley and Swart 1995). The tabulate and rugosan corals which flourished earlier in the Paleozoic Era are not thought to be related by descent to scleractinian corals [see Moore (1956), Hallam (1973), Oliver (1980), and Veron (1995)].

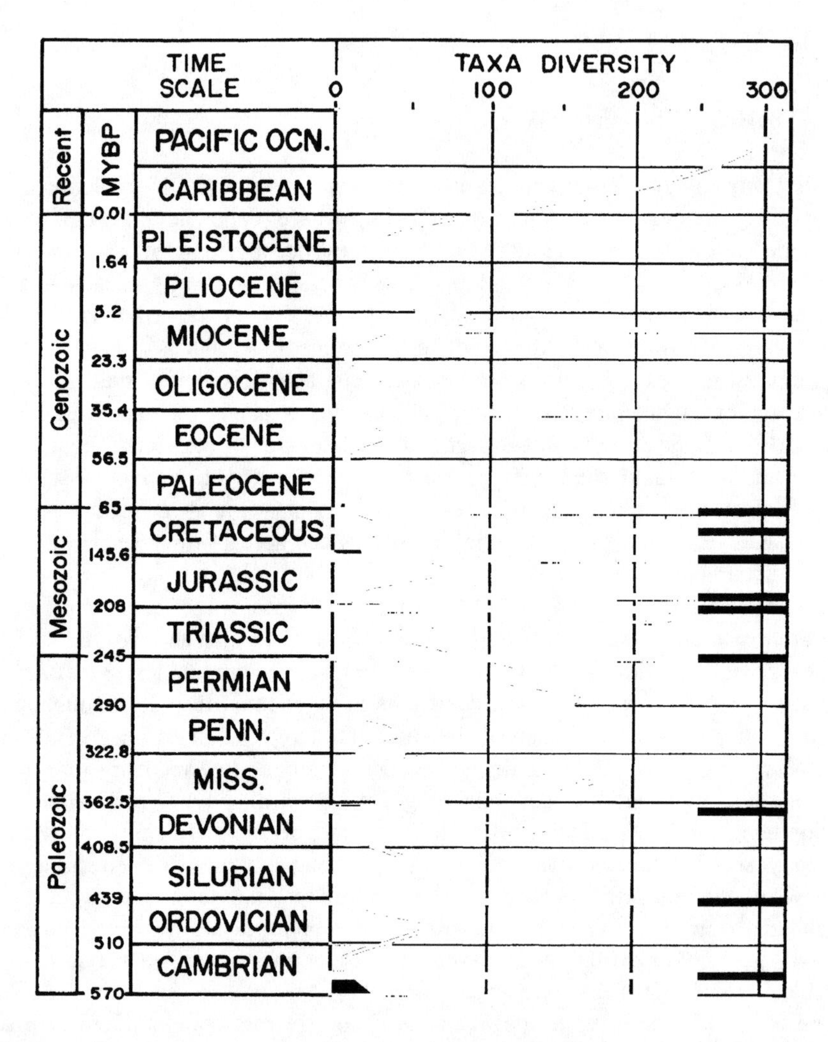

Figure 2.1 Temporal fluctuations in the taxonomic diversity of shallow-water reef communities. Diversity on Recent reefs were derived from data on multiple taxa from Heron Island, Australia (Mather and Bennett 1984) and from Carrie Bow Cay, Belize (Rützler and Macintyre 1982). These included only those taxa with preservable hard parts. Horizontal bars indicate major global extinction events (portion of original redrawn after Kauffman and Fagerstrom 1993). © 1993 by The University of Chicago.

2.2 Patterns of coral diversity

One of the most evident global patterns in species diversity exhibited by corals and many other taxa is the general increase in the number of species which occurs with decreasing latitude (Dobzhansky 1950, Fischer 1960, MacArthur 1965). Although a variety of explanations have been posited to explain this phenomenon (Connell and Orias 1964, Pianka 1966, Krebs 1994), the issue is confounded by a mix of evolutionary and ecological arguments which do not represent clear, mutually exclusive alternatives. In contrast, MacArthur (1965) posed the problem in an unequivocal conceptual context which has strongly influenced community ecology as a predictive science to this day. He noted that the total number of species occurring in tropical habitats might very well be attributed to historical phenomena and still be subject to this nonequilibrial influence. On the other hand, the number of species within tropical habitats may saturate rather quickly as they reach equilibrial levels set by the limiting similarity of species. Species more similar than this limit must occupy different habitats thus contributing to between-habitat diversity. According to this perspective, increases in total diversity are due to habitat specialization (selection) and history has no effect on the number of species within tropical habitats. By removing history as an important explanation of diversity patterns within habitats, MacArthur introduced a predictive, resource-based framework for studying species diversity in ecological communities (MacArthur 1972). This framework further distinguished the field of community ecology from that of biogeography [see section 3.4 and chapters 8 and 9 for more discussion of the influence of history and scale on species distributions and diversity patterns].

The study of within and between-habitat components of species diversity and how they contribute to total species diversity originates with the now classic work by Whittaker (1960) on plant communities in the Siskiyou Mountains of northern California and Oregon. After analyzing plant distributions and associations over elevational gradients, he quantified the number of species within local habitats and related this to the regional total using the notion of between-habitat diversity, the average turnover in species across multiple habitats in the region. This component of diversity accounts for the influence of habitat breadth/specialization over environmental gradients and can be used as a basis for regional comparisons of species diversity (Schluter and Ricklefs 1993a).

As part of his review on patterns of species diversity on coral reefs, Huston (1985a) noted that coral communities occur across a number of gradients in such variables as light, sedimentation, temperature, wave energy, plankton availability, intertidal exposure, storm exposure, and grazing by fish and sea urchins. Although these communities can be quite patchy, much of the influence of these environmental variables is correlated with depth. Consequently, coral reefs located throughout tropical regions of the world exhibit characteristic zonation and diversity patterns (Sheppard 1982, Huston 1985a). In general, the diversity of corals is low near the surface, reaches a maximum at depths of 15-30 m, and then drops again with increasing depth.

In recognition of both speciose and other relatively impoverished, reefal communities in the tropics as well as non-reefal coral communities at higher latitudes, I describe below a range of coral diversity patterns from selected sites scattered throughout the world. Since most of the studies conducted at these sites deal primarily with scleractinian corals rather than with the entire community, I refer to these collections of species as assemblages (after Fauth *et al.* 1996). In what follows, I use the published literature to describe local zonation and species diversity patterns across depth and habitat gradients at a variety of quantitatively sampled sites. In addition, I highlight the average habitat breadth of corals in terms of the number of zones/habitats occupied per species. I include specific examples of common corals with quite broad or more restricted distributions across depth and habitat gradients. Some of these common corals occur at multiple sites distributed over enormous geographic scales, while others have much more limited distributional ranges.

2.2.1 THE PHILIPPINES, MALAYSIA, AND INDONESIA

More scleractinian species occur in the vicinity of the Philippines, Malaysia, and Indonesia than anywhere else in the world (Stehli and Wells 1971, Veron 1995, Figure 2.2).

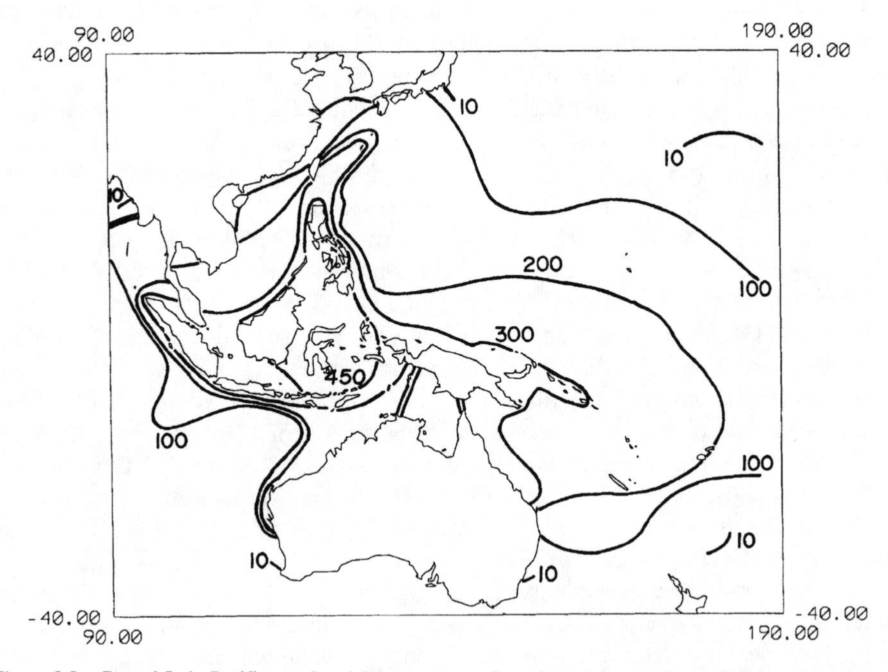

Figure 2.2 Central Indo-Pacific species richness patterns for scleractinian corals predicted on the basis of known distributional ranges and number of species per genus. Contours are given for 10, 100, 200, 300, and 450 species. Coordinates are relative to the equator and east of prime meridian (portion of original redrawn after Veron 1995).

Although there have not been many quantitative surveys of coral diversity patterns across depth gradients in this region, two locations which have been examined include the reefs at Mactan Island (Sy *et al.* 1982) and Apo Island in the central Philippines

(Ross and Hodgson 1982). Sy *et al.* (1982) used 70, 15-m line transects run parallel to shore to survey corals on a fringing reef from the intertidal zone through the reef flat (0-1 m deep), over the reef crest (1-5 m deep), and down to a depth of 30 m on a vertical wall. A total of 204 species were sampled in these three zones. The reef flat samples were numerically dominated by *Pavona decussata* near shore and by *Cycloseris patelliformis* out towards the reef crest, but total coral cover was relatively low. From their graphical representation of the data, one can estimate that there were 3.8 species per transect in this zone. The reef crest was dominated by several species of *Montipora* and *Porites* and there were 16.1 species per transect. The highest diversity levels were found on the wall where there were 21.7 species per transect. The most common species in this zone was *Pavona praetorta* which covered approximately 10% of the substrate. Although a complete set of species distributions across this depth gradient was not provided, Sy *et al.* (1982) did provide this information for eleven of the most common species at this site. Using the three zonal designations as habitats, a mean of 1.6 habitats per species characterizes this group. There were no generalists occurring in all three habitats and five of the most common species occurred over more restricted depth ranges within zones.

Ross and Hodgson (1982) also surveyed a fringing reef in the Philippines, but they used a point intercept method along 300-m line transects in each of seven habitats (back reef 0-2 m, outer reef flat 0-2 m, reef crest 0-3 m, shallow terrace 3-6 m, upper dropoff 6-15 m, lower dropoff 15-25 m, and a scree-buttress slope 25-35 m). Along the 2100 m of sampled reef (under 4200 points), they recorded a total of 197 coral species. With increasing depth and distance from shore, the mean number of coral species per transect in each habitat was 17, 25, 44, 70, 90, 70, and 35 species. The relatively low diversity of the reef flat was attributed to the exposure of these shallow-water species to a variety of environmental stresses. Low diversity in the deepest zone was attributed to low light levels and relatively unstable substrate. On average, each species was found in 1.8 habitats, but this estimate of habitat breadth is not corrected for the presence of rare species. If one considers only the more common species with at least 10 occurrences across all habitats, the mean number of habitats per species was 3.4 (n = 31 species) including eight species which were found in six or all seven habitats (*Acropora formosa*, *Pocillopora danae*, *Seriatopora hystrix*, and five *Porites* spp.). In 29 of these 31 species, all the habitats occupied by a given species were contiguous along the depth gradient. Hence, in spite of the presence of a few generalists, some degree of habitat specialization along this gradient is evident among several coral species (*e.g.*, the eleven most common *Montipora* species which were found in 1.8 habitats per species).

2.2.2 THE GREAT BARRIER REEF

The barrier reef system along the eastern coast of Australia is arguably the largest structure on earth made by living organisms. It is approximately 2000 kilometers long with over 2500 individual reefs spanning a region with almost 350 hermatypic scleractinian species (Veron 1986, Figure 2.2). Habitat complexity is high and the rich variety of species associations led Veron (1995) to suggest that there may be as many as 300 recognizably distinct coral assemblages distributed within and among environmentally homogeneous areas (biotopes) on the Great Barrier Reef. Hence, these coral reef assemblages would appear to be much more patchy and speciose than the three types of epibenthic communities noted in chapter 1.

As a result of a quantitative assessment of coral abundance patterns over depth and wave-exposure gradients on thirteen reefs in the central Great Barrier Reef, Done (1982) distinguished seventeen different types of coral assemblages composed of a total of 108 common species. In this region, mid-shelf reefs are generally most diverse, while atolls out in the Coral Sea and "nearshore silt-affected reefs are almost equally depauperate". Much of this variation is likely to result from the fact that reef flat environments vary extensively across this region. They are virtually absent on many inner reefs, quite extensive in mid and outer shelf reefs, and are submerged in deeper water on the oceanic atolls. As a consequence of this extreme variability, examples based on quantitative surveys of coral diversity patterns over depth gradients on the Great Barrier Reef reflect only local patterns which are not necessarily applicable throughout the region. A stereotypical reef does not appear to exist. Nevertheless, a few local surveys can be used as examples of diversity patterns and as a basis for evaluating depth-related habitat specialization.

Done (1983) reported the distribution of thirty seven coral species over a 40-m depth range through ten topographic zones on a ribbon reef on the outer barrier of the Great Barrier Reef. The habitats in this study included those on the outer slope, over the top of the reef, and back down into a pinnacle slope environment. The shallow mid-reef flat was the least diverse with only seventeen species. Higher diversity characterized deeper zones on the fore-reef and pinnacle slopes (twenty six and thirty species, respectively). A mean of 6.4 zones per species indicates the prevalence of several generalist species on this disturbed, wave-swept reef (*e.g.*, *Acropora humilis*, *A. palifera*, *A. hyacinthus*, *Pocillopora damicornis*, *P. verrucosa*, *Stylophora pistillata*). For these species, Done (1983) emphasized the importance of intraspecific variation in coral morphology among reef zones rather than interspecific habitat specialization. Eight species occurred in all ten zones, while only two species were restricted to two zones and none to a single zone. However, Done (1983) did note the affinity of some species to wave-exposed surf zones (mostly *Acropora* spp.) or to the relatively calm, low-light conditions on the pinnacle slopes.

Surveys conducted between 1980-1983 on Myrmidon Reef, an outer shelf reef, also indicate a general increase in the number of coral species with depth and some degree of zonation. Fisk and Done (1985) determined the mean number of species per quadrat within four, 1 X 10 m belt transects at depths of 0, 1, 8, and 17 m to be 6.2, 9.3, 22.2, and 22.9 species, respectively. These sites were dominated by acroporid and pocilloporid corals including all of the generalist species noted above. Using the assemblage designations defined in Done (1982), they identified the turbulent shallow sites as an *Acropora palifera* / *A. humilis* assemblage and the deepest site as an *Acropora palifera* / *Porites* assemblage. These species associations are typical of reefs on the outer shelf and in the Coral Sea. The 8-m site was an *Acropora* tabulate/branching assemblage which is more common on the slopes of mid-shelf reefs (Done 1982).

Pichon and Morrissey (1981) reported the results of a survey on a mid-shelf reef near Lizard Island in the northern Great Barrier Reef. They examined the abundance of corals across the reef flat at 50-m intervals from the lagoon to the outer edge of the flat and down to a depth of 10 m. Along 31, 30-m line transects (930 m), they found a total of 119 scleractinian species. Much of their sampling effort was restricted to the reef flat at mean low water where they found 11.1 species per transect. There were 38, 36, and 37 species along three deeper transects at 3, 5, and 10 m, respectively. Thirty eight of eighty seven species found on this reef slope were not sampled on the reef flat. Likewise, thirty two of the eighty one species recorded on the reef flat were not found along the reef slope transects. Hence, approximately 60% of the sampled scleractinian species have restricted depth distributions. Based on this survey, Pichon and Morrissey (1981) identified four distinct zones at this site: 1) a diverse reef slope assemblage, 2) a diverse outer reef flat assemblage dominated by such corals as *Acropora humilis*, *A. millepora*, and *Pocillopora eydouxi* (these species did not occur elsewhere in the survey), 3) a barren zone devoid of corals (but large fleshy and turf algae were common), and 4) a less diverse inner reef flat assemblage. Thus habitat specialization on this reef occurs over gradients in depth and distance across the reef flat.

Examples of the depth distributions of corals on inner shelf reefs are reported in two studies from the central Great Barrier Reef (Done 1977, Bull 1982). These reefs are subjected to extensive freshwater runoff and relatively turbid water. Hence, community structure differs considerably between these reefs and those further out on the shelf; there is also much variation among inner shelf reefs. At Magnetic Island, Bull (1982) surveyed two shallow sites to find seventy eight species along 24, 30-m line transects (720 m) through intertidal reef flat, crest, and slope environments. At Cockle Bay, there were 2.3, 10.3, and 11.8 species per transect through the inner reef flat, outer reef flat and crest (all intertidal), and slope (1-4 m deep), respectively. At Geoffrey Bay, there were 6.8 species per transect on the reef flat and crest ($\leq$ 2 m deep) and 22.6 species per transect on the slope (3-9 m deep). *Montipora ramosa*, *Platygyra sinensis*, *Symphyllia recta*, and *Goniastrea aspera* were abundant in shallow habitats, while *Goniopora tenuidens*, *Acropora formosa*, and *A. hyacinthus* were abundant on the reef slope.

On another inner shelf reef at Orpheus Island, Done (1977) used 36, 10-m line transects (360 m) from the shore to a depth of 23 m to find fifty four coral species. Although species identifications were not provided, general diversity patterns can be determined with respect to depth. There was a mean of 7.5 species per 20 m of transect line in shallow water (0-3 m deep). In deeper water, there were 13.8 and 12.0 species per 20 m at depths of 4-10 m and 13-22 m, respectively. The relatively turbid water here favored primarily pocilloporid, acroporid, faviid, and mussid corals in shallow water. Pectinid corals were more prominent in deeper water.

2.2.3 THE INDIAN OCEAN AND RED SEA

Quantitative surveys of coral diversity patterns over depth gradients in this region have been conducted on several oceanic atolls and on a few fringing reefs. One of the best known of these was conducted in the Chagos Archipelago on the slopes of Peros Banhos and Salomon atolls (Sheppard 1980). Using four different transect methods on seaward and lagoonal slopes, the maximum number of species was found at intermediate depths (18-20 m) where there were as many as thirty five coral species in 20 m^2 and one hundred species in 320 m^2 of sampled reef. With increasing depth, the number of species rose to this peak and then steadily declined. Five or fewer species per 20 m^2 were recorded from phototransects at depths of ≤ 2 m and there were fewer than ten species per 20 m^2 in the deepest samples at 48 m and 61 m. In general, there were four conspicuous zones recognized for seaward slopes in this study. In shallow water (less than approximately 12 m), *Acropora palifera, A. humilis*, and *A. reticulata* were numerically dominant. Below this zone to 25 m, there was a diverse assemblage with no dominant species occupying more than 10% of the substrate. At 25-45 and 50-60 m, respectively, the foliose corals *Pachyseris levicollis* and *Agaricialla* sp. were dominant species. Sheppard (1980) noted quite different zones on the lagoonal slopes which were dominated, in order with increasing depth, by *A. palifera, Pavona clavus, A. hyacinthus* (the most numerically dominant species observed for any zone in this study), *Echinopora lamellosa*, and *Galaxea clavus* (dominating extensive areas on the lagoonal floor). Thus, both depth gradients and habitat differences on these atolls result in shifts in species abundance patterns and in the degree to which the coral assemblage is dominated by particular species.

At Gan Island, Addu Atoll in the Maldive Islands, Davies *et al.* (1971) examined the distribution of corals along two belt transects through the reef flat and down the steep lagoonal slope to a depth of 35 m. They found 103 species along a total of 792 m^2 of sampled reef and a mean of 19, 38, 36, 58 species per habitat in mixed coral, *Acropora formosa*, an outer reef flat zones (< 1 m deep), and a deeper zone (0-35 m deep), respectively. Among these four zonal habitats, corals occupied a mean of 1.9 habitats per species. Twelve species occurred in all four habitats, while 52 occurred in only a single habitat. The only habitats noted for having a dominant species or group of species were the *A. formosa* zone on the reef flat and a portion of the deepest zone where *A. formosa, A. convexa, A. digitifera*, and *Echinopora lamellosa* were particularly abundant. Qualitative assessment of coral distributions, diversity, and dominance at other sites around Addu Atoll appear to indicate considerable spatial variability in these patterns (Davies *et al.* 1971).

Barnes *et al.* (1971) reported the results of a detailed quantitative survey from a seaward slope on Aldabra, an atoll located in the western Indian Ocean. Within a total of 35, 1.6 m^2 quadrats (56 m^2) distributed at 10 m intervals from shore to a depth of 44 m, eighty species of hard corals were sampled. There was a maximum of nineteen species per quadrat at a depth of 27 m. Six distinct zones along this transect were recognized together with the dominant benthic forms which included a mix of hard corals, soft corals, gorgonians, and algae. These were 1) a shallow zone (0-6 m) dominated by acroporid and faviid corals, 2) a soft coral zone (6-14 m), 3) a brain coral zone (14-28 m) with abundant calcareous algae (*Halimeda*), 4) an encrusting coral zone (28-38m), 5) a gorgonian zone (38-42 m), and 6) a very depauperate zone dominated by the ahermatypic coral *Dendrophyllia micrantha* (42-44 m). Among all of the eighty species of hard corals, a mean of 2.0 zones (habitats) per species suggests the prevalence of habitat specialization along the depth gradient. After removing nineteen of the rarest species, this measure of habitat specialization remains low at 2.1 zones per species. Species with the highest degree of habitat breadth in this survey occurred in four zones (*Platygyra lamellina, Echinopora gemmacea, Galaxea fascicularis*, and *Lobophyllia hemprichii*).

In the southwestern Indian Ocean, Bouchon (1978, 1981) used both quadrat and line transect samples to study the species composition, diversity, and zonation of coral assemblages over a 40-m depth gradient on fringing reef at Reunion Island. A total of 4350 coral colonies representing 120 species were recorded from reef flat (0.5-1.5 m deep) and reef slope (5-40 m deep) stations. Both methods indicate a maximum number of species at a depth of 30 m. Forty species in 10 m^2 and forty two species along 50 m yielded a combined estimate of fifty four species at this depth. Four stations over 500 m of reef flat represented distinct zones, while seven stations on the slope traversed spur-and-groove (where characteristic reef buttresses are separated by sand channels), lower slope, and unconsolidated calcareous nodule zones. Corals occupied a mean of 2.6 zones per species with twenty one species occurring only on the reef flat and sixty two species only on the reef slope. *Porites lutea* was unusual in that it occurred in all seven zones.

The best known study of depth-related variation in the structure of coral assemblages in the Red Sea was conducted on fringing reefs near Eilat, Israel (Loya and Slobodkin 1971, Loya 1972). A total of 84, 10-m line transects (840 m) were used to survey ninety three coral species at 5-m intervals through back reef, reef crest, and fore-reef environments to a depth of 30 m. The maximum number of species was recorded at the deepest stations where there were 29.7 species per transect. A minimum of 11.3 species per transect was recorded for six, back reef transects. Seven recognized zones over this habitat and depth gradient included back reef, rear reef, and reef flat zones dominated by *Stylophora pistillata*, a reef crest zone dominated by the hydrocoral *Millepora dichotoma*, and three, progressively deeper fore-reef zones dominated by *Echinopora gemmacea*, *Acropora hemprichii*/*A. variabilis*, and *Porites lutea*, respectively. These six dominant species exhibited considerable habitat breadth in that each species occurred in all seven zones. In spite of the fact that the majority of species in this extensive survey of 4404 colonies were relatively rare, only two specialist species were noted by Loya and Slobodkin (1971). *Leptoseris tubulifera* and *L. fragilis* were limited to deeper water at depths of 25-50 m.

2.2.4 THE CENTRAL PACIFIC OCEAN

To the east of the high diversity region in the Philippines, Malaysia, and Indonesia, coral assemblages become increasingly less speciose (Figure 2.2). Within this region of the central Pacific Ocean, quantitative surveys of the number of coral species across a depth gradient include those conducted on the atolls in French Polynesia (Bouchon 1985) and in the Phoenix Islands (Dana 1979). In 1979 and again in 1982, Bouchon (1985) surveyed the fringing and barrier reefs off Moorea over a 30-m depth range. Using 2-5, 10-m line transects at each of nine stations distributed over a shallow fringing reef (three stations, 0.5-1.0 m deep), barrier reef flat (two stations, 1.5-2.0 m deep), and outer reef slope (four stations, 3-30 m deep), there were fifty six and forty seven coral species along a total of 400 m sampled in 1979 and 1982, respectively. The decrease in species richness was attributed to an outbreak of the corallivorous sea star *Acanthaster planci*. The maximum number of coral species was found at the outer edge of the reef flat in 1979 (29 species/50 m) and at 20-30 m on the outer slope in 1982

(28 species/50 m). The minimum occurred at the most inshore station in both years (6 species/50 m). Based on the 1979 samples, the general pattern of higher diversity at intermediate depths (Huston 1985a) does not appear to hold true for this site.

In general, this coral assemblage exhibits intermediate levels of habitat specialization. The dominant corals on the barrier reef flat included *Porites lutea*, *Synaraea rus*, and *Acropora cytherea*. The upper, wave-swept regions of the outer slope were dominated by *Pocillopora verrucosa*, *P. eydouxi*, *P. damicornis*, *Montipora erythraea*, and *Acropora humilis*. At the deeper slope stations, acroporids and pocilloporids remained dominant along with *Porites lutea* and *Synaraea rus*. Among all of sixty species of hard coral, the mean number of stations occupied per species in 1979 and 1982 was 3.4 and 3.0, respectively. Among the thirteen numerically dominant species, there was a mean of 5.5 and 4.7 stations per species in these two surveys. Only *Porites lutea* occurred at all nine stations, while *Montipora circumvallata* and *Pocillopora eydouxi* had the most restricted distributions. The former occurred only at two stations on the shallow fringing reef, the latter only at two, upper slope stations.

At McKean Island (in the Phoenix Islands located near the equator approximately 3000 km to the northwest of French Polynesia), Dana (1979) conducted a brief survey of the corals distributed over a 26-m depth range through seaward terrace and reef slope habitats. Using a point intercept method along transects in three depth intervals (1-10, 10-18, and 18-26 m), a total of thirty six coral species were sampled. As at Moorea, there were more species encountered in shallow water than on the deeper reef slope. There were 25, 22, and 12 species sampled under 127, 167, and 44 points per interval, respectively. This pattern persists even after correcting for unequal sample sizes. Within this assemblage, species occupied a mean of 1.6 depth intervals per species, but the majority of species were uncommon. Among the eight most common species with ten or more occurrences in these samples, there was a mean of 2.2 depth intervals per species. Four of these species occurred in all three intervals (*Porites lutea*, *P. lichen*, *Favia stelligera*, and *Echinopora lamellosa*), while *Cyphastrea microphthalma* and the hydrocoral *Millepora platyphylla* only occurred in the most shallow interval.

2.2.5 THE EASTERN PACIFIC OCEAN

The coral communities of the eastern Pacific Ocean represent some of the most isolated and impoverished in the world. In a recent issue of *Coral Reefs* devoted entirely to this region, Glynn *et al.* (1996) described the reefs at Clipperton Atoll. Using quadrat sampling along multiple transects to a depth of 60 m, a total of only eight hermatypic coral species were reported for this location. Four of these species occurred over large depth ranges. *Pocillopora* sp., *Porites lobata*, and *P. minuta* were common from 3-30 m, and *Pavona varians* was distributed over 5-50 m becoming progressively more abundant in deeper water. Glynn *et al.* (1996) indicated that this atoll is characterized by relatively low habitat diversity possibly due to the limited exchange of lagoonal water with the sea and the absence of a predictable leeward exposure.

Off the Pacific coast of Costa Rica at Caño Island, Guzmán and Cortés (1989) recorded a total of ten coral species along 204, 10-m line transects over an 18-m depth range. Comparably low numbers of species have been reported by Porter (1972) and

Glynn (1976) for quantitative samples taken to the south off the coast of Panama. At Caño Island, the mean number of coral species per transect on the shallow reef flat (0.5-1 m deep), reef slope (1.5-8.5 m deep), and base of the reef (9-18 m deep) was 1.2, 2.4, and 2.7, respectively. Eight of the ten species were present at intermediate depths and seven occurred in the deepest zone. Only three species were recorded on the reef flat. The mean habitat breadth among all species at this site was 1.8 zones per species. Among the most common seven species (occurring in at least 10% of the samples within any single zone), there were 2.1 zones per species. *Porites lobata* and *Pocillopora damicornis* occurred in all three zones, while *Psammacora clavus* was restricted to the reef slope.

2.2.6 THE CARIBBEAN SEA

Some of the best known coral reef communities in the world occur in Jamaica where Tom Goreau pioneered studies on the physiology and ecology of corals (Yonge 1971). Although the Caribbean Sea harbors a relatively impoverished coral fauna [fifty hermatypic scleractinian species (Bright *et al.* 1984)], Jamaican reefs are speciose enough to rival some of the tropical reefs in the Indo-Pacific Ocean (Goreau and Wells 1967). Almost the entire regional scleractinian fauna has been recorded for individual reefs in Jamaica. These corals typically occur in characteristic zones across a gradient from shore to depths exceeding 100 m.

The original description of this zonation pattern comes from a reef near Ocho Rios, where Goreau (1959) surveyed a 125-m wide belt transect through zones designated as the shore (0.5-3 m deep), lagoonal or channel (2-15 m deep), rear (1-3 m deep), reef flat or *Zoanthus* (0.5-3 m deep), *palmata* (0.5-6 m deep), buttress (1-10 m deep), *cervicornis* (7-15 m deep), and *annularis* zones (>15 m deep). Although the eight zones identified by Goreau (1959) included four which were strongly dominated by a single species (*i.e. Zoanthus sociatus, Acropora palmata, A. cervicornis*, and *Montastrea annularis*), the number of species within zones and the considerable depth range of many species across zones is indicative of the extensive habitat breadth characterizing this coral assemblage. The maximum number of all hard (scleractinian and hydrozoan) coral species was reported for the buttress zone at intermediate depths (thirty three species), the minimum in the shallow *Zoanthus* zone (nineteen species). The mean habitat breadth among forty two species was 5.5 zones per species. Fourteen species were found in all eight zones (Goreau 1959).

In an expanded treatment of the zonation of Jamaican coral reefs, Goreau and Goreau (1973) described the general patterns derived from multiple reef locations and a depth gradient including the fore-reef slope dominated by *Agaricia* spp. (30-65 m deep) and the vertical walls of the deep fore-reef (> 65 m deep). The depth ranges of fifty species through eleven zones again indicates broad distributional patterns. The mean was 7.1 zones per species. The most speciose zones were the buttress (2-20 m deep) and fore-reef (15-40 m deep) zones with forty three and forty four species, respectively. Other investigators conducting detailed surveys of coral diversity in the vicinity of Discovery Bay, Jamaica, confirm maximal peaks in the number of species at intermediate depths and large depth ranges for several common coral species (Huston 1985b, Liddell and Ohlhorst 1987).

Species diversity patterns over depth gradients have been reported for several other reefs in this region. For example, Bak (1977) surveyed two sites in Curaçao using continuous quadrats in 5-m wide belt transects to depths of 40-60 m. At one site, the transect extended through seven zones which typify southwest coastal reefs of this island including shore, *Acropora palmata*, barren, upper and lower terrace, drop-off, and upper and lower slope zones. At the second site, zonal distinctions were less well defined. The maximum number of species was reported for the upper slope zone (15-27 m deep) at the first site where twenty eight species were sampled in an area of 125 m². The mean number of species in this zone was 19.2 species per 25 m². A maximum of twenty species in 25 m² of sampled reef was reported for the second site at a depth of only 7 m. The fewest species were surveyed in shallow water near shore (< 2 m deep) where there were only 6-7 species per 25 m².

As in Jamaica, several coral species were distributed over a wide depth range in Curaçao (*e.g.*, *Siderastrea siderea* 1.5-42 m, *Agaricia agaricites* 1-42 m, *Porites astreoides* 0.5-34 m). However, Bak (1977) noted some degree of habitat specialization in corals like *Acropora palmata*, *Diploria clivosa*, and *Siderastrea radians* in shallow, turbulent water and *Agaricia lamarcki*, *A. grahamae*, *Madracis formosa*, and *Mycetophyllia reesi* in deep water. At the site where zonation patterns were well established, thirty six coral species occupied a mean of 3.2 zones per species. Considering only the most common corals at this site (occupying at least 5% of the area within any single quadrat), fourteen species were distributed over a mean of 4.6 zones per species.

In Puerto Rico, Loya (1976a) conducted quantitative surveys of corals over a depth range of 8-20 m in reef flat, slope and patch reef environments exposed to particularly heavy sedimentation and high turbidity. Along a total of 320 m of sampled reef, there were twenty two coral species (1145 colonies). The maximum diversity occurred in an upper fore-reef habitat where a total of twenty species were recorded and there was a mean of 13.2 species per 10-m line transect. Among four designated habitats (a flat reef habitat 8 m deep, upper fore-reef 11-17 m deep, lower fore-reef 18-20 m deep, and a patch reef 20 m deep), the corals occupied a mean of 2.9 habitats per species. Among the eight species with average abundance estimates over 2 colonies per 10 m, the mean habitat breadth was 3.2 habitats per species. *Agaricia agaricites*, *Siderastrea radians*, *Montastrea cavernosa*, and *M. annularis* occurred in all four habitats. *Acropora cervicornis* was restricted to only one (the 8-m site).

In the Virgin Islands, Rogers *et al.* (1984) conducted surveys in St. Croix along the walls of the Salt River submarine canyon using 64, 10-m line transects. They recorded a total of twenty six coral species at depths of 9, 18, 27 and 37 m. A maximum of eighteen species along 80 m of reef was recorded on the east wall of the canyon at a depth of 18 m. Fifteen species were found at 3-4 of these depths including all ten of the most common species, which occupied at least 5% of the live coral cover at any one of eight locations. Due to the abundance of sponges on these walls, total coral cover along the transects was only 5-24%. *Agaricia lamarcki* was the dominant coral at 27 and 37 m where it occupied 43.3-79.9% of the coral cover. *Agaricia agaricites*, *Madracis decactis*, and *Montastrea cavernosa* were dominant at 9 and 18 m (also see Rogers *et al.* 1983).

2.2.7 TEMPERATE CORAL COMMUNITIES

At higher latitudes outside the tropics, corals become increasingly more depauperate (Figure 2.2) as other animal and plant taxa become more important components of epibenthic communities and the capacity for corals to build reefs is lost (Johannes et al. 1983). For example, the single temperate coral species Oculina arbuscula strongly interacts with a variety of algae and a mix of vertebrate and invertebrate consumers in nearshore, hard-bottom communities along the Carolina coast of North America (Miller and Hay 1996). In these communities, competition with algae and latitudinal shifts in temperature, nutrients, and herbivory are likely factors contributing to limitations on coral recruitment, growth, and survivorship. Likewise, in less well-lit environments in this same region, the coral Astrangia astreiformis is a significant, yet minor component of encrusting communities which are competitively dominated by sponges, bryozoans, hydroids, and ascidians (Sutherland and Karlson 1977).

In the western North Atlantic, there is a decrease from forty nine scleractinian species in the Florida Keys and the Bahamas to twenty three species in Bermuda (Bright *et al.* 1984). On this northern most platform, most attention has been directed at the fauna to the north of the island on lagoonal reefs near North Rock and Three Hill Shoals. Wilson (1969) noted the presence of twenty two species on three shallow reefs (1-15 m deep) where most species were distributed throughout habitats on the reef face and top. *Madracis decactis* and three species of *Oculina* were restricted to habitats on the reef face and a deeper zone designated as a reef sediment slope. Dodge *et al.* (1982) used four different survey methods to find fifteen species at three shallow water sites (3-5 m deep). Within the 300-400 m^2 which were sampled at each site, corals only occupied 13-26% of the substrate. *Diploria strigosa* was the most common coral representing 27-50% of the total coral cover. Other abundant corals included *D. labyrinthiformis*, *Montastrea annularis*, *M. cavernosa*, and *Porites astreoides*.

Further south in Florida, Goldberg (1973) conducted quantitative surveys over a 50-m depth range off Boca Raton to find twenty seven coral species co-occurring with thirty nine gorgonian species. Gorgonians (particularly plexaurids) were numerically dominant not only on patch reefs (9 m deep), but also at depths of 16-50 m on an outer reef. In addition, Goldberg (1973) highlighted the abundance of several algae (*e.g.*, *Galaxaura obtusata*, *Dictyota bartayresii*, *Halimeda* spp., and *Udotea flabellum*) as well as some non-coral invertebrate taxa at this site (*e.g.*, the sponges *Xestospongia muta* and *Microciona juniperina*). As at most tropical locations, the number of coral species reached a maximum at intermediate depths. There were fifteen coral species in over 75 m^2 of sampled patch reef at 9 m, nineteen species in 10 m^2 at a depth of 16 m on the outer reef, fourteen species in 30 m^2 at 22-30 m, and only five species in 20 m^2 at 32-44 m. Some common corals were distributed throughout this depth range (*e.g.*, *Montastrea annularis* and *M. cavernosa*). Others had much more restricted ranges (*e.g.*, *Oculina diffusa* was found only on the patch reefs).

One of the most diverse, high-latitude coral assemblages is located in Japan at Miyake-jima (34° 05' N) in the Izu Islands 160 km south of Tokyo. Of the ninety five hermatypic scleractinian species known for this general region (Veron 1993), Tribble and Randall (1986) encountered fifty seven species along belt transects covering 445 m^2

at nine stations over a depth range of 3-27 m. These stations were located in shallow boulder fields and along a submarine lava flow where the percentage of substrate covered by corals was low (1.4-36.8%). A maximum of thirty seven species were sampled in 60 m^2 at the base of the lava flow (15 m deep). The most common species (*Favia speciosa*, *F. valenciennesi*, *Echinophyllia aspera*, *Acanthastrea echinata*, and *Goniastrea* spp.) were widely distributed across this range of depths and habitats and no common species was noted to have a restricted distribution. Although the authors emphasized the remarkable diversity of this high-latitude assemblage, this site is subjected to severe disturbances as well as a cool temperature regime which limits reef building by these corals.

At the other latitudinal extreme in the Pacific Ocean, the coral assemblage at Lord Howe Island occurs in one of the southern most coral reef communities in the world (Veron and Done 1979). Almost all of the sixty five hermatypic scleractinian corals known for this island (Veron 1986) also occur on the Great Barrier Reef. At Lord Howe Island, these corals occur as co-dominants with over 200 species of algae (Veron and Done 1979). The reef building activities of corals are restricted to the western side of the island at depths less than 15-20 m. In a semi-quantitative survey at approximately 100, 40-m^2 sites around the island, Veron and Done (1979) characterized the coral assemblages into three primary habitat types: 1) lagoonal (with passages and hole environments), 2) reef flat and outer slope, and 3) non-reefal assemblages. Algae and several generalist coral species dominated sites within the two reefal categories (*e.g.*, *Acropora horrida*, *A. palifera*, *A. hyacinthus*, *A.* sp., *Goniastrea favulus*, *Pocillopora damicornis*, *Porites lichen*, *Seriatopora hystrix*, and *Stylophora pistillata*). Dominant species at the non-reefal sites included *Acropora palifera* and *A. clathrata*. Non-reefal sites were exposed to strong wave action, yet almost all species known for Lord Howe Island occurred there.

When Veron and Done (1979) evaluated only the sixty six reefal sites sampled at Lord Howe Island, they found three principal assemblages characterized by differences in habitat, depth, and wave exposure: 1) exposed and/or shallow, 2) lagoonal, and 3) deep outer slope or passage sites. A fourth grouping included sites from a mixture of lagoonal, passage, and outer slope sites. There were only four dominant species representing more than 10% of the coral cover at a site in this analysis. *Acropora palifera* dominated sites at depths of 0-8 m on the outer reef slope and exposed reef flat. *Goniastrea favulus* dominated very shallow (0-1 m), exposed sites. *Porites lichen* dominated deeper sites (12-15 m) in passage and hole environments or more protected sites in shallow water (4-6 m). Lastly, *Pocillopora damicornis* dominated shallow (0-2.5 m) sites located mostly in the lagoonal habitat.

To the south of Lord Howe Island, epibenthic communities along the Australian coast become even more dominated by algae and non-coral invertebrates. For example in South Australia, there are only three hermatypic scleractinian corals (*Plesiastrea versipora*, *Coscinaraea mcneilli*, and *C. marshae*) in shallow water, over fifty ahermatypic species in cryptic habitats or deeper water (Shepherd and Veron 1982), and an exceedingly rich collection of sponges, hydroids, bryozoans, and ascidians (Butler 1991, Shepherd and Thomas 1982) representing the encrusting communities discussed in chapter 1. As noted previously, corals play a diminishing role in epibenthic communities as the physical and biological attributes of the environment begin to favor other taxa at high latitudes.

2.3 Diversity of non-coral taxa

Corals provide a structural and trophic base for many other taxa within coral communities. In order to convey just how diverse these taxa can be, I use two sources covering the coral reefs around Heron Island on the Great Barrier Reef (Mather and Bennett 1984) and Carrie Bow Cay in Belize (Rützler and Macintyre 1982). Kauffman and Fagerstrom (1993) used these same references to estimate total taxonomic diversity in living coral communities (Figure 2.1). I also use five additional sources covering the fishes (Randall *et al.* 1990) and scleractinian corals (Veron 1986) of the Great Barrier Reef and Coral Sea and the general biota of reefs in the western Atlantic Ocean (Chaplin and Scott 1972, Colin 1978, Kaplan 1982). Although these last three sources are not comprehensive taxonomic works identifying most of the described species for each taxon, they do provide an indication of the wide variety and relative diversity of some of the taxa in coral communities.

In the vicinity of Heron Island, Mather and Bennett (1984) specifically noted 175 species of scleractinian corals representing more than 40% of the total number of species known for the Great Barrier Reef (Table 2.1). They also covered several other higher taxonomic groups of aquatic organisms which, when compared with corals, collectively comprise between one and two orders of magnitude more species. Although many groups have not been well studied, several exceptionally rich taxa (*e.g.*, Pisces, Crustacea, Polychaeta, and Mollusca) are known for this southern region of the Great Barrier Reef. Likewise, coral communities in the western Atlantic harbor rich assemblages of fishes, arthropods, and other taxa (Table 2.1).

Intensive sampling of small areas of coral reef and even single coral heads have yielded amazingly high diversity estimates for non-coral species. For example, Grassle (1973) reported finding 103 infaunal polychaete species in a single head of *Pocillopora damicornis* sampled on the reef slope at Heron Island. Furthermore, he noted that this class of invertebrates represented approximately 67% of more than 2100 macro-invertebrates in the sample. Although polychaetes are well known for their diverse feeding modes, such high species richness at local spatial scales continues to provide an intriguing focus for ecological research (*e.g.*, Grassle and Grassle 1994).

The assemblage of crustaceans associated with *P. damicornis* is also noted for its diversity (see section 8.2.2 for regional comparisons). A single coral head may be inhabited by up to twenty five species. Some of these are obligatory symbionts with specialized feeding adaptations, while others are much more general in their habitat and trophic requirements (Abele 1979, 1984). Some of the symbiotic species (*e.g.*, alpheid shrimp and xanthid crabs) are also known to protect pocilloporid corals from predation by the sea star *Acanthaster planci* (Glynn 1976; see section 3.3.3). At Uva Island off the Pacific coast of Panama, a collection from 119 heads of *P. damicornis* contained 4724 individual decapod crustaceans representing fifty two different species (Abele 1984). The most common species was the xanthid *Trapezia corallina* which feeds only on coral mucous. At nearby Pearl Island, there were fifty seven decapod species in a smaller sample of 1107 individuals from 35 coral heads. The porcellanid crab *Pisidia magdalenensis* was the most abundant of these species. This generalist filter feeder occurs in a variety of different habitats (although it did not occur in the samples collected at Uva Island).

Besides the diverse assemblages associated with specific coral taxa, there are also rich collections of species which occur in structurally complex habitats found in coral communities. For example, in a study of the genus *Conus* conducted at eleven sites in the Maldive Islands and Chagos Archipelago, Kohn (1968) found up to seventeen sympatric species occurring together in shallow, subtidal habitats characterized by complex topography, eight species on smooth, intertidal, limestone benches, and only two on shallow, subtidal sand. There were a total of forty three species representing 66% of the regional fauna sampled at these sites. Analysis of the diet of these carnivorous gastropods did not support the hypothesis that there was relatively more trophic specialization in the more diverse assemblage. Instead, the most specialized diets were found among gastropods inhabiting the intertidal limestone bench.

In general, *Conus* spp. [and many other taxa of predatory gastropods (Taylor 1978, 1984)] are more specialized as predators than they are with respect to microhabitats (Kohn 1971), so species co-exist sympatrically within reef habitats as they exploit different food resources. In a study of intertidal bench habitats at Eniwetok Island in the Marshall Islands and One Tree Island on the Great Barrier Reef, Kohn and Leviten (1976) once again highlighted the importance of topographic complexity in promoting the species richness of predatory gastropods. Even though this habitat is topographically quite simple, depressions in the smooth horizontal limestone and algal turf acted as significant refuges from the physical stress associated with intertidal exposure. Both numbers of individuals and species of gastropods were enhanced by these substrate features.

As indicated in Table 2.1, the diversity of fishes in coral communities is very high. At quite local scales, as many as 150-200 fish species have been collected in individual rotenone samples taken on reefs in the central Indo-Pacific region (Goldman and Talbot 1976). Many fishes have evolved specialized feeding modes, are strongly influenced by the physical structure of coral reef habitats, and are involved in a number of species-specific associations (see reviews in Sale 1991a). Fish feeding activities have obvious direct effects on their food resources and also contribute to indirect effects such as the mediation of algal-coral interactions and the generation of sediment-based habitats (Choat and Bellwood 1991). Habitat-related associations appear to be strongest among smaller fish for whom shelter is very important. Larger fish are more mobile and more prone to travel beyond the limits of coral communities (Williams 1991). Important habitat attributes for smaller species include substrate type, topographic relief and water depth, yet larger fish which are typically found in deeper water are also known to selectively utilize sheltered sites.

TABLE 2.1 Number of species/genera/families (s/g/f) for various higher taxa of aquatic organisms found in coral communities. Estimates are derived from general field guides, multiple studies conducted at Heron Island and Carrier Bow Cay, and two comprehensive sources covering the fishes and scleractinian corals of the Great Barrier Reef. These estimates of taxonomic richness are provided only as a basis for relative comparisons among taxa.

Taxa	Heron Island[1]	Great Barrier Reef/Coral Sea	Carrie Bow Cay[4]	Caribbean Sea/ Florida/Bermuda /The Bahamas
Scleractinia	175 s	422 s[2]	42 s	52 s[6]
Algae	61 s		74 s	
Foraminiferida	28 s			
Porifera	23 s			45 s[6]
Hydrozoa	79 s		107 s	
Actiniaria				15 s[5]
Zoanthidea	8 s		4 s	8 s[5]
Antipatharia	2 g			5 s[6]
Alcyonaria	36 g		36 s	
Platyhelminthes	38 s			
Nemertea	6 s			
Polychaeta	11 f			14 f[5]
Oligochaeta	33 s			
Sipuncula	10 s		8 s	
Echiura	2 s			
Crustacea	40 f			>17 f[6]
	>300 s			
Stomatopoda				>60 s[5,6]
Caridea	200 s			
Anthuridea			14 s	
Pycnogonida			31 s	
Bryozoa	25 g			
Mollusca	96 f			>15 f[6]
Echinodermata	39 f			
	121 s			77 s[5]
Ophiuroidea	17 s		36 s	
Crinoidea	26 s		4 s	7 s[5]
Tunicata	200 s			
Pisces	109 f			89 f[7]
	>850 s	>1234 s[3]		>600 s[7]
Reptilia	8 s			

[1] Mather and Bennett (1984), [2] Veron (1986), [3] Randall *et al.* (1990), [4] Rützler and Macintyre (1982), [5] Kaplan (1982), [6] Colin (1978), [7] Chaplin and Scott (1972).

Numerous studies have focused on how the number of fish species varies across environmental gradients on coral reefs. Given the wide variety of families and trophic guilds represented among the fishes, many of these studies have focused on one or a small number of such groups rather than on the entire fish fauna. Furthermore, the scale of these analyses spans individual reefs, local regions, and oceanic basins. Consequently, I present below just a few examples of how fish diversity on individual reefs varies along environmental gradients.

Bouchon-Navaro (1981) surveyed the distribution of butterflyfishes (Chaetodontidae) at Moorea in French Polynesia over the same nine stations mentioned earlier for coral surveys by Bouchon (1985). Three stations were located on a shallow fringing reef (0.5-1.0 m deep), two on a barrier reef flat (1.5-2.0 m deep), and four on an outer reef slope (3-30 m deep). At each station, counts of butterflyfishes were conducted within a pair of 5 X 200-m, belt transects situated parallel to shore (except at the deepest station where only a single transect was used). Overall, there were a total of nineteen species encountered (878 fish) within 17,000 m^2 of reef habitat. Quantitative estimates of the mean number of species per 1000 m^2 include 2, 5, and 14 species on the shallow fringing reef, 11 and 12 species on the reef flat, and 9, 14, 11, and 9 species on the outer slope. Among all species, there was a mean of 5.4 stations per species. Among the thirteen most common species represented by at least twenty individuals in these surveys, the mean was 6.1 stations per species. Six of these common species were quite broad in their distributions occurring at 7-8 stations. Although Bouchon-Navaro (1981) noted several habitat preferences and dominance patterns linking particular species with a specific station or reef zone, only one of the common species (*Chaetodon quadrimaculatus*) was restricted to fewer than four stations and a single reef zone. This species was found only at the two shallowest stations on the outer slope.

Green (1996) surveyed the distribution and abundance of wrasses (Labridae) at three sites along the semi-exposed northeastern coast of Lizard Island on the northern Great Barrier Reef. Visual census techniques within 3-m wide belt transects were used to quantify the number of wrasses per 750 m^2 in six habitat zones: inner reef flat (0.0-1.5 m deep), outer reef flat (0.0-1.5 m deep), reef crest (0.0-3.0 m deep), reef slope (2.0-20.0 m deep), reef base (5.0-20.0 m deep), and sand flat (6.0-20.0 m deep). A total of sixty-four labrid species were recorded in five surveys conducted in 1990-1993. Detailed information from one such survey in February 1992 indicates that among thirty eight recorded species (1804 individual fish), seven of the eight most abundant species exhibited strong, depth-related habitat associations. "*Halichoeres* spp., *Stethojulis bandanensis*, *Thalassoma hardwicke*, and *T. jansenii* were most abundant in the shallow habitats, while *Coris schroederi*, *H. melanurus*, and *T. lunare* were most abundant in the deep habitat zones". Furthermore, Green (1996) concluded that the entire assemblage of wrasses within each habitat zone was unique. "Only one species, *Labroides dimidiatus*, occurred in similar densities in all habitat zones". However, when one combines the data from all five surveys, most of the abundant species noted above occurred in most habitats. There was a mean of 4.8 habitats per species. Some of this habitat breadth can be attributed to ontogenetic shifts in habitat use as juveniles grow to adult size. In particular, Green (1996) noted these shifts in *L. dimidiatus* whose young recruits were more common in reef slope and base habitats. Adults were more common in shallower habitats. Likewise, juveniles of *T. lunare* moved from reef slope to deeper habitats as adults.

Williams and Hatcher (1983) examined the total fish fauna on an inshore (Pandora), mid-shelf (Rib), and outer shelf (Myrmidon) reef in the central region of the Great Barrier Reef. On each reef, they used five explosive charges to sample approximately 750 m^2 of reef slope habitat at depths of 5 and 9 m. Following the coral assemblage designations in Done (1982), the reef at Pandora was a *Galaxea* assemblage (both depths), at Rib an

Acropora tabulate/branching assemblage (both depths), and at Myrmidon an *Acropora palifera/A. humilis/A. hyacinthus* assemblage at 5 m and an *Acropora palifera/Porites* assemblage at 9 m. Overall, these quantitative samples contained 28,537 individuals representing 45 families and 323 species. Just as Done (1982) had found for corals, the greatest number of species occurred on a mid-shelf reef. The mean number of species per sample was 38, 70, and 56 species for collections at Pandora, Rib, and Myrmidon, respectively. Williams and Hatcher (1983) attributed this pattern to the high habitat heterogeneity on mid-shelf reefs, but also considered inshore-offshore differences in the aquatic environment, supply of larvae, and reef size as contributing factors. Furthermore, they noted that the general pattern exhibited by all fishes did not represent the distribution of species among reefs for several families and trophic guilds. Five of the nine most common families in these samples exhibited exceptional distributional patterns as did six of fourteen trophic guilds (Table 2.2). They emphasized that their results indicate "major changes in community structure in the same reef habitat at a single latitude within one reef system". Over this environmental gradient, large differences in the number of species and trophic structure of fish assemblages were observed. Such variability at this scale warrants caution in interpreting community differences among reefs from widely separated regions (Williams and Hatcher 1983)!

Russ (1984a,b) surveyed the distribution and abundance of three families of fishes [Acanthuridae (surgeonfishes), Scaridae (parrotfishes), and Siganidae (rabbitfishes)] in this same central region of the Great Barrier Reef to highlight both cross-shelf and among-zone variation in these primarily herbivorous taxa. Censuses were conducted at three, relatively small, inshore reefs (Pandora, Phillips, and Lorne) at depths of 3-8 m. At three mid-shelf (Rib, John Brewer, and Lodestone) and three outer shelf (Myrmidon, Dip, and Bowl) reefs, censuses were conducted within windward reef slope (12-15 m deep), windward reef crest (2-3 m deep), reef flat (2-3 m deep), lagoonal (3-7 m deep), and back reef (< 12 m deep) zones. Within each zone or inshore reef, four replicate, 10-m wide belt transects covering 3000-4000 m^2 each were surveyed.

Using cluster analysis on the abundance data for fifty one species, Russ (1984a) determined that the largest difference was between the fish assemblages on inshore reefs with only twelve recorded species and those on mid- and outer reefs with forty six and forty nine species, respectively.

The second major distinction generated by this analysis was that between the deep (reef slope and back reef) and shallow (reef crest, flat, and lagoonal) zones. Three of five trophic guilds (large croppers, small croppers, and scrapers) representing thirty eight species were more abundant in the shallow zones (Russ 1984b). However, this general pattern did not hold true for individual families or for the other two trophic guilds (sand and fine sediment suckers) and there was considerable variation among replicate reefs within shelf locations. For example, surgeonfishes and parrotfishes were more abundant in reef crest and lagoonal zones than on reef flats and slopes. Rabbitfishes were more abundant in lagoonal and back reef zones than elsewhere. Likewise, fine sediment suckers were more abundant in windward and leeward zones regardless of depth, and sand suckers more common in lagoonal and back reef zones. Thus, Russ (1984a,b) also emphasized the high degree of variability exhibited by these three families within and among reefs in this region.

TABLE 2.2 Number of species in the nine most common families and in fourteen trophic guilds of fish sampled in the central Great Barrier Reef by Williams and Hatcher (1983). The trophic guilds are categorized into four broad feeding groups which are cross-referenced using Sorokin (1993: Table 6.5).

Family	Pandora Reef	Rib Reef	Myrmidon Reef	Total
Pomacentridae[1,2,3,4]	17	43	30	57
Labridae[1,3]	17	29	27	46
Chaetodontidae[1,3]	9	15	20	28
Gobiidae[1,2,3,4]	11	11	0	21
Serranidae[2,3,4]	11	14	7	22
Apogonidae[2,3,4]	11	10	5	17
Holocentridae[2,3]	2	6	13	16
Scaridae[1,3]	3	11	9	14
Acanthuridae[1,2,3]	1	3	9	10
Miscellaneous	28	59	43	92
Total	110	201	163	323
Trophic Guilds				
Algal grazers[1]	13	30	28	48
Suckers	0	1	3	3
Small croppers	6	15	12	21
Large croppers	4	4	5	11
Scrapers	3	10	8	13
Planktivores[2]	14	47	41	64
Zooplankton	5	24	27	38
Omnivores	8	20	9	21
Gelatinous	1	3	4	4
Algal	0	0	1	1
Invertebrate feeders[3]	69	103	89	185
Motile	56	77	64	146
Sessile	3	5	7	10
Sessile omnivores	4	10	9	14
Coral feeders	6	11	9	15
Piscivores[4]	14	21	5	26
Confirmed	2	8	1	11
Facultative	12	13	4	15
Total	110	210	163	323

Because the butterflyfishes (Chaetodontidae) may best illustrate among-zone distributional variation among fishes on individual reefs, Williams (1991) compared their distribution and abundance on three reefs spanning much of the Indo-Pacific faunal province. He used data from quantitative surveys conducted at One Tree Island in the southern Great Barrier Reef (Fowler 1990), Moorea in French Polynesia (Bouchon Navaro 1981), and Aqaba in the Red Sea (Bouchon-Navaro 1980). In comparing the habitat breadth across reef zones at these three locations, Williams (1991) found that species with restricted or quite broad distributions at one location were not necessarily distributed in a similar fashion at other locations (Table 2.3). At One Tree Island and Aqaba, the most abundant species were widely distributed across zones and most rare species were quite restricted in their distributions. At Moorea, however, there was no clear relationship between distribution and abundance. Six species with restricted distributions at One Tree Island exhibited much broader distributions at Moorea and only four of thirteen species occurring at both locations were similarly distributed (Williams 1991). Thus there is considerable intraspecific variation in habitat breadth among reefs when evaluated at this large geographic scale. For further consideration of the diversity patterns of fish assemblages at larger scales, I refer the reader to Sale (1980, 1991b) and section 8.2.1.

TABLE 2.3 Distribution across reef zones and mean abundance of butterflyfishes at One Tree Island (from Fowler 1990), Moorea (from Bouchon-Navaro 1981), and Aqaba (from Bouchon-Navaro 1980). The index W is a weighted measure of habitat breadth calculated as described by Clark (1977); $W_{max} = 1.0$. Abundance data are given as mean densities per zone (analysis from Williams 1991).

	One Tree Island		Moorea		Aqaba	
	W	Mean Density	W	Mean Density	W	Mean Density
Chaetodon rainfordi	0.60	62.1				
C. plebius	0.88	16.7				
Chelmon rostratus	0.47	15.7				
Chaetodon flavirostris	0.83	11.9				
C. trifasciatus	0.40	10.4	0.42	6.1		
C. melannotus	0.53	10.3			0.26	1.0
C. auriga	0.33	5.7	0.23	1.2	0.07	0.4
C. lineolatus	0.48	4.1				
C. trifascialis	0.10	1.3	0.10	3.4		
C. aureofasciatus	0.34	0.9				
C. ulietensis	0.06	0.9	0.10	0.8		
C. speculum	0.54	0.7				
C. ephippium	0.08	0.6	0.27	0.4		
C. citrinellus	0.10	0.6	0.14	5.6		
C. baronessa	0.07	0.5				
C. vagabundus	0.02	0.4	0.53	2.3		
C. pelewensis	0.05	0.3	0.17	6.1		
C. lunula	0.02	0.3	0.49	2.2		
C. bennetti	0.31	0.2	0.13	0.0		
C. ornatissimus	0.05	0.2	0.71	3.7		
Forcipiger spp.	0.05	0.2	0.33	3.7		
C. unimaculatus	0.05	0.2	0.52	3.3		
C. reticulatus			0.33	8.7		
C. quadrimaculatus			0.06	1.1		
C. pauscifasciatus					0.83	-
C. austriacus					0.76	-
C. fasciatus					0.22	-

2.4 Overview

Coral communities are extremely diverse. The most species-rich areas of the Indo-Pacific include individual reefs with more than 200 hermatypic scleractinian coral species and approximately 1-2 orders of magnitude more described species of other animal taxa. While local diversity patterns vary with several environmental factors, much of this variation occurs along depth and habitat gradients. Maximum coral diversity is typically found at intermediate depths (15-30 m) in fore-reef environments. Non-coral taxa can exhibit similar patterns, but they may also be strongly influenced by species-specific associations, structural complexity of the environment, and geographic variation in a number of other environmental factors.

Although many species are very widely distributed geographically as well as locally across depth and habitat gradients, there is considerable variation in species abundance patterns at several spatial scales. At the local scale, some species can be very abundant as they dominate particular zones or habitats within much wider distributional limits. There are also species in coral communities exhibiting quite restricted distributional limits (*e.g.*, narrow habitat breadth in *Montipora* spp. in the Philippines). However, overall estimates of habitat breadth among corals and fishes suggest that most species do not have narrow distributional limits nor is habitat specialization particularly strong. There is also considerable variation in species composition and abundance patterns among reefs within regions as well as across quite large spatial scales. Among broadly distributed species, significant variation in habitat associations occurs across their distributional ranges.

3 STABILITY

3.1 What is stability?

Stability is "the ability of a system to return to an equilibrium state after a temporary disturbance" (Holling 1973). Hence, at the simplest level, the study of the stability of ecological communities involves the description of the equilibrium condition and the response of the community to disturbances. This approach has strong theoretical foundations backed by the mathematical analysis of convergent dynamics around locally stable equilibria. An equilibrium is said to be locally stable "if the system returns to the equilibrium after being displaced... by a sufficiently small but otherwise arbitrary amount" (Yodzis 1989). Yet natural communities are sometimes exposed to quite large disturbances representing major displacements/perturbations. A locally stable equilibrium is said to be globally stable, only if the system converges to this point after being displaced to any other condition no matter how large the displacement (Lewontin 1969). If displacements are followed by convergence to multiple, locally stable equilibria, a globally stable equilibrium does not exist (Lewontin 1969, Sutherland 1974). In such cases, the eventual response of a community to disturbance will depend on the initial state of the community, the size and direction of the disturbance (*i.e.*, the community displacement), and what are called the basins (or domains) of attraction about the local equilibria. Knowlton (1992) has argued that there are several attributes of coral communities which should favor the existence of multiple equilibria. If true, major disturbances should commonly result in phase shifts in community structure among local equilibria (see section 3.3 and chapter 7).

The evaluation of stability by ecologists has generally tended to extend beyond consideration of local and global equilibria to include the interrelated notions of persistence, resilience, and resistance (see Holling 1973, Sutherland 1981, Connell and Sousa 1983, Dayton *et al.* 1984). Persistence merely refers to the existence of a community in some state over some period of time. Highly persistent communities may or may not be highly stable depending on the history of potential perturbations [*e.g.*, species invasions and larval recruitment (Sutherland 1981) or loss of a keystone species (see below)]. Resilience refers to the recovery of a community towards some local equilibrium condition following a perturbation which actually changes community structure. In theory, this recovery process is highly dependent on the size and shape of the basin of attraction (Holling 1973). Highly resilient communities recover from perturbations rapidly yet they may be easily perturbed or, at the other extreme, quite resistant to perturbations. Resistance refers to the inertial properties of a community which tend to minimize the impact of perturbing forces on community structure. Highly resistant communities should be quite persistent, yet may or may not be resilient once community structure has been disrupted.

In order to determine if a natural community is stable and whether there is a single global equilibrium or multiple local equilibria, one must make several decisions. One must decide which community variables to measure (*e.g.*, the number of species and trophic levels, the density of selected species, patch size and distributional limits, biomass, productivity, etc.). One must also choose an appropriate perturbation or set of perturbations. The scale at which perturbations operate should match the scale at which one measures the community. Lastly, one must select the appropriate stability criteria (Connell and Sousa 1983, Sutherland 1981, 1990b). By restricting their stability criterion to resilience, Connell and Sousa (1983) concluded that there was little evidence to support the view that multiple equilibria exist in natural communities. They noted that many studies appear to have been conducted over too short a period of time relative to the scale of disturbances and species generation times. Sutherland (1990b), on the other hand, argued that "there is abundant evidence for the existence of multiple stable points" in that different species assemblages in the same community are resistant to invasion by other species. "Understanding how assemblages resist change is equally as important as understanding how they respond to perturbations causing significant change" (Sutherland 1990b).

3.2 Are stability and diversity related?

The study of the stability of ecological communities dates back to early investigations of succession and the notion of the stable climax [*e.g.*, Tansley (1935), Clements (1936)] and later theoretical treatments of the relationship between community stability and diversity (*e.g.*, MacArthur 1955, May 1973). The former tend to focus on the self replacement of dominant species in the community and the resistance of the community to environmental disturbances (see chapter 4 for further discussion of succession). The latter emphasize the inherent structure of the community and trophic interactions among species. MacArthur (1955) proposed that community stability increases as the number of trophic links between species in a food web increases. Hence, quite simple communities with relatively few trophic links should be less stable than highly diverse communities with many trophic links. Alternative energy pathways in complex food webs were proposed to favor more constant population sizes with reduced fluctuations thus promoting community resistance. In the absence of alternative trophic links, fluctuations in the size of a population at one trophic level were predicted to cause relatively large fluctuations at other trophic levels thus destabilizing the community. As a consequence of formally evaluating the mathematical stability of several types of model communities, May (1973) argued that there is no general theoretical basis to support the hypothesis that communities with many species should be more stable than less speciose communities. Furthermore, these models indicated that the number of trophic links (*i.e.*, trophic complexity) in a community "is no guide to community stability". Although more stable model communities can be generated by introducing time lags into the effects of trophic interactions (May 1973) and greater detail in the nature of these interactions (De Angelis 1975), the validity of these notions should be evaluated using natural communities (McNaughton 1977).

The first direct, empirical test of the relationship between diversity and stability used nutrient enrichment as a perturbation to different aged, successional old fields in central New York (Hurd *et al.* 1971, Hurd and Wolf 1974, Mellinger and McNaughton 1975). The species diversity and net productivity of primary producers and arthropod consumers were monitored in control and treatment plots in six- and seventeen-year old fields. Stability was evaluated in terms of the resistance of three trophic levels (producers, herbivores, and carnivores) to modification by enrichment. These authors found support for a relationship between stability and diversity, but only when viewing the responses at each trophic level separately. The plant assemblage in the younger fields was less diverse and less resistant to the perturbation than that in the older fields. Among the consumer trophic levels, stability was not positively related to either high diversity or to increasing successional age. Increased diversity among the plants and herbivores resulted in decreased stability at the next higher trophic level. Thus, MacArthur's prediction regarding trophic complexity and stability was not supported by this evidence.

McNaughton (1977) further explored the relationship between diversity and stability in studies of the effects of grazing and variable rainfall on plant assemblages on the Serengeti Plains. Although more diverse plant assemblages experienced a greater change in diversity in response to grazing by the African buffalo (H' fell from 1.783 to 1.302 as opposed to no significant change in less diverse assemblages), green plant biomass was less susceptible (falling only 11.3% as opposed to 69.3%). The green plant biomass of more diverse assemblages was also less susceptible to variation in rainfall. The coefficient of variation in green plant biomass was significantly lower in diverse assemblages subjected to periodic showers (Figure 3.1).

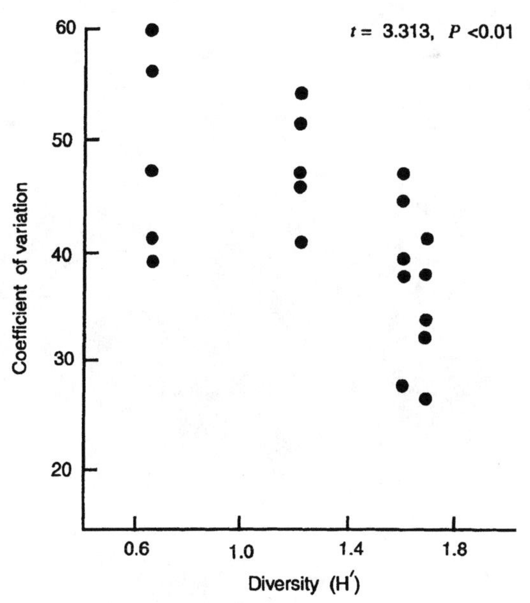

Figure 3.1 The instability of green plant biomass (given as the coefficient of variation) of all species in four adjacent stands in the Serengeti National Park. The data were generated from five replicate plots in each stand sampled at weekly intervals over a 5-wk period and are presented here as a function of the maximum species diversity (H') observed per stand over the study period (redrawn after McNaughton 1977). © 1977 by The University of Chicago.

This relative resistance to change resulted from the differential response of species to intermittent rainfall and drought. More diverse assemblages included species with a wide range of adaptations to variable rainfall. Just as was found in central New York, these studies confirmed that the biomass of diverse plant assemblages is more resistant to perturbations than that of less diverse assemblages. Since these diversity and stability attributes characterize a single trophic level, MacArthur's notion of trophic complexity does not appear to apply here. Instead, the more general notion that "species diversity stabilizes ecosystem functional properties" is supported (see review in Johnson *et al.* 1996). Additional studies in the Serengeti ecosystem have verified the relationship between "vegetation stability" and plant species diversity (McNaughton 1985). However, McNaughton (1985) highlighted the importance of coevolution in this "grazing ecosystem" rather than emphasizing some theoretical notion of a diversity-stability relationship. Strong plant-herbivore interactions as well as facilitative and competitive interactions among herbivores have contributed to the organization of a trophic web "which gives the ecosystem coherence and continuity in space and time... Understanding trophic processes ultimately will depend upon understanding the role of evolution in shaping them" (McNaughton 1985).

Recently published work on diversity-stability relationships in a North American grassland ecosystem solidly supports the notions that species-rich plant assemblages tend to enhance ecosystem functional properties (*i.e.*, primary productivity and nutrient utilization), stabilize total plant biomass, and promote drought resistance (Tilman 1996, Tilman *et al.* 1996). However, the biomass of individual plant species was found to be more variable (*i.e.*, unstable) in plots with more plant species (Tilman 1996, Figure 3.2).

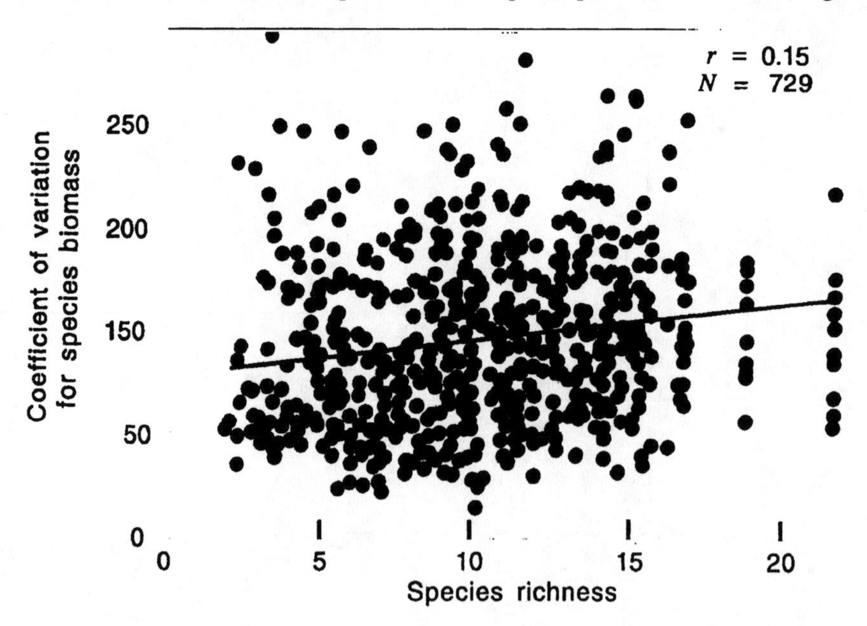

Figure 3.2 The interannual coefficient of variation in plant biomass of individual species as a function of the mean species richness per plot from 1984 through 1994 (redrawn after Tilman 1996).

Tilman (1996) explained this seemingly contradictory result by invoking the differential response of species to disturbance and the competitive release experienced by disturbance-resistant species. During two of the eleven years this ecosystem was under study, the third worst drought in 150 years drastically reduced total plant biomass. However, among the twenty seven most abundant species, only one (*Solidago graminifolia*) was severely reduced to less than 5% of its previous biomass. The biomass of eleven other species was reduced to 10-50% of previous levels. Remarkably, four species (*Asceplias syriaca*, *A. tuberosa*, *Andropogon gerardi*, and *Sorghastrum nutans*) actually increased 100-700% during the drought. This interplay of disturbance and competition acts to stabilize total plant biomass in species-rich assemblages and again emphasizes the role of evolution in shaping the individual response of species to variable environments.

3.3 Keystone species

While considering the potential influence of the complexity of food webs on community stability, Paine (1969) noted that some predatory species can greatly modify the structure and persistence of a community. At the same time, other species of similar trophic status have no comparable impact. The former class of predators are called keystone species. By implication, such species serve a major stabilizing role in communities. In rocky intertidal communities, for example, the sea star *Pisaster ochraceus* strongly controls prey population densities, species composition, and zonation patterns (see section 1.4.1). Numerous other predatory species have much weaker effects or none at all. Paine (1969) also considered predatory gastropods (the tritons *Charonia* spp.) as a potential keystone species in coral reef communities. These predators feed preferentially on the sea star *Acanthaster planci* and have become quite rare in many coral communities in recent years. Speculation that recent plagues of *Acanthaster* have resulted from the loss of the keystone species have been widely circulated. The strong impact of keystone species on community structure and stability and the much weaker influence of other predators led Paine to conclude that trophic complexity contributes minimally to community stability. Once again, MacArthur's hypothesis was not supported and attention to specific biological interactions involving keystone species was emphasized.

Keystone species (*e.g.*, certain predatory fish, mammals, echinoderms, gastropods, and crustaceans) have now been identified in a wide variety of aquatic communities (see reviews in Menge *et al.* 1994, Power *et al.* 1996), yet not all communities are controlled by them. Sometimes predation is important in organizing community structure, but it involves multiple species of predators each having relatively weak effects on lower trophic levels. Menge *et al.* (1994) referred to this situation as "diffuse predation" citing earlier work by Menge and Lubchenco (1981) and Robles and Robb (1993). They recommended that "any general synthesis of the conditions under which predation controls community structure should consider both causes of variation in interaction strength and in predation mode (keystone or diffuse)". Although some communities may not be organized by predation at all, the structure of most aquatic communities appears to be controlled by either keystone or diffuse predation (Menge *et al.* 1994). In coral communities, complex trophic structure involving diffuse herbivory and predation may be important (see sections 3.3.1 and 6.4).

In recognition of the fact that some nonpredatory species in ecological communities have large effects on community structure which are disproportional to their abundance, it has been recommended that the notion of the keystone species be expanded to include them (Power *et al.* 1996). Such large effects can be quantitatively estimated using a standardized measure of the importance of a species. Community importance "is the change in a community or ecosystem trait per unit change in the abundance of the species" (Power *et al.* 1996). Thus a variety of species with diverse effects on a community may be considered keystone species. Examples of nonpredatory keystone species include 1) plants which flower or produce fruit at critical times of the year when resources are severely limited, 2) soil cyanobacteria and endolithic algae which fix nitrogen and support snail populations which, in turn, breakdown rock and create soil in desert environments, and 3) beavers, gophers, leaf cutter ants, and badgers as they physically modify the soil and/or flow of water through their habitat. Regardless of the trophic status or community function of a keystone species, the central issue is the large stabilizing effect it has on the community. Major shifts in the abundance of keystone species, whether by experimental removals/additions or natural processes, should destabilize community structure.

3.3.1 KEYSTONE SPECIES IN CORAL COMMUNITIES: *DIADEMA ANTILLARUM*

One of the best documented examples of a keystone species in a coral community is the sea urchin *Diadema antillarum.* It is widely distributed throughout the tropics in the Caribbean Sea and Atlantic Ocean. It is primarily herbivorous in that it feeds mostly on algae and marine grasses (Ogden *et al.* 1973), but is also known to eat a variety of benthic invertebrates [*e.g.*, corals (Bak and van Eys 1975, Sammarco 1980, Carpenter 1981), hydrocorals (Bak and van Eys 1975), encrusting gorgonians (Karlson 1983), and bryozoans (Jackson and Kaufmann 1987)]. In coral communities with many herbivorous species, this urchin is unlikely to act as a keystone species (Hay 1984). However, in coral communities which have been decimated by overfishing (*e.g.*, Jamaica, Haiti, and the Virgin Islands), trophic structure has been simplified with a loss of predatory and herbivorous species resulting in a shift from diffuse herbivory in these communities to a situation in which *D. antillarum* became the keystone herbivore. This urchin's role in stabilizing overfished coral communities has become evident following the region-wide mass mortality of *D. antillarum* in 1982-1984 (Lessios 1984, 1988, Hughes 1994a). These events provide an excellent historical example supporting the trophic complexity-stability hypothesis (MacArthur 1955). Although coral communities persisted as overfishing reduced trophic complexity, they became more vulnerable to perturbations.

Prior to the mass mortality, several studies (conducted primarily in Jamaica and the Virgin Islands) documented the effects of grazing by *D. antillarum* in coral communities using experimental manipulations of urchin density. For example, Sammarco *et al.* (1974) found that the removal of urchins resulted in shifts in algal species composition, increased algal biomass and species richness, and reduced algal species evenness. Elevated urchin densities reduced algal cover/biomass as well as the density and diversity of coral recruits (Sammarco 1980, 1982). Carpenter (1981) even noted that algal species richness and primary production were maximized at intermediate densities of *D. antillarum.* This result implicated the keystone status of this urchin in controlling the structure of algal assemblages in much the same manner as *Pisaster ochraceus* controls the structure of rocky intertidal communities (Paine 1966, 1969). Severe grazing pressure at high densities clearly reduced all measures of algal abundance and diversity. In the absence of urchins, macroalgae (*e.g.*, *Halimeda incrassata*, *Padina santae-crucis*, *Laurencia obtusa*) became more common (even though algal turf species were the preferred food of *D. antillarum*) and algal species richness was relatively low. This suggests some level of competitive superiority of macroalgal species over turf species. The higher algal species richness observed at intermediate urchin densities can clearly be attributed to increased herbivory and possibly to reduced interspecific competition (Carpenter 1981). In a later set of experiments, Carpenter (1986) again noted the competitive superiority of macroalgal species in the absence of herbivory and the important role *D. antillarum* plays in controlling algal biomass, productivity, and the relative abundances of algal turf, macroalgae, and encrusting coralline algae. As in many other epibenthic communities, encrusting coralline algae are favored when grazing by herbivores is intense (see section 1.4).

The mass mortality of *D. antillarum* affected populations distributed throughout its range killing 85.44-99.99% of the sea urchins at study sites located in Panama (Lessios 1988), Curaçao (Bak *et al.* 1984), Jamaica (Hughes *et al.* 1985), the Virgin Islands (Levitan 1988, Carpenter 1990), Barbados (Hunte *et al.* 1986), and elsewhere. The immediate consequences of this epidemic included increases in algal abundance especially on overfished reefs. In Jamaica, substantial increases in the abundance of macroalgae in accordance with earlier urchin removal studies were reported by Hughes *et al.* (1987) and Morrison (1988). By September of 1986, back and fore-reef sites at multiple locations near Discovery Bay, Jamaica, all exhibited high algal cover (typically 85-95%) and biomass estimates of 110-1730 g dry weight per m^2 (Hughes *et al.* 1987). At five sites in the shallow water of Lameshur Bay on Saint John, U.S. Virgin Islands, Levitan (1988) documented initial algal biomass increases of 3000% to approximately 180 g dry weight per m^2 by 1984, but then a reduction to approximately 40 g dry weight per m^2 by 1987.

Speculation regarding the cause of this latter reduction in algal biomass has included compensatory responses of other herbivorous species and increased metabolic needs of the large remaining urchins (Lessios 1988). The former notion is supported by evidence from St. Croix where Carpenter (1988) reported nearly a 300% increase in the algal grazing intensity by herbivorous fishes [most notably acanthurids (surgeonfishes), pomacentrids (damselfishes), and scarids (parrotfishes)] only two months after the mass mortality event. On the shallow fore-reef near Discovery Bay, Jamaica, Morrison (1988) found that the grazing intensity by herbivorous fishes twelve months after the mass mortality increased 180% over pre-mass mortality levels. This increase includes a spectacular rise of 1300% by the redband parrotfish (*Sparisoma aurofrenatum*) and a rise of 140% by the predominant herbivorous fish, the striped parrotfish (*Scarus iserti*).

The notion that algal biomass decreased in 1984-1987 because of increased sea urchin herbivory (even though densities remained quite low) is supported by the evidence from Lameshur Bay that urchin biomass dramatically increased by over 4000% (from 1.02 to 41.99 g live weight per m^2) during in this post-mass mortality interval (Levitan 1988). Urchin test diameters grew at rates of 11-14 mm per yr as maximum diameters approached 100 mm and the median individual weight increased from 20 to 200 g. Although both urchins and herbivorous fishes may have contributed to the reduced algal biomass observed in Lameshur Bay, such reductions were not evident on the overfished reefs of St. Croix (Carpenter 1990) or Jamaica (Hughes *et al.* 1987).

In fact, following the mass mortality event in Jamaica, the abundance of algae continued to increase into the 1990s as the abundance of corals continued to decline (Hughes 1994a, 1996). Phototransects, taken at depths of 7, 10, and 15-20 m at Rio Bueno and 35 m at Pinnacle I off Discovery Bay, clearly indicate that the percent cover of macroalgae at all depths increased from 1983 (2-34%) through 1990-1993 (60-94%). The most rapid increases occurred at 7 m (Figure 3.3). Coral abundance dropped from approximately 40-75% in 1977 to 20-40% in 1981 due to the devastation caused by Hurricane Allen (see Woodley *et al.* 1981). Following the mass mortality and increased dominance of macroalgae at these sites, coral abundance dropped even farther to 4-18% by 1993. This general reduction in coral abundance from the 1970s to the 1990s was further documented using 10-m sites sampled at multiple locations around the island. In the 1970s, coral cover was approximately 24% in the Port Royal Cays near Kingston and 40-80% at eight sites along the north coast from Negril in the west to Port Antonio in the east (Hughes 1994a). By 1993, coral abundance ranged from 4-19% at these sites.

Hughes (1994a) attributed the dramatic shift from a coral-dominated benthic assemblage to one dominated by macroalgae to a combination of natural and human disturbances to Jamaican coral reefs spanning several decades (*i.e.*, overfishing, Hurricane Allen, and the mass mortality of *D. antillarum*). Citing earlier work by Munro (1969, 1983), Hughes noted that 1) fish populations had already been greatly reduced by the late 1960s, 2) estimates of fishing effort in 1973 were far in excess of sustainable yields, 3) the catch per unit effort decreased by 50% from 1970-1985 despite increased fishing effort, and 4) current fishing is heavily reliant on small, pre-reproductive individuals. Munro (1983) also indicated that by the early 1980s, "negligible numbers of fishes survive long enough to spawn in the most heavily exploited areas" like those along the north coast of Jamaica. This intensive level of fishing greatly reduced or eliminated both predators of *D. antillarum* and herbivorous competitors. Thus "the unusually high abundance of *D. antillarum* on overfished reefs such as Jamaica's was almost certainly a result of the over-exploitation of reef fisheries" (Hughes 1994a), a position supported earlier by Hay (1984). This situation shifted *D. antillarum* to a major keystone species status not anticipated by those of us working on these reefs at that time. As a consequence of the 1983-1984 mass mortality event, these communities moved towards algal domination and "the classic zonation patterns of Jamaican reefs, described by Goreau and colleagues just two or three decades ago, no longer exist" (Hughes 1994a) (see section 2.2.6).

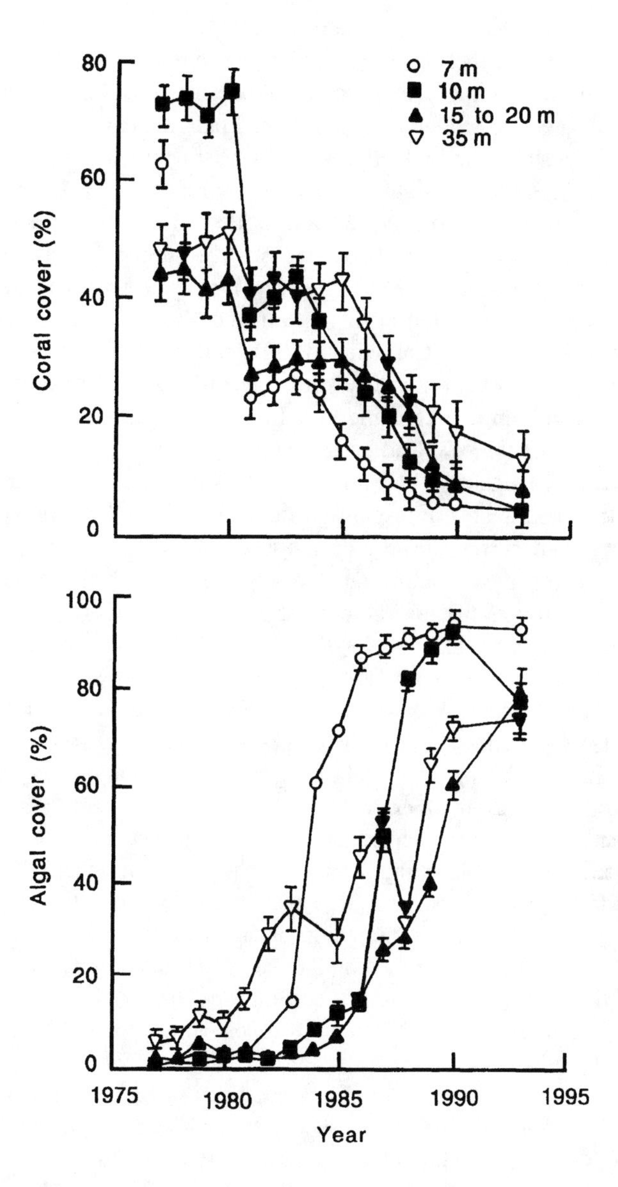

Figure 3.3 The percent cover of macroalgae and corals in 10-20, 1-m2 quadrats sampled near Discovery Bay. The sites were located at Rio Bueno (7, 10, and 15-20 m deep) and at Pinnacle I (35 m deep) off Discovery Bay (redrawn with permission from Hughes, T.P. 1994a. Catastrophes, phase shifts, and large-scale degradation of a Caribbean coral reef. Science 265, 1547-1551). Copyright 1994 American Association for the Advancement of Science.

The assessment that the algal domination of Jamaican coral reefs should be attributed primarily to overfishing, Hurricane Allen, and the mass mortality event has been criticized for its failure to consider alternative explanations. These alternatives include "nutrification by increased runoff, sewage discharge, and fertilizer use" (Hodgson 1994). Additional criticism was leveled at the assessment because it did not support monitoring programs and management strategies focused on linkages between the land and sea (Hodgson 1994). The statement by Hodgson (1994) that "nutrient enrichment is usually required to support high rates of algal productivity on reefs" was countered by Hughes (1994b) by citing evidence from St. Croix. Following the mass mortality there, the shift favoring macroalgae was associated with a short-term reduction in algal primary productivity (Carpenter 1988). However, long-term increases in the primary productivity per unit area occurred at this site in the Virgin Islands as algal biomass increased 300-440% by 1986 (Carpenter 1990). Much of this algal biomass was comprised of herbivore-resistant species thus leading to increased algal detritus, reduced energy transfer through "grazing-based food webs" and the export of "organic matter to adjacent communities" (Carpenter 1990).

More convincingly, Hughes (1994b) noted that algal blooms occurred in numerous pre-mass mortality experiments in response to reductions in urchin density without any changes in background nutrient levels. Furthermore, the extremely rapid initiation of algal blooms following the mass mortality events on overfished reefs also supports the view that overfishing, trophic simplification, and greatly reduced densities of *D. antillarum* are largely responsible for the observed shifts in community structure. These rapid changes occurred over large stretches of the coast in Jamaica at sites mostly in rural areas and in clear water not likely to have been suddenly enriched by nutrients in 1983 (Hughes 1994b). Ogden (1994) provided supportive evidence for the importance of overfishing in destabilizing coral communities in noting that of fourteen sites in the Caribbean where researchers had been asked to assess 1993 coral cover and a variety of other local factors (*i.e.*, "fishing, nutrient pollution, sedimentation, storms, diseases, and coral bleaching"), "the five reporting no change in coral cover (Belize, Bermuda, Cayman, Saba, and St. Lucia) were located within parks, reserves, or areas where control of fishing is the major management tactic".

Following the massive shift from coral to algal domination on overfished reefs, concern over the recovery of urchin populations has grown. The density of *D. antillarum* in Lameshur Bay, St. John (Karlson and Levitan 1988) and Jamaica (Hughes 1994a, Figure 3.4) has shown no sign of recovery towards pre-mass mortality levels. A counter example noted for urchin populations in Barbados (to 17.9-57.4% of pre-mass mortality densities by 1985) seems to be quite exceptional (Hunte and Younglao 1985). Given the region-wide mass mortality event, the potential for recruitment to most local populations has been severely damaged. This is evident in spite of the fact that surviving urchins have grown to enormous sizes and have greatly increased their gamete production (Levitan 1989). Reduced fertilization success of gametes spawned by urchins at very low densities has offset increased gamete production to yield negligible changes in the number of zygotes produced per female (Levitan 1991). Thus urchin recruitment rates in most populations appear to have paralleled the large reductions in urchin densities (*e.g.*, Bak 1985, Karlson and Levitan 1988) and current net reproductive rates (i.e, the average number of female offspring per female per generation) would appear not to be high enough to promote recovery.

Furthermore, as pointed out by Hughes (1994a), even a full recovery of *Diadema* to pre-mass mortality densities would leave Jamaican reefs "reliant once more on a single dominant herbivore and vulnerable to a recurrence of disease". In addition, Hughes (1994a) noted that future hurricanes are likely to contribute to the present algal domination of these reefs rather than recovery because algal recruitment following major storms is much more rapid than that of corals. A full recovery of the entire coral community to some more stable structure would then require more herbivorous species. Hughes (1994a), like Munro (1969, 1983) before him, strongly emphasized the need to control overfishing to achieve this goal. Other management foci may also be required to control the numerous other threats to coral communities (noted above), but these are likely to be fruitless in the absence of significant trophic complexity.

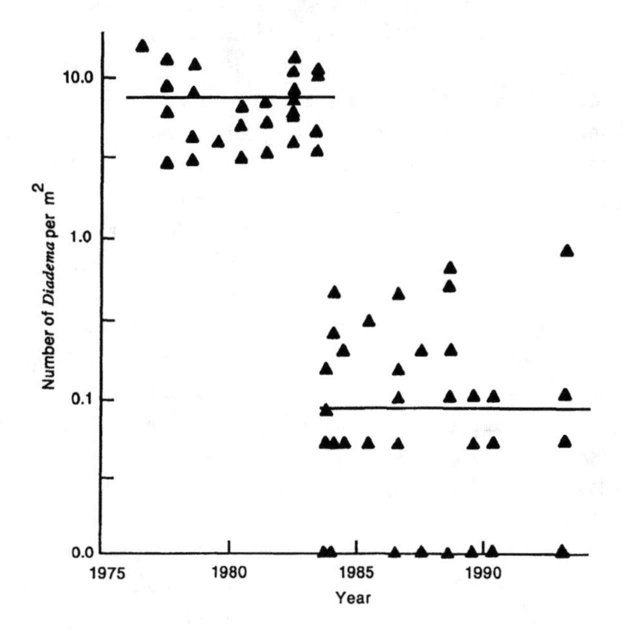

Figure 3.4 Density of *Diadema antillarum* at 14 sites along >100 km of the north coast of Jamaica. The pre-mass mortality mean of 9.0 urchins per m^2 dropped to 0.9 urchins per m^2 in 1983 with no sign of recovery through 1993 (redrawn with permission from Hughes, T.P. 1994a. Catastrophes, phase shifts, and large-scale degradation of a Caribbean coral reef. *Science* **265**, 1547-1551). Copyright 1994 American Association for the Advancement of Science.

3.3.2 KEYSTONE SPECIES IN CORAL COMMUNITIES: DAMSELFISHES

Damselfishes (Pomacentridae) are one of the most abundant of all the families of coral reef fishes [*e.g.*, fifty seven species in the central region of the Great Barrier Reef Table 2.2 and Williams and Hatcher 1983)]. This taxon is comprised of approximately 320 species worldwide (Randall *et al.* 1990) representing a variety of trophic guilds exploiting both benthic and planktonic food resources (Table 2.2). Most damselfishes are small and quite territorial (especially algal grazers). In addition to defending food resources, they also very "zealously defend" their eggs which are attached to the bottom by adhesive strands (Randall *et al.* 1990). Territorial interactions with herbivorous fishes and sea urchins, predatory fishes, and virtually all other intruders result in direct and indirect effects on the structure of the benthic assemblage and the community as a whole.

On patch reefs dominated by the staghorn coral *Acropora cervicornis* in Discovery Bay, Jamaica, Williams (1979) examined interactions between the threespot damselfish *Stegastes* (*Eupomocentrus*) *planifrons* and the urchins *Diadema antillarum*, *Echinometra viridis*, and *Lytechinus williamsii* to find that this herbivorous damselfish is extremely aggressive when confronting *D. antillarum* as compared to the other two urchin species. In experiments in which pairs of different species of urchins were placed within damselfish territories, *D. antillarum* was usually attacked first at rates in excess of twenty two attacks per min. *D. antillarum* responded to these attacks by moving out of the territories, while the other urchin species were removed through a series of "lifts and drops". Williams (1979) also reported spatial segregation patterns for *D. antillarum* and *E. viridis* populations relative to damselfish territories. The former urchin occurred primarily at outer edges of coral patches relatively far from central damselfish territories. The latter urchin occurred in closer proximity to these territories. Thus the damselfishes were inferred to interfere with direct competitive interactions between the herbivorous urchins and to influence their distribution patterns.

Sammarco and Williams (1982) also added *Diadema antillarum* to territories defended by *S. planifrons* on patch reefs in Discovery Bay. The agonistic behavior of > 100 damselfishes included spine nipping and territorial displays which were restricted to daylight hours. Most behaviors persisted throughout 5-min observation periods and usually resulted in an urchin alarm response and movement out of the territory. Sammarco and Williams (1982) concluded that the damselfish behavior greatly reduces urchin abundance within territories thereby influencing urchin distribution patterns, algal abundance, and coral recruitment patterns (as reported by Sammarco 1980, 1982). At such high densities where > 50% of the reef substrate is comprised of damselfish territories, these effects can be quite pervasive.

Williams (1980) directly evaluated the impact damselfish can have on urchin densities by repeatedly removing damselfish from patches of *A. cervicornis*. The mean density of *D. antillarum* rapidly increased in removal patches to 2.5 times that in controls after only three days. Although the mean density of *E. viridis* also increased in removal patches to twice that in controls, this increase was not statistically significant. Williams (1980) noted the similarity in the ability of damselfish to remove "competitors from its territory according to their ability to invade" and the effects of keystone predators in selectively removing prey, reducing the effects of competitive exclusion, and stabilizing competitive interactions. This represents an early suggestion to expand the notion of the keystone species as later recommended by Power *et al.* (1996). It also represents an interesting example of two interacting keystone species in a single community.

In Kaneohe Bay, Oahu, Hawaii, the yelloweye damselfish *Stegastes fasciolatus* has also been identified as a keystone species because of its strong effects on algal diversity (Hixon and Brostoff 1983, 1996). Territorial interactions between damselfishes and other herbivorous fishes (specifically parrotfishes and surgeonfishes) indirectly influence the structure and diversity of the algal assemblage within territories. In a one-year long experiment, Hixon and Brostoff (1983) documented the response of the algal assemblage on settlement plates to three different conditions. Outside territories, intense grazing by fishes eliminated erect filamentous algae leaving a less diverse, fourteen-species assemblage of prostrate forms dominated by the coralline alga *Hydrolithon reinboldii* (H' = 1.1). Plates left exposed to grazing by damselfishes within territories for one year developed a highly diverse assemblage composed of mostly filamentous algae (H' = 2.2). On plates placed within exclosure cages within territories (thus eliminating grazing by the damselfishes), algal diversity peaked after six months (H' = 2.0) and then dropped (H' = 1.8) as the red alga *Tolypiocladia glomerulata* dominated the assemblage of seventeen mostly filamentous and foliose species. Thus the lowest grazing intensity within cages favored a competitive dominant, intermediate grazing within territories (the uncaged condition) favored algal coexistence in the most diverse assemblage, and intense grazing outside territories favored a less diverse assemblage of resistant species.

In a more thorough analysis of the above experiment using data from seventeen sampling times over the year, Hixon and Brostoff (1996) examined the effects of grazing and territoriality on the successional patterns of the algal assemblage. They interpreted the high algal diversity exhibited within territories as evidence that the damselfishes were "decelerating succession" as they "prolonged the high-diversity midsuccessional stage". Damselfish grazing activities prevented the domination of the algal assemblage by competitive dominants. Intense grazing by parrotfishes and surgeonfishes outside territories destroyed algal holdfasts and was interpreted to "deflect" the successional process to favor crustose coralline algae. Thus the different grazing regimes influenced the endpoint of temporal succession and it was the presence or absence of the keystone species (*i.e.*, damselfishes) which controlled the rate and trajectory towards these alternative states. "Given that damselfish territories can cover well over 50% of shallow reef tracts (Sammarco and Williams 1982, Klump *et al.* 1987), can account for 70-80% of the primary productivity on fore-reefs (Brawley and Adey 1977), and have numerous indirect effects on local reef benthos disproportionate to the abundance of damselfish (review by Hixon 1997), the keystone label seems particularly appropriate" (Hixon and Brostoff 1996).

Experiments introducing corals into damselfish territories also indicate how their behavioral interactions with other species can affect species composition and zonation patterns within coral communities. In the Gulf of Panama, shallow reef flat environments are dominated by branching pocilloporid corals (*Pocillopora* spp.). Massive colonies of *Pavona gigantea* predominate in slope-base environments (6-10 m deep) (see section 2.2.5). Over this depth range, the damselfish *Stegastes* (*Eupomocentrus*) *acapulcoensis* maintains algal territories which it vigorously defends. Wellington (1982) placed small coral fragments of *P. damicornis* and *P. gigantea* into caged and uncaged conditions within and outside territories at depths of 1, 4, and 7 m. After 100 days, *P. damicornis* experienced the highest mortality and grew most slowly at 7 m when exposed to predation by pufferfishes and parrotfishes (under uncaged conditions). In contrast, *P. gigantea* did not suffer from exposure to these predatory fishes. Instead, it experienced the highest mortality and grew most slowly when exposed to damselfish after having been placed in close contact with the algal mat within territories. These results indicate that damselfish territories provide a refuge for *P. damicornis* from predatory fish and promote the establishment of branching corals in shallow water [such coral refuges have also been suggested for western Atlantic species (Sammarco 1980, Sammarco and Williams 1982)]. In Panama, the interaction favoring branching pocilloporid corals, in turn, provides additional shelter which is required for the establishment of new damselfish territories. Thus these interactions reinforce the observed zonation patterns on these reefs (Wellington 1982).

3.3.3 KEYSTONE SPECIES IN CORAL COMMUNITIES: *ACANTHASTER PLANCI*

Since the 1950s, population explosions of the corallivorous sea star *Acanthaster planci* have devastated coral communities throughout the Indo-Pacific province (see reviews by Endean 1973, 1974, Moran 1986, Birkeland and Lucas 1990). This predator is considered a keystone species (Power *et al.* 1996) because it is the only corallivore known to consume coral on such unprecedented scales with major consequences to community structure. These include both direct and indirect effects involving interactions which are noted to differ at high (> 150 individuals per hectare) and low densities of *A. planci*. These densities are known to fluctuate as much as "six orders of magnitude within a year or two" (Birkeland and Lucas 1990). Furthermore, there is some supportive evidence for the notion that outbreaks of *A. planci* favor alternative communities dominated by persistent benthic macroalgae, sponges, and alcyonacean corals (Birkeland and Lucas 1990). Perhaps the explosive population growth of *A. planci* on coral reefs represents destabilization of these communities due to the loss of some other keystone species (as suggested by Paine 1969), but there is currently no consensus regarding the cause of these outbreaks nor how long such outbreaks have been occurring. Nevertheless, shifts in the stability of coral communities and changes in the status of keystone species are likely consequences of perturbations to community structure. The current status of *A. planci* as a keystone species is justified whether or not it served this role in the past.

Given the fact that the biology of *A. planci* and its role in coral communities have received so much attention, I focus here on just a few ecological studies illustrating the diversity and magnitude of its effects which sometimes reach "over areas on the scale of square kilometers" (Birkeland and Lucas 1990). To begin, I note that the first report of a population outbreak by *A. planci* has been attributed to S. Shirai for the Amami Islands of Japan in 1956. These islands are located near the middle of a 2000 km stretch of islands (from the Izu Islands in the north to the Yaeyama Islands in the south) where "the most extensive continuous series of outbreaks recorded to date" occurred in 1969-1989 (Birkeland and Lucas 1990). On the Great Barrier Reef, population explosions began as early as 1962 (Endean 1974) followed by extensive outbreaks over large distances in 1969 and again in 1979. By 1987, 28% out of hundreds of surveyed reefs on the Great Barrier Reef exhibited some degree of the effects of *A. planci* (Birkeland and Lucas 1990).

The coral communities of Guam represent one of the better studied examples of how *A. planci* affects coral communities. Prior to 1967, *A. planci* densities on the reefs of Guam were apparently quite low. Chesser (1969) reported a massive population explosion of this coral predator which occurred rapidly over a period of 2.5 years. The direct effect of corallivory was the death of 90% of the living coral at multiple sites along 38 km of the northwest coast. "At some localities, with population densities as high as one animal (*A. planci*) per square meter of reef", Chesser (1969) estimated average feeding rates which would completely kill all the coral in one month. Following this large-scale coral mortality, algal cover increased thus favoring herbivorous parrotfishes and surgeonfishes. Most other fishes vacated these "dead reefs" (Chesser 1969) as did *A. planci* until a second, small outbreak occurred in 1982 (Colgan 1987). At Tumon Bay and Tanguisson Reef, the coral assemblages were surveyed by J. Randall in 1969-1974 and at the latter site again in 1980-1981 by M. Colgan. In reef front (1-6 m deep), submarine terrace (6-16 m deep), and seaward slope (16-33 m deep) zones, the percent cover of corals was 49.1%, 59.1%, and 50.1%, respectively, prior to the peak infestation in 1969 (Pearson 1981). One year later, percent cover had dropped to 20.9%, 0.9%, and 0.5%, respectively, with the loss of 29-44% of the coral species (Pearson 1981). Species of *Acropora* and *Montipora* suffered particularly high mortality as they are the preferred prey of *A. planci* (Colgan 1987).

Although it was anticipated that recovery from such massive mortality would require quite long periods of time, this was not the case. These reefs appeared to be much more resilient to this perturbation than previously suspected (Colgan 1987). "In terms of generic and species diversity, recovery was almost complete by 1971" (Pearson 1981). Overall coral cover had increased fourfold and the number of coral colonies had almost doubled on the submarine terrace and more than tripled on the seaward slope. By 1980-1981, preferred prey species had recruited and grown faster than nonpreferred species especially in the reef front and submarine terrace zones (Colgan 1987). Over the 12-year interval, these two zones exhibited major shifts in species composition, increases in species richness, yet no additional major changes in species diversity (Colgan 1987, Table 3.1). In contrast, the heavily devastated seaward slope continued to recover with large increases in species richness and diversity, and nonpreferred prey remained the dominant corals.

In the eastern Pacific, the effects of *A. planci* on the coral community have been shown to be modified by the presence of symbiotic crustaceans (Glynn 1976). On two reefs in the Gulf of Chiriquí in the Contreras and Secas Islands, feeding by this sea star resulted in declines in coral cover and species diversity especially in lower reef zones dominated by nonpocilloporid corals. Field observations in 1973 indicated that these corals (*e.g.*, *Porites panamensis*, *Pavona varians*, and *Psammocora stellata*) were eaten more often than pocilloporid corals than one would predict based on their relative abundances (Glynn 1976). Nevertheless, pocilloporid corals represent a large fraction of the diet because these corals are much more common on the reefs as a whole. Glynn (1976) demonstrated that crustacean symbionts (*i.e.*, *Trapezia ferruginea* and *Alpheus lottini*) living in association with *Pocillopora damicornis* actively repulsed attacks by *A. planci* or delayed the feeding activities of the sea star. Furthermore, symbiont removal experiments substantiate 1) that *A. planci* was more successful feeding on *P. damicornis* without symbionts than when symbionts were present, and 2) a preference for *Pavona gigantea* and *P. varians* over *P. damicornis* only when the symbionts were present. These symbionts were most common in the shallow reef flat zone where they clearly reduced the negative impact of *A. planci* on large branching pocilloporid corals. Additional evidence indicates that the defensive behavior of these symbionts is a species-specific response which is stimulated by chemical cues from *A. planci* (Glynn 1980). These interactions mediate the direct effects of this predatory sea star on these eastern Pacific reefs.

TABLE 3.1 Abundance of coral colonies, species richness, and species diversity (H') on reefs off Guam one and twelve years after a major outbreak of *Acanthaster planci.* Samples were collected in reef front (RF), submarine terrace (ST), and seaward slope (SS) zones sampled in 1970 and again in 1981 (36-42 m^2 per sample). Rank abundance is given for some of the most common fifteen species along with the percentage of the total number of colonies present in each sample. Preferred prey include species of *Acropora* and *Montipora.* Nonpreferred prey include *Favia, Galaxea, Goniastrea, Leptastrea, Porites,* and *Pocillopora* (Colgan 1982, 1987). *Pavona* was considered a neutral prey (after Colgan 1987).

1970			1981		
		RF			
1	*Galaxea fascicularis*	15.21%	1	*Acropora surculosa*	14.79%
2	*Goniastrea retiformis*	10.92%	2	*Galaxea fascicularis*	13.48%
3	*Favia stelligera*	6.17%	3	*Pocillopora setchelli*	10.23%
6	*Pocillopora setchelli*	5.06%	4	*Goniastrea retiformis*	6.08%
	Acropora surculosa[1]	?	6	*Favia stelligera*	4.96%
Number of colonies:		632			987
Number of species:		60			66
Species diversity:		4.87			4.71
		ST			
1	*Pavona varians*	12.19%	1	*Leptastrea transversa*	19.96%
2	*Favia stelligera*	6.25%	2	*Montipora elschneri*	7.50%
3	*Galaxea fascicularis*	5.94%	3	*Montipora verrilli*	6.84%
6	*Leptastrea transversa*	5.00%		*Pavona varians*[1]	?
12	*Montipora verrilli*	3.13%		*Favia stelligera*[1]	?
	Montipora elschneri[1]	?		*Galaxea fascicularis*[1]	?
Number of colonies:		320			1306
Number of species:		41			73
Species diversity:		4.81			5.00
		SS			
1	*Leptastrea transversa*	15.92%	1	*Leptastrea transversa*	13.20%
2	*Porites lutea*	11.39%	2	*Porites lutea*	9.52%
3	*Leptastrea purpurea*	8.23%	3	*Porites lobata*	6.35%
	Porites lobata[1]	?	15	*Leptastrea purpurea*	1.78%
Number of colonies:		159			1576
Number of species:		26			73
Species diversity:		4.04			5.10

[1] Rank abundance <15 or species was absent

Unlike the reefs of Guam and many other locations in the Indo-Pacific province, densities of *A. planci* in the Contreras Islands have remained relatively constant (Glynn 1982). Periodic outbreaks have not occurred. In 1971-1979, Glynn (1982) estimated sea star densities at 7-27 individuals per hectare. Furthermore, the rate at which new individuals immigrated to this reef during a two-year period following the removal of all *A. planci* in 1979 was quite low (0.7-2.2 individuals per hectare per month). This immigration rate and observed rates of mortality due to predation on *A. planci* are sufficient to explain the stable low density of *A. planci* observed over this decade (Glynn 1982). The predators in this case were the painted shrimp *Hymenocera picta* and the amphinomid polychaete *Pherecardia striata* (Glynn 1977, 1982). Apparently, the branched pocilloporid reefs at this site not only shelter the previously mentioned symbionts (and the damselfishes noted in section 3.3.2), but also these predatory shrimp and polychaetes. Thus explosive population growth of *A. planci* and its potentially devastating effects are limited in this coral community.

The indirect effects of outbreaks of *A. planci* on coral reef fish assemblages occur as a consequence of shifts in the abundance of food resources for herbivorous and corallivorous species. Birkeland and Lucas (1990) suggested that "the best quantitative documentation (of this type of effect) is from Tutuila, the main island in American Samoa". In late 1978 and 1979, an outbreak of *A. planci* killed a large proportion of corals at multiple sites around the island. However, the corals were "largely untouched" at a site in Sita Bay (Birkeland *et al.* 1987). Data from this Sita Bay site were used as a "control" with which to contrast the response at two heavily affected sites at Fagatele Bay and Cape Larsen. At the exposed sites, major increases in the relative proportion and absolute abundance of herbivorous fishes were observed in 1985 in response to high algal abundance (Table 3.2). A small increase observed in Sita Bay was not significant (Birkeland and Lucas 1990). By 1988, herbivorous fishes decreased towards pre-outbreak levels as the coral community was recovering from this perturbation. The response of corallivorous fishes was equally dramatic. These predators virtually disappeared from the exposed reefs and recovery was not evident even after ten years (Table 3.2).

From 1969-1989, repeated outbreaks of *A. planci* have occurred in the Ryukyu Islands (Birkeland and Lucas 1990). Two years after a 1981-82 outbreak had killed all the coral at a site off Iriomote Island, Sano *et al.* (1987) contrasted living and dead reefs in the area. Coral cover representing primarily branching acroporids was more than 80% on the living reef, while corals were still totally absent on the dead reef. On this latter reef, coral skeletons showed obvious signs of erosion and the reef exhibited low structural complexity. There were significantly fewer fish species on the dead reef (62 vs. 43 species per 100-m^2 transect) due to the total absence of corallivores. There were no significant differences in number of species of herbivores, omnivores, or benthic invertebrate feeders on these two reefs. However, the number of individual herbivorous and omnivorous fishes per transect was significantly lower on the dead reef. By 1986, an outbreak had occurred on the living reef which reduced coral cover there to 50% and the dead reef was further eroded to low-relief rubble. The fish assemblage on this rubble included only 22 species (10 benthic invertebrate feeders) as all trophic groups were represented by significantly fewer species and individuals than were present on this reef in 1984. These extremely devastating effects of *A. planci* were only evident after four years.

TABLE 3.2 The absolute abundance of herbivorous and corallivorous fishes per 200-m^2 transect and the percentage of these fishes among all adult fishes (excluding data for the surgeonfish *Ctenochaetus striatus* which recruited heavily during 1985) at three sites on Tutuila, American Samoa, surveyed before and after a major 1978-1979 outbreak of *Acanthaster planci* (after Birkeland and Lucas 1990, and S.S. Amesbury unpublished data). Reprinted with permission from Birkeland, C. and Lucas, J.S. (1990) *Acanthaster planci: major management problem of coral reefs*, CRC Press, Boca Raton. Copyright CRC Press, Boca Raton, Florida,© 1990.

Site	Year	Herbivorous fish		Corallivorous fish	
		individuals per transect	%	individuals per transect	%
Sita Bay	1977-78	71	13.3%	28	5.3%
	1985	75	20.7%	17	4.7%
	1988	60	30.8%	13	6.7%
Fagatele Bay	1977-78	26	7.3%	29	8.1%
	1985	92	35.5%	0	0.0%
	1988	50	18.5%	2	0.7%
Cape Larsen	1977-78	56	11.7%	14	2.9%
	1985	134	39.2%	2	0.6%
	1988	26	13.9%	3	1.6%

Surprisingly, not all outbreaks of *A. planci* have resulted in major shifts in the abundance of trophic groups of fishes. Williams (1986) examined the fish assemblages occurring in reef slope environments on several reefs in the central Great Barrier Reef in 1980-1984 to highlight their large degree of spatial and temporal variability. In addition, he noted the response of fishes on three mid-shelf reefs exposed to high densities of *A. planci* in 1983 and 55-90% reductions in coral cover (Rib, John Brewer, and Lodestone reefs). Among 187 pairwise comparisons of pre- and post-outbreak abundance estimates of 74 species on these reefs (13 surgeonfishes, 16 parrotfishes, 10 wrasses, 24 damselfishes, and 11 butterflyfishes), only five species exhibited highly significant differences. Two obligatory corallivorous species (*Chaetodon rainfordi* and *C. plebius*) were less abundant (among six such obligatory species) on reefs after exposure to *A. planci*. In contrast, another butterflyfish (*C. melannotus*), a damselfish (*Stegastes fasciolatus*), and a surgeonfish (*Naso lituratus*) increased on one or more of the reefs after the outbreak. None of 54 similar comparisons on a "control" reef (Davies), where no outbreak had occurred, yielded highly significant differences. Thus most species were not affected by this outbreak of *A. planci.*

It would appear that the keystone status of *A. planci* like that of many such species is variable from place to place due to a variety of controlling factors. In general, this status is confirmed when outbreaks (or local extinctions, population crashes, and experimental removals) result in major destabilization of community structure. When factors like predation control populations of *A. planci* (Glynn 1982), keystone status for some community attributes may lie elsewhere. When population sizes of *A. planci* are maintained at seemingly equilibrial levels due to extrinsic factors influencing larval transport and recruitment into local communities (*i.e.*, recruitment limitation), perhaps no keystone species acts to control community structure. In any case, our challenge is to understand the relative importance of strong biological interactions, the local population dynamics of the major interacting species, and larger-scale climatic, geological, and evolutionary processes influencing observed community structure. Integrating these multi-scale effects into this framework is the ultimate objective.

3.4 Transient dynamics of nonequilibrial systems

The notions that trophic complexity and keystone species stabilize community structure around single or even multiple equilibrial conditions are both based on a deterministic view that this structure is controlled by predictable causal relationships. In a broad sense, this view encompasses top-down as well as bottom-up control mechanisms involving interactions among resources, competitors, and consumers. This deterministic perspective has generated equilibrial explanations for observed community structure in a variety of different ecological communities. Furthermore, these explanations, being largely based on local biological interactions and resource availability, have been subjected to widespread experimental evaluation (Hairston 1989). However, we now know that communities can be buffeted about by frequent perturbations, unique historical events, and environmental changes which can potentially maintain them in a state of perpetual nonequilibrium.

How then can we understand the dynamics of ever changing communities? Chesson and Case (1986) reviewed classical equilibrial theories along with four new directions for the development of a nonequilibrial theoretical framework. Classical theories generally invoke continuous competition in constant environments. Some equilibrial theories also add predation or spatial heterogeneity as explanations for how more species can coexist than predicted on the basis of competition alone. The former allow more species to coexist locally in the presence of predation and reduced competition. The latter permit coexistence of potential competitors in spatially segregated patches. Perhaps, the two most relevant nonequilibrial theories which apply to coral communities are those invoking 1) climate change where average environmental conditions shift over ecologically relevant time scales (Davis 1986, Graham 1986, Van Devender 1986) or 2) slow competitive displacement of ecologically similar species (Hubbell and Foster 1986). These two new directions differ from most other theories in that they allow for strong effects of history on community structure. As noted in chapter 1, historical events and cyclical climatic phenomena with multi-year periods can influence the structure of a variety of aquatic communities. Coral communities should not be exceptional in this regard.

Davis (1986) noted that directional climate changes over a wide range of temporal scales (10-1000 yrs) are the rule, so natural communities are never exposed to constant environmental conditions except for very short periods of time. For forest communities, this is true for both high-latitude temperate forests as well as rain and sclerophyll forests in the tropics. Furthermore, species respond to environmental change in unique ways which can differ in direction, magnitude, and timing (Davis 1986, Van Devender 1986). Shifts in the direction and rate of change (*i.e.*, trajectory) of community structure should occur with environmental changes as they affect species abundances, distributional limits, trophic structure, and community stability (persistence, resilience, and resistance). At longer temporal scales, cyclical climatic variation (reviewed in Webb and Bartlein 1992) has resulted in periodic phase shifts between alternate terrestrial communities (Davis 1986, Van Devender 1986) just as has been observed in marine plankton assemblages (section 1.3, Figure 1.2). However, these periodic shifts in community structure do not represent simple geographic displacement of entire communities with fixed attributes. After analyzing a 35,000-year record of the plant and animal remains found in packrat middens in the Chihuahuan desert, Van Devender (1986) noted that "many animal species occurred during the late Pleistocene in the same geographical area that they inhabit today, but were living in different habitats". Therefore, the community was "extensively reshuffled" as the environment changed (Van Devender 1986). Graham (1986) also noted massive and repeated reshuffling of North American mammalian assemblages during the Quaternary Period.

In shallow marine environments, these periodic climate changes have resulted in major sea level changes, fluctuations in water temperatures, and a variety of species removals, additions, and substitutions within local communities. Using the well-studied Pleistocene and Recent marine molluscs from the west coast of North America as an example, Valentine and Jablonski (1993) noted consistent anomalous associations of species. Pleistocene associations are compositionally quite distinct from those living today even though many species have persisted throughout this time period. They emphasized species-specific responses to climate change, yet also noted that this reshuffling of community structure resulted in few extinctions in these seemingly stable communities. A broader review of fossil communities in general led them to conclude that although climate changes can cause extinctions and opportunities for speciation, these communities as a whole appear to be flexible, loosely integrated entities which are "quite resilient" to perturbations when considered over geologic time scales [see Roy *et al.* (1996) for a recent review of periodic Pleistocene climates and their biological implications].

Jackson (1992) also noted long-term stability after comparing coral dominance and depth-related zonation patterns on Recent reefs in Jamaica (data from Liddell *et al.* 1984) and Pleistocene reefs in Barbados (data from Mesolella 1967, 1968). In spite of short-term disturbances, spatial variation across disturbance gradients, and rapid sea level changes, living coral assemblages in Jamaica prior to the mass mortality of *Diadema antillarum* were remarkably similar to Pleistocene reefs dated at 120 ky and 200-600 ky (Figure 3.5). Jackson (1992) viewed short-term fluctuations in community structure as noise in the long-term dynamics of coral communities operating on these millenial scales. Hence, the current disruption of Jamaican coral reef communities may represent a geologically temporary condition rather than an alternative equilibrial state (due to a phase shift) or a nonequilibrial response to a changing environment (see sections 2.2.6 and 3.3.1). The conserved patterns exhibited over these time and spatial scales strongly support this equilibrial perspective, but nonequilibrial changes in community structure may also be operating. Since Jackson (1992) focused only on three dominant species on Recent and Pleistocene reefs (*Acropora palmata*, *A. cervicornis*, and *Montastrea annularis*), the potential for some degree of species reshuffling in the rest of the community or shifts in trophic structure remains. However, the stability of the distribution and abundance patterns of these three dominant species is indeed impressive. This equilibrial view can be contrasted with nonequilibrial perspectives invoking disturbances at shorter (see chapter 7) and longer (section 2.1, Kaufmann and Fagerstrom 1993) time scales. Given that fossil assemblages themselves represent long-term averages of species occurrences, the stability noted by Jackson (1992) represents a time-averaged view which masks phenomena operating at much shorter time scales. Therefore, much more needs to be learned regarding the dynamical processes controlling these observed recurrent patterns in coral communities.

Hubbell and Foster (1986) suggested another direction for the development of nonequilibrial theories to explain observed community structure. Their study of a 0.5 km^2 -plot of forest on Barro Colorado Island in Panama indicated that species richness was high (> 300 species of woody plants) and the "biotic neighborhood" in which trees interact was complex and unpredictable. They argued that "evolution will reflect the temporal and spatial average of the selective conditions created by a suite of ever-changing" diffuse competitors. Furthermore, the selective environment experienced by seedlings and saplings on the forest floor should lead to convergent generalized abilities to tolerate shaded conditions. Thus species-specific competitive interactions are weak relative to diffuse competition. Species turnover at local scales involves a slow process of unpredictable competitive displacements of ecologically similar species. Under these conditions, the systematic elimination of any species from the community due to differential competitive abilities is unlikely. Hubbell and Foster (1986) acknowledged the existence of different guilds and divergent adaptive strategies among tropical tree species, but noted that they "see evidence for only a dozen or so guilds, each of which is composed of many morphologically and phenologically similar tree species". Under this nonequilibrial scenario, the richness of the assemblage of tree species varies stochastically as it is influenced by largely unpredictable large-scale phenomena (*e.g.*, climate, historical events, immigration from regional sources, extinction, and speciation).

Although once appreciated as primary determinants of species diversity patterns, such large-scale unpredictable phenomena were not incorporated into the early development of predictable equilibrial explanations for the structure of ecological communities (Schluter and Ricklefs 1993a). However, ecologists have recently once again begun to realize their importance (*e.g.*, Karieva *et al.* 1993, Ricklefs and Schluter 1993, Giller *et al.* 1994) and have begun to integrate the effects of niche differentiation and biotic interactions with the historical/geographical perspective (see chapters 8 and 9 for further discussion of these ideas and recent evidence from coral communities).

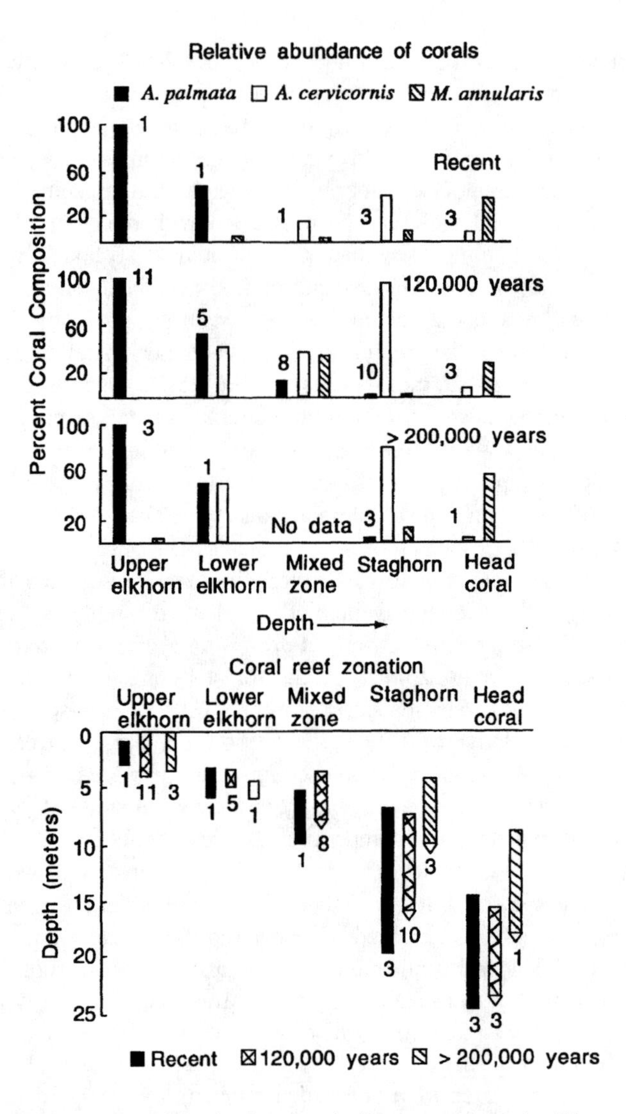

Figure 3.5 Relative abundance of the dominant corals (*Acropora palmata* in elkhorn, *A. cervicornis* in staghorn, and *Montastrea annularis* in head coral zones) and depth distributions of five zones in Recent and Pleistocene Caribbean reef communities. Number of sampled reefs in Jamaica (Recent) and Barbados (Pleistocene) are indicated. Downwardly pointing bars indicate minimum estimates due to the lack of exposure of deeper communities in some outcrops (redrawn after Jackson 1992).

3.5 Overview

Although coral communities are quite diverse, they do not appear to be very stable on ecological time scales. The structure of some communities has been radically disturbed following large fluctuations in the abundance of such keystone species as *Diadema antillarum* and *Acanthaster planci.* In contrast, other coral communities seem to be stabilized by the presence of damselfishes. The degree of stability of coral communities is quite variable in space and time. While some communities may persist for long periods of time, others are frequently disturbed and perpetually changing. Furthermore, persistent communities need not be particularly stable. The resilience of these communities is also substantially variable. Recovery from comparable perturbations may be quite rapid in a some places and not occur at all at other places. The major phase shift of Jamaican coral communities from coral to algal domination followed the regional mass mortality of one species. In this regard, the community was unstable (not resistant) being highly vulnerable to major reductions in the density of *Diadema antillarum.* The keystone species status of this and other species in coral communities appears to depend on a variety of local and regional phenomena. At the local scale, keystone species status and trophic structure may be linked. For example, overfishing may significantly reduce trophic complexity and the importance of diffuse predation and herbivory. As a consequence, some keystone species might be eliminated from the community or others species might be elevated to keystone species status. At the regional or global scale, climate fluctuations may result in "reshuffling" of local communities as individual species respond to variation in the environment in different ways. New species associations within habitats should have a number of repercussions on biological interactions which can potentially influence the stability of the community.

4 SUCCESSION

4.1 Successional mechanisms on ecological time scales

Ecological succession refers to the temporal progression of a community through a series of predictable species replacements leading to a stable, self-perpetuating condition called the climax (Ricklefs 1990). Hence, the notions of predictable causality, stability, and equilibria are joined with that of a changing community. Successional changes in community structure are typically considered over relatively short ecological time scales. Thus, the focus is often on the relevant ecological processes driving the succession of a community. Much longer term changes in community structure are viewed here as being driven primarily by orbital forcing (see sections 1.3 and 3.4) or geological rather than ecological processes. Orbital forcing directly alters climate, the environmental conditions favoring particular successional equilibria, and therefore must deflect successional trajectories as well. Major disturbances during succession are generally considered to be extrinsic perturbations which oppose the successional process. However, frequent disturbances (*e.g.*, storms, fires, flooding, etc.) can represent important predictable features of the selective regime resulting in adaptations which may actually contribute to succession and community development (Horn 1981).

The distinction between processes occurring on ecological and geological time scales is less clear when the organisms in question are quite long-lived and their skeletal remains are well preserved in the fossil record. This is certainly the case when dealing with coral communities. Some corals live for centuries and many tend to leave "massive, cemented, biogenic structures" which are much more likely to remain in place than the remains of most other organisms (Aronson 1994). Thus some successional sequences may be well preserved in the fossil record and provide evidence for particular successional processes. For example, the remarkable similarity of dominance and zonation patterns on Pleistocene and Recent coral reefs in the Caribbean (section 3.4, Figure 3.5, Jackson 1992) may support the view that the end-points of succession in these communities represent predictable equilibria. However, the effects of relatively rapid successional processes are likely to be obscured in the fossil record as individual assemblages can only present a time-averaged perspective of species replacements. Thus successional studies on the mechanisms creating these patterns in coral communities may very well include evaluation of both living and fossil communities.

The fundamental mechanisms influencing successional sequences have been postulated as formal hypothetical models of succession (Connell and Slatyer 1977) and as assembly rules (Lawton 1987). Both of these contributions highlight the role of biological interactions (competition, herbivory, predation, disease, and mutualisms) in directing the rate and direction of successional changes. These changes are posited to occur in the absence of significant shifts in the physical environment. Again, such shifts are typically viewed as being driven by processes extrinsic to the ecological community. Although here I do not pursue a broader perspective incorporating geologically important processes (*e.g.*, calcification, lithification, accretion, bioerosion, etc.) with the biological control of ecological succession in coral communities, some integration of these processes should eventually occur. However, as noted by Connell (1987), most experimental evidence documenting successional mechanisms in marine hard substrate communities comes from intertidal rather than subtidal communities. In general, the mechanisms operating in coral communities are not well documented.

The specific successional mechanisms presented by Connell and Slatyer (1977) were identified as part of "facilitation", "tolerance", and "inhibition" models. Each model describes the colonization of open space following some disturbance. In the facilitation model, only early successional species can colonize open space. These species modify the environment making it more suitable for late successional species, but less so for early successional species. Sequential colonization occurs with environmental modification "until resident species no longer facilitate the invasion and growth of other species". In the tolerance model, early successional species modify the environment but do not influence the rates of recruitment and growth of later colonists. However, these early successional species are less tolerant of these environmental modifications and thus are eliminated from the community as succession proceeds. "The sequence of species is determined solely by their life-history characteristics". In the inhibition model, colonization of open space is followed by environmental modifications which make it less suitable for both early and late successional species. Early colonists inhibit invasion by subsequent colonists and suppress the growth of those already present. The successional process can only occur as generally short-lived, early colonizing species die or are damaged and then are displaced by longer-lived, late successional species. Under the facilitation and tolerance models, the demise of early colonists is associated with competition with later successional species in a modified environment. In the inhibition model, early colonists die due to a variety of causes other than competition (*e.g.*, extreme physical conditions and interactions with natural enemies) and late successional species are more resistant to these factors.

Lawton (1987) used these same three models as part of a set of general rules for the predictable assembly of a community during succession. Although he specifically directed his attention to terrestrial systems, one might use these assembly rules as predictions for coral communities. To begin with, biological interactions involving corals may comprise important assembly rules. The interspecific competitive interactions implicit to the facilitation and tolerance models involve differential resource exploitation by early and late successional coral species. The inhibition model invokes interference competition as well as the potential influence of natural enemies on the successional process. Additional assembly rules predict species turnover among non-coral species during succession because of their preferences for specific successional assemblages, habitat stratification, and other changing aspects of coral distribution and abundance patterns. Although Lawton added a null, random colonization model to describe successional sequences with no interspecific interactions, several assembly rules invoke the specific effects of these interactions. Interspecific competition among non-coral species is predicted to result in competitive exclusion and species turnover during succession. Furthermore, the rate and direction of successional changes are predicted to be influenced by interactions with herbivores, carnivores, mutualists, and pathogens. All these interspecific interactions form a web of interactions which, as a whole, are predicted to affect the assembly of a community during succession.

Lawton (1987) noted a growing body of theoretical literature which predicts at least three of these "community-wide" successional patterns. First is the prediction that with the addition of species to a community during succession, trophic structure converges to some constant proportion among trophic levels (*e.g.*, the ratio of the number of predator to prey species) (*e.g.*, see Mithen and Lawton 1983). Second is the prediction that species persistence increases and species invasions become increasingly less successful in speciose, late successional communities (*e.g.*, see Drake 1990, Law and Morton 1996). Third is the prediction that initial random differences in the order in which species invade a community can result in the development of alternative climax communities (*e.g.*, Drake 1990). Such priority effects indicate that historical events can be important in determining the structure of some successional communities [a result noted previously by Sutherland (1974), Horn (1981), and others]. Thus, "distinguishing between the effects of chance and determinism on community assembly undoubtedly constitutes one of the most important and difficult problems in contemporary ecology" (Lawton 1987).

4.2 Succession in coral communities

Although much of successional theory goes well beyond what has actually been demonstrated for coral communities (and most other natural communities), there has been considerable effort to establish some of the basic patterns. However, much remains to be done in order to evaluate the underlying mechanisms generating these patterns and to discriminate among alternative explanations. In what follows, I focus primarily on coral assemblages in describing empirical evidence for species replacements, differential colonizing ability, and species interactions during succession. Some of the early evidence suggested that species replacements do not occur and therefore the applicability of successional theory was uncertain. More recent evidence indicates that in some coral assemblages (particularly species-rich assemblages), species replacements do occur. Some of these replacements result from highly variable competitive relationships which may or may not be predictable depending on the actual mechanisms involved. It is also noted that most coral communities are so frequently disturbed that the notion of an undisturbed climax is probably not applicable. As an alternative, local disturbance regimes are considered as part of the environmental setting in which succession occurs. The role of life history phenomena on species colonization during recovery from disturbances is also considered as a critical component of successional processes. In particular, I emphasize the contrast in colonizing ability among assemblages of adult and juvenile corals. Lastly, I consider how interspecific interactions other than competition among corals may influence coral succession.

4.2.1 CORAL SPECIES REPLACEMENTS

Pearson (1981) reviewed published studies on the recovery of coral communities following a variety of disturbance events. Although most of these were primarily descriptive and did not address specific predictions regarding successional processes, Pearson noted the potential importance of a number of relevant factors which influence the recruitment and survival of corals. These include life history characteristics (*e.g.*, larval dispersal, availability, and settlement as well as coral fragmentation and regeneration), micro-habitat availability and diversity, and biological interactions with competitors and consumers. Furthermore, the effects of these factors can be modified by the nature, extent, and frequency of disturbances. In general, the rate at which coral communities recover from disturbance is quite variable (estimates range from 1-100 years). Most of these assemblages experience disturbances so frequently that only the faster successional processes are likely to reach anything approximating undisturbed climax conditions. For example, in one study of algal succession in a shallow reef environment in the Ryukyu Islands, climax conditions were reached in 10 months as thin crustose species were replaced by thicker, competitively superior species (Matsuda 1989). When disturbances intervene so frequently that undisturbed conditions are rarely if ever achieved, one can view temporal changes in the community from a perspective which includes disturbances as intrinsic processes in the community. Rather than merely resetting the successional clock, disturbances can modify successional trajectories as species are differentially favored based on their tolerance of disturbed conditions and regenerative capabilities.

One early study of the temporal changes in a coral community following major disturbances was conducted off the island of Hawaii where Grigg and Maragos (1974) compared the coral assemblages occurring on six lava flows dated 1868-1969. These dates permitted the authors to estimate the recovery times for the convergence of these assemblages with "controls" occurring on nearby substrate adjacent to each flow. The six sites were distributed over both windward (east and southeast) and leeward (west) exposures so comparisons based on relative disturbance could also be evaluated. At each site, ten, randomly selected, 1-m^2 quadrats were surveyed for corals on each flow and on "control" substrates at a depth of 8 m. The number of coral species, areal coverage, and species diversity estimates for these sites all indicate the general lack of recovery at windward sites on the youngest lava flows (Table 4.1). Grigg and Maragos (1974) estimated that a recovery time of approximately 20 years is required at these sites, while more than 50 years may be required at less exposed, leeward sites. They noted that *Montipora verrucosa* was present at all sites on control substrates, but totally absent on the youngest (1955-1969) lava flows with windward exposures (Table 4.1). This may indicate the influence of recruitment processes on these assemblages on decadal or longer time scales, although alternative explanations have not been excluded. *Porites compressa* was one of only three coral species found on the youngest 1969 lava flow, yet this coral was absent on all other flows. *Pocillopora meandrina* was particularly common on younger lava flows, but it was also very common on older flows and all control substrates (Table 4.1). Likewise, all species colonizing lava flows dated 1955-1969 were also present on older flows or control substrates. Hence, it would appear that coral species tend to accumulate over time and that a successional sequence of coral species replacements may not actually take place in this assemblage.

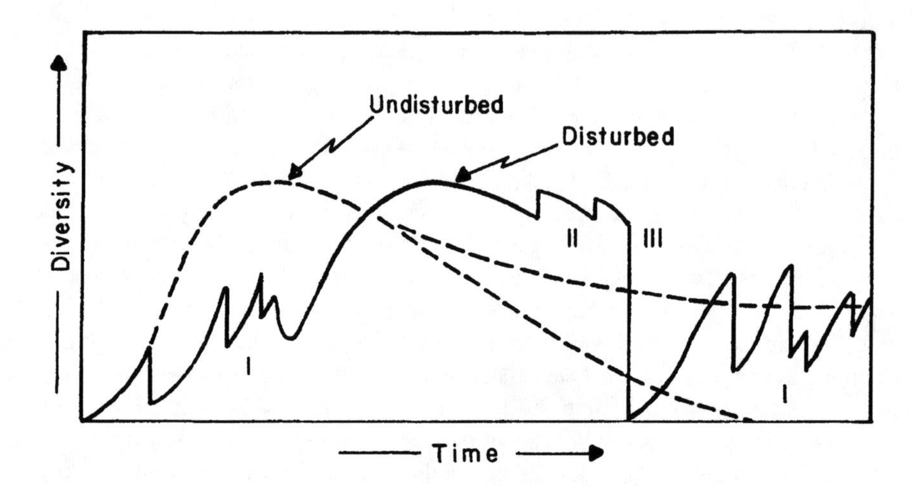

Figure 4.1 Theoretical model of coral community succession under undisturbed (broken line) and disturbed conditions (solid line). Disturbances can either decrease (I and III) or increase (II) species diversity depending on the state of the community and the frequency and magnitude of disturbances (redrawn after Grigg 1983).

Nevertheless, the effects of disturbance, recruitment, and competition have been postulated by Grigg (1983) to contribute to the succession of Hawaiian coral communities (Figure 4.1). Under undisturbed conditions, the community is predicted to gradually accumulate species, reach a peak in species diversity, and then become dominated by one or a few species. As an example, Grigg (1983) noted that at intermediate depths in Hawaii *Porites lobata* can occupy almost 100% of the substrate. This coral species was identified as "the most tolerant of wave stress" among Hawaiian corals. Analyses using southwesterly sites off 14 Hawaiian islands indicate a significant negative relationship between coral cover for the entire assemblage and degree of exposure to "tradewind generated sea and long period swell" (Grigg 1983). This same relationship holds for individual species (*i.e.*, the wave-tolerant *P. lobata* and competitively dominant *P. compressa*). Large islands generally provide more shelter thus promoting the best development of seaward reefs. Differences among these reefs were "interpreted primarily as differences in successional age". Further evidence from the dated lava flows (Table 4.1, Grigg and Maragos 1974) indicate much lower species diversity on older control substrates dominated by *P. lobata* than on the adjacent 1926 flow at that leeward site. Thus, reduced species diversity due to competitive interactions may require several decades to be achieved. However, local episodic recruitment events can also result in spatial monopolies as can differential tolerance among species to particular disturbances. Hence, the disturbance regime at a site can influence succession by differentially favoring a subset of species and by deflecting trajectories towards alternative end-points of succession.

TABLE 4.1 Species richness (S), diversity (H'), and areal cover (cm^2 X100) of corals within 10 m^2 sampled on six lava flows (F) and adjacent control (C) substrate in Hawaii. The date of each lava flow and a rank of the exposure of each site to windward seas are indicated (after Grigg and Maragos 1974).

Date:	1868		1926		1950		1955		1960		1969	
Exposure:	4		5		6		2		1		3	
	C	F	C	F	C	F	C	F	C	F	C	F
S:	6	9	11	8	6	6	5	4	7	4	5	.3
H':	1.0	.8	.5	1.3	.7	.9	.9	.5	1.2	.1	.7	.4
Species												
1	51	67	586	176	220	80	78	35	59	<1	248	<1
2	<1	-	-	-	1	-	-	-	-	-	62	<1
3	-	-	1	<1	-	-	-	-	<1	<1	-	-
4	109	98	79	91	24	52	137	163	29	12	19	<1
5	12	<1	<1	53	19	4	11	-	6	-	<1	-
6	-	<1	-	-	-	-	<1	<1	<1	-	-	-
7	5	5	3	30	<1	1	-	-	-	-	-	-
8	-	<1	6	9	-	2	4	-	<1	-	-	-
9	-	-	<1	-	-	-	-	-	-	-	-	-
10	3	<1	3	<1	6	<1	-	-	-	-	-	-
11	-	<1	2	-	-	-	-	-	16	-	-	-
12	-	<1	1	<1	-	-	-	<1	-	<1	-	-
13	-	-	2	-	-	-	-	-	-	-	-	-
14	-	-	-	-	-	-	-	-	-	-	<1	-
Total Cover:	181	172	684	360	270	140	230	198	111	12	330	<1

Species designations: 1 = *Porites lobata*, 2 = *P. compressa*, 3 = *P. brighami*, 4 = *Pocillopora meandrina*, 5 = *Montipora verrucosa*, 6 = *M. flabellata*, 7 = *M. patula*, 8 = *Leptastrea purpurea*, 9 = *L. bottae*, 10 = *Pavona varians*, 11 = *P. explanulata*, 12 = *Cyphastrea ocellina*, 13 = *Psammocora stellata*, 14 = *Leptoseris incrustans*

In another study of the recovery of Hawaiian coral assemblages, Dollar and Tribble (1993) examined changes in the abundance of corals over a 20-year period at a leeward site on the west coast of Hawaii. Although this site is generally protected from trade-wind generated seas, it experiences recurrent storms which can have catastrophic effects. In 1973, the "pre-storm" coral assemblage covered 51.6% of the bottom and 94.4% of the coral cover was represented by massive colonies of *Porites lobata* and branching colonies of *Pocillopora meandrina* and *Porites compressa* (Table 4.2). Moderate disturbance to this assemblage due to a north Pacific storm in 1974 had little negative effect on coral cover. Due to the regeneration of colony fragments, there was even some "expansion of the range of *Porites compressa*" following this storm. A severe 1980 storm drastically reduced coral cover of all species (Table 4.2). Even massive colonies of *P. lobata* were broken and overturned by this storm, yet some flat encrustations were "relatively unaffected". Between 1980 and 1992, one hurricane and two north Pacific storms caused little apparent damage as there were only small increases in coral cover. By 1993, coral cover was reduced again by two storms and the coral assemblage remained quite far from recovering to the "pre-storm" conditions of 1973. Dollar and Tribble (1993) estimated hypothetical recovery times of 40-70 years for this assemblage in the absence of disturbance. They emphasized that recovery rates depend on the site-dependent frequency of severe disturbances and the adaptive capabilities of dominant species. Like Grigg (1983), they viewed the succession of coral assemblages as a cycle of damage and recovery processes resulting in temporal changes in species diversity (Figure 4.1). However, there continues to be no evidence from the Hawaiian coral assemblage that the recovery process involves a series of predictable species replacements.

TABLE 4.2 Mean percent cover of corals surveyed in 1973-1993 along 675 meters of line transect at a site on the west coast of Hawaii at depths of 3-40 m in boulder, bench, slope, and rubble zones. Ranges (in parentheses) represent estimates of the mean total coral cover within the four zones. A severe Kona storm preceded the survey in 1980. Hurricane Iwa struck in 1982 and Hurricane Iniki in 1992 after the survey. Five, less severe, north Pacific storms contributed to some disturbance at this site over this 20-year period. Species designations as in Table 4.1 (reprinted with permission after Table 1 in Dollar and Tribble 1993). © Springer-Verlag 1993.

	Percent cover				
	1973	1974	1980	1992	1993
Species					
1	17.8	17.6	9.1	10.5	8.4
2	22.5	18.6	0.5	1.5	0.6
4	8.4	6.5	0.4	1.8	1.6
All corals	51.6	45.8	10.5	15.0	11.8
	(3-75)	(4-67)	(5-20)	(4-26)	(4-19)

The slow recovery of disturbed coral assemblages in Hawaii can be contrasted with that of corals colonizing substrate after a 1988 eruption of Gunung Api, a volcano in the Banda Islands of Indonesia which had been dormant for ninety seven years (Tomascik *et al.* 1996). In 1986, seven sites around these islands were surveyed by Sutarno (1990) who reported a total of eighty six coral species and mean of 32.9 species per 100-m line transect. At one site on the northeast coast near surveys conducted after the eruption, there were a total of forty one coral species and a total coral cover of 36.7%. Massive colonies of *Porites lutea* and *Diploastrea heliopora* were dominant species at this site contributing to the 24.3% cover for all massive forms; soft corals covered 18.5% of the substrate. In 1993, Tomascik *et al.* (1996) surveyed the coral assemblages at three devastated sites which differed in terms of the type of disturbance caused by the eruption, local environmental conditions, and substrate. On the northeast coast, one location was completely covered by an andesitic lava flow to a depth of 50 m, whereas a second was located adjacent to the flow and suspected to have experienced considerable thermal stress during the eruption. A site on the southwestern coast was "smothered by large quantities of pyroclastic deposits".

After only five years, a diverse assemblage of 124 coral species had colonized the andesitic lava flow and occupied 61.6% of the substrate (Table 4.3). There was a mean of 25 coral species per 20-m line transect at this site which is quite comparable to pre-eruption estimates of coral richness. However, the dominant species after five years of colonization were tabulate acroporids and other branching corals rather than massive and soft corals (Table 4.3). Although the absolute abundance of corals at the other two sites was not as high, acroporids and other branching corals were also the most prevalent type of corals. If one assumes that these assemblages will ultimately be dominated by massive and soft corals as described for the pre-eruption assemblages by Sutarno (1990), then there must be species replacements during the recovery process. Tomascik *et al.* (1996) noted that 45% of the acroporid species colonizing the andesitic lava flow (*i.e.*, 20 species) are reported to be rare or uncommon species in Indonesian waters. Their prevalence on post-eruption substrates clearly indicates their successful colonizing abilities. However, their longer term persistence or gradual elimination from these assemblages still needs documentation.

TABLE 4.3 Mean species richness (number of species per 20-m line transect), total coral cover, and relative cover of six types of corals at three sites on Gunung Api five years after the 1988 eruption (reprinted with permission after Figures 2 and 3 in Tomascik *et al.* 1996). © Springer-Verlag 1996.

Location:	northeast	northeast	southwest
Substrate:	andesitic lava	carbonate reef	unstable aggregate of pyroclastic deposits
Mean coral richness:	25	10	5
Total coral cover:	62%	8%	4%
		Relative cover	
Acroporid corals:	69%	47%	3%
Other branching corals:	19%	25%	47%
Encrusting corals:	6%	18%	29%
Poritid corals:	1%	4%	12%
Other massive corals:	4%	4%	7%
Soft corals:	0%	1%	1%

In the previous chapter, the recovery of coral assemblages following a 1967-1969 population explosion of *Acanthaster planci* was considered for reefs on the north-western coast of Guam (section 3.3.3). Although some reefs lost as much as 99% of the coral cover and 29-44% of the coral species (Pearson 1981), coral diversity had recovered after only one year and major shifts in species composition were evident after twelve years (Table 3.1). Colgan (1987) attributed most of this shift to the re-establishment of species of *Acropora* and *Montipora* in reef front and submarine terrace zones. In 1970, these two genera accounted for only three of the most common twenty two species in these two zones (Colgan 1987). By 1981, four species of *Acropora* and seven species of *Montipora* were among the twenty eight most common species. The most notable increases included *Acropora surculosa* in the reef front and *Montipora elschneri* and *M. verrilli* in the submarine terrace zones. Notable decreases occurred in this latter zone where the three most common species in 1970 (*Pavona varians*, *Favia stelligera*, and *Galaxea fascicularis*) were not among the fifteen most common species in 1981 (Table 3.1). Thus species replacements may very well have been occurring over this time interval as preferred prey species rapidly recruited and grew on these reefs faster than nonpreferred prey species (Colgan 1987).

TABLE 4.4 Number of transitions among nine benthic categories[1] recorded at three reef sites near Heron Island between 1962 and 1989 at approximately two year intervals. A total of ten, 1-m² quadrats were sampled using a square grid of 40 X 40 points. Columns represent the condition under each point in one sample, while rows represent the condition in the next sample. *N* is the total number of transitions from each category at a site (after Tanner *et al.* 1994).

	A	B	C	D	E	F	G	H	I
				Exposed reef crest (5 m², low intertidal)					
A	1110	3	45	19	0	16	9	15	31
B	16	336	13	23	1	3	4	7	11
C	64	7	1873	34	4	22	22	18	67
D	34	3	35	695	3	25	6	18	37
E	2	4	0	1	20	0	0	1	6
F	14	7	29	32	1	45	3	5	16
G	2	0	13	3	0	4	106	2	9
H	12	0	22	18	2	3	3	146	19
I	529	235	1167	638	44	210	97	196	1230
N	1781	595	3196	1464	75	328	250	408	1425
				Exposed pools (2 m², -0.5 m deep)					
A	29	4	3	4	0	5	6	0	36
B	6	1018	70	79	3	67	11	44	617
C	17	69	442	5	5	66	6	27	398
D	17	74	76	636	0	31	32	40	344
E	0	4	0	0	5	3	0	1	18
F	1	9	26	0	2	98	10	19	380
G	0	2	7	11	0	16	47	1	107
H	3	14	14	11	3	20	4	68	214
I	40	577	494	260	21	340	94	182	3823
N	113	1770	1130	1006	39	646	210	382	5937
				Protected reef crest (3 m², low intertidal)					
A	566	17	106	47	0	15	18	3	717
B	34	115	17	6	1	3	0	0	170
C	105	11	1589	121	5	18	55	8	1106
D	78	6	126	649	3	12	22	4	474
E	2	2	17	6	268	1	8	0	170
F	14	13	23	6	0	7	11	0	85
G	24	1	43	21	2	11	245	0	389
H	3	2	3	1	0	0	0	17	49
I	770	199	1400	622	42	143	361	44	9003
N	1598	366	3325	1478	321	211	720	76	12150

[1] A = encrusting acroporids and *Montipora* spp., B = tabular acroporids (predominantly *A. hyacinthus*), C = bushy acroporids, D = staghorn acroporids, E = soft corals, F = algae, G = massive corals including *Porites* spp., H = pocilloporids (predominantly *P. damicornis*), I = free space

Using survey data from Tanguisson Reef collected in 1970, 1971, 1974, and 1980, Colgan (1982) identified several stages of the successional process occurring during the recovery of the coral community on Guam. Recovery began with a short period in which bare substrate was covered with crustose and filamentous algae. These ephemeral forms were heavily grazed by herbivorous fishes. By 1970 and 1971, the reef had entered a stage characterized by small coral colonies which had settled as planulae, a larval form common to many cnidarians. By 1974, additional recruitment by planulae and regeneration of coral fragments resulted in a proliferation of several coral growth forms on the reef and increased topographic complexity. The small encrusting growth forms which had dominated in 1971 were in the process of being replaced by massive, cespitose, and other growth forms (Colgan 1982). With this "growth form differentiation" and the continued growth of coral colonies, there was a reduction in available space on the reef and inhibition of coral settlement by planulae. Colgan (1982) noted a reduction from 7 new colonies per m^2 in 1971 to 1.25 new colonies per m^2 in 1974. Also as a consequence of colony growth, competitive interactions among corals on the reef became more evident. Hence, several mechanisms involving life history attributes and biological interactions are implicated as important components of the succession of the coral community.

Since 1962, Connell and his colleagues have been studying the dynamics of the coral community at Heron Island on the southern Great Barrier Reef (Connell *et al.* 1997). This project is the longest continuous investigation of corals anywhere in the world and it has generated a number of important contributions dealing with coral life histories, recruitment, disturbance, and competition. Here I focus on data collected from three sites which were used to evaluate transitions among eight categories of benthic species and free space from 1962 through 1987 (Tanner *et al.* 1994). I have recalculated the number of transitions observed over this time period based on the reported transition probabilities. These transitions represent a simplified set of shifting abundance patterns among seventy two coral and nine algal species recorded within ten, 1-m^2 quadrats at Heron Island. Overall, there were approximately seventeen thousand transitions representing the combined effects of colonization, growth, and mortality (Table 4.4). More than twenty four thousand transitions represented self-replacements or no change.

Species replacements appear to be a common feature of this assemblage in spite of high levels of disturbance from frequent storms. If one assumes that transitions from free space represent colonization events, transitions to free space represent disturbance-related mortality events, and transitions among the other benthic categories represent species replacements, then the relative frequencies of these events averaged over 27 years were 32.5%, 51.7% , and 15.8%, respectively. Although some species replacements may represent cases in which an intervening disturbance event removed one species and then was followed by the colonization of another, others represent the outcome of interspecific interactions among corals (Connell 1973, 1976). Tanner *et al.* (1994) noted that virtually all possible transitions occurred in this assemblage. Only the transition from soft corals to encrusting acroporids and *Montipora* spp. was not among the 2655 observed species replacements (Table 4.4). Among the most common species replacements were reciprocal transitions between categories A and C, B and C, B and D, and C and D (see designations in Table 4.4). This suggests that no single type of coral represents a clear competitive dominant in this assemblage because the outcome of interspecific interactions appears to be quite variable. For example, in the exposed pools on the outer reef, transitions between tabular acroporids (predominantly *Acropora hyacinthus*) and the categories representing bushy and staghorn acroporids occurred at approximately the same frequency (Table 4.4).

Using the observed frequencies of transitions, Tanner *et al.* (1994) simulated coral succession at the three sites as a first order Markov process. These simulations provide an approximation of the process of recovery from disturbances and the end-point of succession under some simplifying assumptions. It was assumed that past history provided a good basis for predicting the future and that this past could be best characterized using average values for the entire observational period. Therefore transition probabilities were assumed not to vary with time. It was also assumed that all predictions could be based on the previous state of the community without regard to any other historical considerations. Although Tanner *et al.* (1996) found evidence that this assumption is violated in this coral assemblage, such historical effects did not greatly influence predicted community dynamics and structure. Only the prevalence of highly persistent soft corals was greatly affected by adding historical information to the first order model. Lastly, it was assumed that equilibrium can be reached in the presence of severe disturbances rather than in the complete absence of disturbances. Projections for the three sites at Heron Island indicate that equilibrial conditions can be reached in approximately 20-25 years (Tanner *et al.* 1994), an interval far longer than the average 5.3-6.0 years between severe cyclones (Tanner *et al.* 1996, Connell *et al.* 1997).

The equilibrial structure of the benthic assemblage at the three sites around Heron Island were predicted to be dominated by bushy and encrusting acroporids (on both the exposed and protected reef crests) or by tabular and staghorn acroporids (in the exposed pools) (Figure 4.2). Although no categories were predicted to be totally absent from any site, five of the eight "living categories" were predicted to occupy < 1% of the substrate at one or more sites. Free space was predicted to be quite common (50-74%) as a consequence of the frequent storm disturbance to these shallow sites. Given that the scale of local disturbance due to severe cyclones is large enough to completely obliterate most of the benthos within a 1 m^2-quadrat, one should probably view these predictions at a larger scale. The percent cover estimates at equilibrium represent the relative proportion of benthic categories occurring across multiple patches of comparable habitat at a scale of approximately 10-100 m. At Heron Island, 0.1-10 km encompasses the full extent of all the habitats on the protected and exposed reefs (Connell *et al.* 1997).

Tanner *et al.* (1994) noted that their predicted equilibrial abundance patterns were "sensitive to only a small number of transitions". That is there are only a few transitions which had a large effect on equilibrial structure. Their predictions were most sensitive to transitions to or from free space (disturbance or colonization events) at all three sites. However, there were also significant site-dependent differences in the importance of particular transitions. In the exposed pools, transitions from space occupied by tabular and staghorn acroporids were especially important. Species replacements of the former by bushy and staghorn acroporids were common, while the latter was commonly replaced by tabular acroporids (Table 4.4). On the protected and exposed reef crests, replacements to and from bushy acroporids were very important. These were commonly replaced by encrusting and staghorn acroporids, but the reciprocal transitions were also common (Table 4.40). Further discussion of the significance of variable competitive interactions and the predictability of species replacements due to competition is presented in chapter 5.

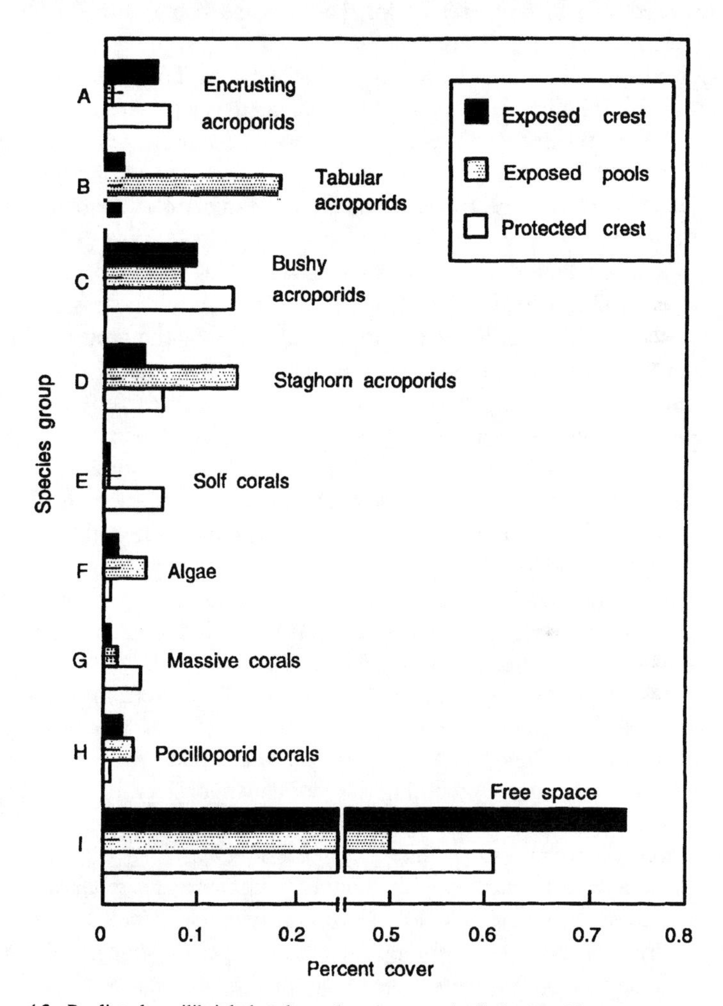

Figure 4.2 Predicted equilibrial abundance (percent cover) of nine benthic categories at three sites at Heron Island. Letter designations are as in Table 4.4 (redrawn after Tanner *et al.* 1994).

4.2.2 DIFFERENTIAL COLONIZING ABILITY

At Heron Island, the colonization of free space resulted in over twice as many transitions as did species replacements (Tanner *et al.* 1994). When examined at approximately two-year intervals, most colonization events involved the four groups dominated by acroporid species and there was over an order of magnitude range in the number of colonizing transitions recorded for all eight groups (Table 4.4). Tanner *et al.* (1994) evaluated the importance of colonization of each category during the recovery process by first quantifying the damping ratio, a dimensionless term indicating the relative rate of recovery. The damping ratios were higher for the assemblages at the two exposed sites (1.66 and 1.74 for the crest and pools, respectively) than at the protected crest site (1.21). These correspond with exponential rates of recovery which are approximately 2.5 times faster at the exposed sites. The most important colonizing transitions contributing to recovery were evaluated using a sensitivity analysis on the damping ratios. At the exposed crest, exposed pools, and protected crest, the largest sensitivities were attributed to colonization by encrusting acroporids, staghorn acroporids, and soft corals, respectively. Although the number of colonizing transitions recorded for these three groups were not among the largest observed (Table 4.4), they represent important constituents of the equilibrium assemblage at each site (Figure 4.2). Yet Tanner *et al.* (1994) also noted that these groups were not necessarily the most abundant groups at equilibrium. For example, the soft corals at the protected crest were among four groups predicted to be common at equilibrium (Figure 4.2). Because of their slow rate of colonization (Table 4.4), soft corals had the largest effect on the time required to reach equilibrium at this site. Likewise, staghorn corals were not predicted to be as abundant as tabular corals in the exposed pools (Figure 4.2) and their colonizing transitions were only 56% of those observed for tabular corals (Table 4.4.). Nevertheless, the sensitivity of the damping ratio was over three times higher for staghorn corals than for tabular corals.

The wide range of colonizing abilities among the major components of coral communities has been recognized for many years. At Heron Island, the mean recruitment rates of all corals over thirty years have been estimated at 1.7-12.7 colonies per m^2 per year (Connell *et al.* 1997). During 1962-1970, Connell (1973) determined this rate to be 5 colonies per m^2 per year, but there was more than an order of magnitude variation among individual species (0-24 colonies per m^2 per year). Such recruitment estimates represent composite measures combining the influence of larval settlement, coral fragment regeneration, and post-settlement juvenile survival. The 1962-1970 data indicate that more common species like *Acropora squamosa*, *A. digitifera*, *Porites annae*, and *P. lutea* at Heron Island exhibited high recruitment rates relative to the mean as did some less common species like *Pocillopora damicornis*, *Montipora foliosa*, and *M. hispida*. High recruitment rates among uncommon species suggests that they are replaced during succession as they experience disproportionately higher rates of mortality and become increasingly less abundant. For example, *P. damicornis* was common among colonizing species particularly in quadrats on the exposed reef crest (Connell 1973), but was not an abundant component of the predicted equilibrial assemblage at that site (Figure 4.2). Tanner *et al.* (1994) reported frequent transitions from pocilloporid corals to space dominated by acroporid corals at this site (Table 4.4). Further study of the mechanism governing the demise of this species during succession should be explored.

Species may occur in equilibrial assemblages more than, less than, or equal to the expectations based on relative colonizing ability. This last possibility is consistent with the null, random colonization model which predicts successional sequences in the absence of interspecific interactions (Lawton 1987). This model represents an extreme "supply-side" perspective in which abundance patterns are totally determined by factors controlling the availability of colonists. Greater than expected abundances of species with relative low colonizing ability (and low abundances of good colonizers) in equilibrial assemblages is compatible with 1) the facilitation model (in which early successional species modify the environment so as to favor competitively superior late successional species), 2) the tolerance model (in which late successional species are favored by environmental modifications but life-history attributes are not influenced by interspecific interactions), and 3) the inhibition model (in which longer-lived, late successional species are favored as early colonists die from physical extremes or attacks by natural enemies) (Connell and Slatyer 1977). The observation that good colonizing species are even more abundant in equilibrial assemblages than expected would be consistent with the current notion that there are no necessary life history tradeoffs between r- and K-selected traits (Begon and Mortimer 1986, Stearns 1992). Local selective regimes may favor unique combinations of attributes contributing to competitive success, longevity, tolerance of physical extremes and natural enemies, and colonizing abilities which are not compatible with simple predictions contrasting early versus late successional species.

The evidence from Heron Island and elsewhere provides mixed support for predictable relationships between relative colonizing ability and abundance in equilibrial assemblages. While some species (*e.g.*, *P. damicornis*) exhibit a tradeoff between rapid colonizing ability and traits promoting competitive success and tolerance to disturbances in late successional assemblages (Connell 1987), others do not. But if one focuses on good colonizing species (which appear to be replaced over successional time) and the poor colonizing species which replace them, one may be able to discriminate among the alternative mechanisms driving ecological succession in coral assemblages. One might use the rare or uncommon acroporid species, which can rapidly colonize lava flows in Indonesia, to determine why they do not persist in local assemblages (see Tomascik *et al.* 1996, section 4.2.1)]. Alternatively, one might select *Stylophora pistillata*. This species is known to colonize substrate rapidly in the Red Sea and is quite common in disturbed reef zones in shallow water. However, it is thought to be "competitively excluded" in deeper water when space is limited (Loya 1976b, section 2.2.3). This association between disturbance and high colonization rates is an important consideration in that coral recruitment is known to be inhibited by high coral and even macroalgal cover (*e.g.*, Hughes 1985, Connell *et al.* 1997). Consequently, successional rates (*i.e.*, species replacements) are likely to be quite low under undisturbed conditions.

In the Caribbean Sea, several studies over the years have focused on variation in recruitment rates and contrasted the patterns exhibited by assemblages of juvenile and adult corals. Bak and Engel (1979) conducted such a comparison for coral assemblages in the Netherlands Antilles. Although they found that the abundance of juveniles of some species actually "paralleled that of larger colonies", large discrepancies were noted for other species. For example, very few (<0.5%) of the 1183 juvenile colonies sampled in their surveys were noted for such common species in adult assemblages as

Acropora palmata, *Montastrea annularis*, *M. cavernosa*, *Madracis mirabilis*, and *Siderastrea siderea* (Bak 1977 and section 2.2.6). In contrast, *Agaricia agaricites* represented 60.6% of the juvenile corals overall dominating all depth zones (3-37 m deep) on reefs sampled in Curaçao and Bonaire. The difference in the abundances of these corals in juvenile and adult assemblages indicates that some species replacements may occur over successional time as species experience differential rates of mortality and recruitment. In an effort to examine relative rates of mortality, Bak and Engel (1979) monitored a cohort of 390 juvenile corals over 6-months to note 124 deaths or disappearances, but 55% of the mortality was attributed to unknown causes and *A. agaricites* actually experienced lower mortality (27%) than did other coral species (38%). The major known causes of mortality to juvenile corals included interactions with other spatial competitors, predation by parrotfishes and sea urchins, burial by sediments, and substrate collapse, but the relevance of these observations in terms of successional mechanisms is unclear.

In Jamaica, Rylaarsdam (1983) conducted a more detailed comparison of the relative abundances of coral species in adult and juvenile assemblages (Table 4.5). The most common adult corals at three sites (11-15 m deep) included *Acropora cervicornis* (1-49%), *Agaricia agaricites* (17-32%), and *Montastrea annularis* (7-32%) with some colonies estimated to have been over 50 years old. Assemblages of small corals at these sites (< 5 cm in diameter and no more than 5-10 years old) were dominated by *Agaricia agaricites* (58-63%), *Leptoseris cucullata* (13-18%), and poritid corals (8-16%). Even younger assemblages of corals recruiting to experimental substrates over 9-20 months were strongly dominated by agariciid corals (66-96%). Mortality to species with high recruitment rates was attributed to a variety of causes including overgrowth by algae and the colonial foraminiferan *Gypsina* as well as grazing by *Diadema*. Furthermore, Rylaarsdamm (1983) contrasted the extreme rarity of both acroporid and faviid corals (including *M. annularis*) in juvenile assemblages with their prevalence among adult corals (Table 4.5). These corals are favored once colonies escape strong size-dependent mortality on these reefs, but "routine processes" are apparently insufficient to account for their high abundance as adults. Rylaarsdam (1983) speculated that the explanation for this pattern may involve longer term processes such as slow competitive overgrowth and damage caused by rare, severe storms. Massive corals like *M. annularis* can be favored under some disturbance regimes, yet they also succumb to overtopping by acroporid corals and toppling during storms. Rapid growth rates and the ability to disperse by fragmentation and regenerate following storms can favor branching forms like *A. cervicornis*.

In another study conducted in Jamaica, Hughes (1985) monitored recruitment by coral larvae and fragments into 12-m^2 quadrats on a steep wall at depths of 10 and 20 m. Half of these quadrats were cleared of all epifauna, while the remainder (with living coral occupying 33-97% of the substrate) served as controls. During 1977-1984, most larval recruitment into cleared quadrats was attributed to *Leptoseris cucullata* (231 colonies), *Agaricia agaricites* (190 colonies), and *Tubastraea aurea* (98 colonies). The two agariciid species reached control densities in only four years. Recruitment by fragments was observed for *Madracis mirabilis* (76 colonies), *Porites furcata* (57 colonies), and *Acropora cervicornis* (15 colonies). No recruitment was observed for such locally common species as *Colpophyllia natans*, *Montastrea annularis*, *M. cavernosa*, and *Siderastrea siderea*. Once again there appear to be shifts in the abundance of corals over successional time as some good colonizing species are replaced by poor colonizing species, yet the mechanisms favoring this group of largely massive corals remains open to speculation. Hughes (1985) noted that adult colonies in the controls had an strong inhibitory effect on larval recruitment, yet no such differences in the rates of asexual recruitment were observed between controls and cleared quadrats. Processes which inhibit recruitment are likely to slow the rate at which species are replaced. However, it is unclear whether species replacements generally depend more on episodic recruitment and disturbance events or on more constant features of the local environment.

TABLE 4.5 Relative abundances of corals in adult (A) and juvenile (J) assemblages at three fore-reef sites off Discovery Bay, Jamaica, sampled in 1976 (after Rylaarsdam 1983).

Species	West Fore-reef				East Fore-reef	
	Site 1 11 m deep		Site 2 14 m deep		Site 3 14 m deep	
	A	J	A	J	A	J
Acropora cervicornis	49	-	1	-	21	-
Agaricia agaricites	17	59	32	58	20	63
Leptoseris cucullata	-	16	-	18	-	13
Siderastrea siderea	4	-	3	-	6	-
Porites astreoides	2	8	5	13	2	5
Porites furcata	6	5	12	2	-	1
Other poritid corals	-	2	-	1	-	2
Colpophyllia natans	6	-	6	-	-	-
Montastrea annularis	7	1	21	-	32	-
Montastrea cavernosa	-	-	6	-	14	-
Stephanocoenia michelini	-	3	-	3	-	7
Millepora spp.	1	2	7	-	1	-
Others	9	2	7	8	4	10

As a last example of comparisons between assemblages of juvenile and adult corals, I use a recent study conducted on nine offshore reefs along 38 km of the Florida Reef Tract (Chiappone and Sullivan 1996). These included high-relief spur and groove reefs (2.7-10.5 m deep), relict reef flats (4.5-7.8 m deep), and relict spur and groove reefs (7.2-18.6 m deep). Over a twelve-month period in 1993-1994, 916 juvenile corals were found within an area totalling 450 m^2 on these reefs. Sixteen coral species (5-12 per reef) comprised these juvenile assemblages. The highest densities of juvenile corals (>0.45 corals per m^2) were observed for *Agaricia agaricites*, *Porites porites*, *P. astreoides*, *Siderastrea siderea* and *Montastrea cavernosa*. For each of these species, Chiappone and Sullivan (1996) reported a significant positive correlation between juvenile and adult ("non-juvenile") densities. Unlike the studies noted above, this result does not support the views that species replacements occur as juvenile assemblages age nor the existence of an apparent tradeoff between colonizing ability and success in adult assemblages.

In fact, "parental abundance and composition may be a direct function of juvenile abundance". Post-recruitment processes would appear to be less important than the physical environment controlling recruitment and the disturbance regime at these sites. The disturbed nature of these assemblages is indicated by the low mean coral cover across all nine reefs (only 4%) and the fact that juvenile corals represented a very large proportion (20-52%) of all colonies sampled in these surveys (Chiappone and Sullivan 1996).

4.2.3 METAMORPHIC INDUCTION

Although there has been speculation in the literature that observed abundance shifts in coral communities recovering from disturbances represent succession (Pearson 1981, Connell 1987), we are only just beginning to appreciate the specific mechanisms controlling these processes. One area receiving considerable attention is directed at studying how planktonic larvae are chemically induced to metamorphose into benthic adults (Chia and Rice 1978, Morse and Morse 1996). For example, seawater extracts from the corals *Porites compressa* and *P. lobata* in Hawaii have been shown to induce metamorphosis in the corallivorous, nudibranch gastropod *Phestilla sibogae* (Hadfield 1978, Hadfield and Schever 1985). The entry of this predator into the benthic assemblage from the water column requires the presence of these specific prey species. Thus trophic complexity and the reassembly of this component of the coral community are promoted by the chemical induction of metamorphosis in predators by their prey.

Acanthaster planci provides another example in which larval metamorphosis into a benthic consumer is induced by benthic organisms. Juvenile sea stars of this notorious corallivore (see section 3.3.3) are known to spend a period of approximately six months following metamorphosis feeding on coralline and epiphytic algae (Lucas and Jones 1976, Moran 1986). In a study of recruitment on reefs in Fiji, Zann *et al.* (1987) found juvenile *A. planci* attached to the undersurfaces of coral blocks and rubble. In all cases, this substrate was encrusted with coralline algae, primarily *Porolithon onkodes.* Johnson *et al.* (1991) conducted laboratory experiments which documented high rates of settlement and metamorphosis of *A. planci* larvae onto coral rubble and the coralline alga *Lithothamnium pseudosorum.* Their studies indicate that this process is mediated by bacteria associated with the algal surface.

Among corals, the chemical control of settlement, metamorphosis, and recruitment has advanced to the stage where signal molecules are now being evaluated for species-specific induction (Morse and Morse 1996). Using three species of agariciid corals (*Agaricia tenuifolia, A. agaricites humilis, A. agaricites danai*), Morse *et al.* (1988) provided evidence that the former two species recruit preferentially to substrates covered by crustose coralline algae, yet differ in terms of recruitment to exposed or shaded surfaces. Metamorphosis in all three species was induced by crustose coralline algae, yet differed in the degree of specificity to particular algal species. *A. agaricites humilis* was most restricted in that only some species of crustose coralline algae induced metamorphosis. *A. agaricites danai* was the least restricted species. Morse *et al.* (1988) clearly differentiated the chemical characteristics of the inducer in this study from those of a water-soluble peptide previously isolated from crustose coralline algae and known to induce metamorphosis in gastropods. Thus multiple inducer substances occur in these algae. Further chemical studies (Morse and Morse 1991, Morse *et al.* 1994) have isolated the inducer of agariciid metamorphosis and resulted in the development of artificial substrates suitable for field experiments on the chemical basis of settlement in corals. These have yet to be linked to successional processes in coral communities, yet they offer an intriguing opportunity for the future. The maintenance of high diversity in coral communities may very well depend on such highly specific biological interactions.

4.2.4 INFLUENCE OF CONSUMERS

The assembly rules identified by Lawton (1987) included the influence of consumers on the rate and direction of community succession. The empirical evidence from coral communities is beginning to support this view, but much more needs to done. Given that many aspects of coral succession are slow, most current evidence appears to focus on herbivore-alga interactions because algae colonize substrate rapidly and have the capacity to exhibit species replacements much faster than corals. Sousa and Connell (1992) reviewed experimental results on the effects of herbivores on succession in marine systems using the three models of succession from Connell and Slatyer (1977) as the conceptual framework. The possible effects of herbivores on succession include acceleration, deflection, deceleration, cessation, and none (the null case). Sousa and Connell (1992) noted a rich variety of successional mechanisms in marine systems and the general prevalence of inhibition of late successional species by early colonists. Under the inhibition model, succession occurs only as a consequence of death or damage to these early colonists. The empirical evidence from coral communities includes the documentation of preferential feeding by the damselfish *Stegastes* (*Eupomocentrus*) *planifrons* on a variety of early successional algal species in the Caribbean (Irvine 1983). This herbivore accelerated the successional process favoring replacement by mid-successional red algae. Furthermore, "by selectively weeding out" late successional species of perennial algae (*e.g.*, *Dictyota* and *Jania*), damselfish also can slow the successional process by maintaining it at an intermediate stage.

In chapter 3, damselfishes were highlighted as keystone species in coral communities because they regulate algal diversity within territories through herbivory and interactions with other herbivores and they influence coral recruitment and zonation (section 3.3.2). Some of the best evidence for their effects on succession comes from the study by Hixon and Brostoff (1983, 1996) in which deceleration of algal succession was noted in the presence of damselfishes in Hawaii. The mid-successional assemblage dominated by red algal species persisted in the presence of damselfishes, but was replaced by late-successional macroalgae when damselfishes were excluded by cages within territories. Furthermore, an alternative successional trajectory leading to substrate domination by crustose coralline algae was favored under intense grazing by parrotfishes and surgeonfishes outside damselfish territories.

The evidence for how damselfishes influence the succession of corals is less clear probably due to the long time required for this process to occur. We know that coral recruitment can be affected by the intense grazing of sea urchins and the interaction between sea urchins and damselfishes (Sammarco 1980, 1982). In the Gulf of Panama, where coral zonation patterns appear to be partially maintained by damselfishes, Wellington (1982) speculated that coral succession would eventually lead to domination by *Pocillopora damicornis* even in deeper water where *Pavona gigantea* is normally dominant. Although this putative process is quite slow, he noted that large colonies of *P. gigantea* can, like *P. damicornis*, also provide refuges for damselfishes and thus promote the establishment of their territories and some degree of coral mortality. This, in addition to the transport and establishment of fragments of *P. damicornis* in deeper water, may lead to the eventual demise of *P. gigantea* in late successional stages as the typical zonation pattern observed on these reefs disappears (section 3.3.2).

Another way fishes may influence succession in coral communities involves their interaction with planktonic larvae [note the prevalence of planktivory among the most common fishes of the central Great Barrier Reef (Table 2.2)]. In a broad survey of a variety of benthic invertebrates, Lindquist and Hay (1996) contrasted the palatability of the larvae from brooding and broadcast-spawning species to the bluehead wrasse *Thalassoma bifasciatum* and the bicolor damselfish *Stegastes partitus*. These two Caribbean fishes are quite common generalist predators which feed on benthic and planktonic prey. In chemically rich taxa of prey (*e.g.*, sponges and gorgonians), broadcast spawners had more palatable larvae than did brooders. In contrast, the larvae from three species of Caribbean corals which brood their young (*Agaricia agaricites*, *Porites astreoides*, and *Siderastrea radians*) were "readily consumed by fishes". Since hard corals are not known to be chemically rich in secondary metabolites (Faulkner 1993), their larvae are likely to be particularly susceptible to planktivory. Thus in "consumer-rich" coral communities, the contrast in the palatability of highly susceptible and chemically defended larvae may have large, yet currently unexplored effects on relative rates of recruitment and the rate and direction of succession. After all, intense planktivory on some palatable species "can reduce the number of larvae enough to affect the distribution and abundance of adults" (Lindquist and Hay 1996).

Empirical investigations into larval mortality due to planktivory have typically been limited because most species have small larvae and their densities in the plankton are usually quite low. However, synchronous spawning of corals on the Great Barrier Reef [first documented by Harrison *et al.* (1984)] results in much higher than usual densities of eggs and zygotes. Planktivorous fishes such as the damselfishes *Acanthochromis polyacanthus* and *Abudefduf bengalensis* have been shown to switch from their normal omnivorous diets to feeding primarily on coral spawn (Westneat and Resing 1988). Likewise, the large tadpole larvae of ascidians have been shown in some cases to incur significant mortality from fishes in coral communities. Olson and McPherson (1987) observed 133 larvae of *Lissoclimun patella* at Davies Reef in the central Great Barrier Reef following their release from benthic colonies. The majority of these larvae (72%) were eaten by territorial damselfishes, 8% were captured by adult corals or zoanthids, 12% settled successfully on the bottom, and 8% were lost. Olson and McPherson (1987) contrasted these palatable larvae with those of *Didemnum molle* whose larvae are typically rejected by fish. I also note the study of *Diplosoma similis* in Hawaii where Stoner (1990) documented low rates of larval mortality due to damselfishes. The palatable larvae of this colonial ascidian suffered much higher mortality after they settled and metamorphosed into benthic juveniles. Given these species-specific differences, the general prevalence of damselfishes in coral communities throughout the world, and their aggressive behaviors, it would appear that planktivory on the larvae of benthic species may represent a major selective force shaping not only the palatability of larvae, but also spawning times, larval behaviors, settlement patterns, and recruitment into the benthic community. Variation in the magnitude of this source of mortality may very well contribute to significant differences in species composition and succession within and among coral communities.

4.3 Overview

Succession refers to the temporal progression of species replacements occurring in a developing community over ecological time scales. Evaluation of the evidence from coral communities indicates that predictable species replacements actually do occur. Following outbreaks of *Acanthaster planci* on reefs on Guam, predictable shifts in the abundance of coral species have been documented. These indicate the importance of life history attributes and strong biological interactions on the successional process. At Heron Island, coral communities have been studied long enough to generate a predictive successional model of species transitions. Although severe cyclones are very frequent at this location, the benthic assemblage is predicted to reach an equilibrial condition after 20-25 years. This equilibrium represents the balance among colonization events, disturbances, and species replacements. Local disturbance regimes can strongly influence the successional process. The biological mechanisms influencing the direction and rate of successional changes typically include facilitative and inhibitive interactions among resident species and potential colonists (*e.g.*, interspecific competition, facilitation or inhibition of colonization by resident benthic organisms, and planktivory on the larvae and eggs of colonizing species by fishes). Although alternative successional mechanisms have not been studied well enough in coral communities, we do know that a rich variety of processes are involved, the patterns of species replacements can be quite complex, and the entire process is highly variable from place to place.

5 INTERSPECIFIC COMPETITION

5.1 General considerations

In the preceding chapters, interspecific competition has been emphasized as an important process in coral and other marine epibenthic communities. Considerable empirical evidence has documented the influence of competition on species replacements, temporal successional sequences, species diversity, distributional limits, and the densities of interacting species. In addition, these effects vary along environmental gradients in physical conditions (environmental stress, disturbance, light, depth, flow regime, etc.), larval supply, and intensity of consumption by herbivores and predators. Thus the predicted consequences of interspecific competition must account for variation not only in competitive interactions, but also in these other factors [*e.g.*, as suggested by Menge and Sutherland (1987) and Roughgarden (1989), see section 1.4.1]. Although such an integrated perspective can result in complex patterns due to interactions among factors, it is important that our predictive framework incorporate these conditional relationships.

Before dealing in more detail with competition specifically in coral communities, it is appropriate to consider some general notions regarding competition in ecological communities. In an early review, Miller (1967) noted the influential contributions of Volterra (1926), Lotka (1932), and Gause (1935) to the development of what is called the competitive exclusion principle. This principle was most simply stated by Hardin (1960) as follows: "complete competitors cannot coexist". That is to say that populations of distinctly different taxa which utilize very similar (or identical) resources cannot coexist indefinitely when resources are in limited supply. In such cases, the sympatric population which is most efficient in converting resources into newborn individuals is predicted to exclude less efficient populations. Thus the focus of many empirical studies has been directed at understanding the mechanisms responsible for either competitive exclusion or the coexistence of competitors.

In another historical review, Jackson (1981) noted several contributions to the study of interspecific competition which are often ignored by current ecologists. Most notable were those by the British plant ecologists A.G. Tansley, E.J. Salisbury, and A.S. Watt. Tansley (1914) stressed the importance of experimental studies to reveal how interspecific competition contributes to community structure. Subsequently, Tansley and Adamson (1925) experimentally demonstrated how different levels of grazing by rabbits can modify the outcome of plant competition and thus deflect the successional process. Watt (1947) highlighted the importance of integrating the interactive effects of plants, animals, micro-organisms, soil, and climate on plant community dynamics. Competition alone does not dictate the successional sequence. Lastly, Salisbury (1929) evoked the role of evolved life history and morphological attributes in promoting the competitive success of species in ecological communities.

Jackson (1981) strongly emphasized how Salisbury's ideas were decades ahead of his time. The blending of competition theory with the evolutionary perspective generated by the "modern synthesis" to form the foundations of what we now call niche theory occurred much latter this century. Much of this body of theory describes how interspecific competition may influence resource utilization patterns in natural communities. The notions of niche breadth, niche overlap, resource partitioning, limiting similarity, and character displacement are central tenets of niche theory. All of these were based on the premise that communities are structured primarily by interspecific competition [see Schoener (1989) and Yodzis (1989)]. Predictions from niche theory include how active competition influences current resource utilization and habitat associations as well as how competition in the past has caused shifts in these attributes (Connell 1980, Jackson 1981, Roughgarden 1983, Schoener 1989).

In two other major reviews, Schoener (1983) and Connell (1983) independently evaluated the published experimental evidence documenting the occurrence of interspecific competition in nature. Using a set of 164 field studies published over more than five decades, Schoener (1983) reported that interspecific competition was detected in 90% of these studies. Each study employed "proper" controls and included manipulations of the abundance of one or more hypothetical competitors. Schoener (1983) found no significant differences among habitats as competition was detected in 91%, 94%, and 89% of the studies from 22 freshwater, 36 marine, and 106 terrestrial habitats, respectively. Connell (1983) assembled a sample from 72 papers describing 527 controlled experiments which were published in 1974-1982 in six major ecological journals. Overall, 46% of these studies documented interspecific competition and Connell noted "consistently higher frequencies of competition" among marine organisms.

When Connell (1983) restricted the analysis to the 15 studies reporting results from a single experiment in which only one species was monitored following the experimental manipulation, the frequency of interspecific competition was 93% and quite comparable to Schoener's result. Connell suggested that these higher values do not represent unbiased estimates of the frequency of interspecific competition in natural assemblages. Instead he suggested that they may reflect biases influencing the selection of target species and editorial policies favoring the publication of positive results. Possibly contributing to this bias is the fact that interspecific competition was strongly asymmetrical in many of these published studies (Connell 1983, Schoener 1983). Competitive release was often evident in inferior species upon removal or reductions in the density of the competitive dominant. The reciprocal experiment (when conducted) had no or little effect on the dominant species. In either case, however, there is ample evidence supporting the position that interspecific competition is common in nature.

Traditionally biological interactions between competing species have been categorized as either interference or exploitative competition (Elton and Miller 1954, Park 1954, Figure 1.8). The former includes cases in which one species directly interferes with another's access to some common resource (*e.g.*, aggressive territoriality and the defense of feeding and/or nesting sites). The latter is restricted to cases in which species interact only indirectly by differentially exploiting some common resource like food. Miller (1967) noted that interference competition appears to be "especially well developed" among vertebrates with highly evolved behaviors. In contrast, lower metazoans seem to interact more through exploitative competition. As indicated below, this generalization appears not to apply to metazoans in coral communities.

Because of confusion especially over interspecific competition for space (which is often a major limiting resource in epibenthic communities), Schoener (1983) proposed six different kinds of competition (Table 5.1). Consumptive competition clearly represents a form of exploitation as the consumer utilizes a resource (*e.g.* food, water, or a nutrient) and makes it unavailable to other consumers. Chemical and encounter competition are clearly forms of interference. Chemical competition includes interactions in which allelochemicals are used by one species to interfere with the use of resources by others. Encounter competition involves interactions between mobile organisms which result in some harm to a competitor (*e.g.*, loss of foraging time, loss of resources, injury, death). The remaining three categories cover various kinds of competition for space. They are somewhat ambiguous in that they include aspects of both exploitative and interference competition. For example, overgrowth competition often involves direct contact among encrusting organisms (see section 1.4.3) whereby one species interferes with another's access to food or light, yet one can also view this as exploitation of space. Likewise, territorial competition involves active behaviors which can interfere with feeding and nesting opportunities, but occupants of territories clearly are exploiting space as well. Lastly, pre-emptive competition, a form of competition involving primarily sessile organisms, includes cases in which such organisms occupy space and thereby make it unavailable to competitors. Thus pre-emption of space may be thought of as an exploitative mechanism, yet it may also involve the use of allelochemicals and aggressive behaviors as these organisms directly interfere with one another. Schoener (1983) suggested that his proposed categories need not be mutually exclusive of one another as some competitive interactions clearly involve more than a single mechanism.

TABLE 5.1 The distribution of six competitive mechanisms among ecosystems [freshwater (F), marine (M), and terrestrial (T)] and taxa [plants (P) and animals (A)] used in field competition experiments. One example of "an herbivorous fungus" is included among the terrestrial animal experiments (after Schoener 1983).

	Consumptive	Pre-emptive	Over growth	Chemical	Territorial	Encounter	Unknown
FP	0	0	1	1	0	0	0
FA	13	1	0	1	1	5	2
MP	0	6	4	1	0	0	0
MA	9	10	6	0	7	6	0
TP	28	3	11	7	0	1	9
TA	21	1	0	1	11	15	6
Sum	71	21	22	11	19	27	17

There are some interesting contrasts in the distribution of the six proposed mechanisms among the ecosystems and taxa in Schoener's review. Pre-emptive competition is common among marine plants and animals, yet apparently occurs at low frequency in freshwater and terrestrial ecosystems (Table 5.1). Overgrowth competition was the second most important form of interspecific competition among marine ecosystems, uncommon in freshwater ecosystems, and common among terrestrial plants, but not animals. Among the marine field experiments included in the reviews by Schoener (1983) and Connell (1983), there were only a few examples from coral communities. Schoener (1983) noted four experiments with herbivorous fishes in which territorial competition was noted. Two others documented consumptive and territorial competition in interactions involving sea urchins and damselfishes. These six experiments were published by Low (1971), Robertson *et al.* (1976), Sale (1974), and Williams (1978, 1981). The review by Connell (1983) included only the two studies from coral communities published by Robertson *et al.* (1976) and Williams (1981). These contributions will be considered more fully below.

5.2 Interspecific competition in coral communities

Here I present major examples of interspecific competition in coral communities using three different species assemblages. First, competition among scleractinian corals and other sessile taxa is considered with an emphasis on the richly diverse set of biological adaptations associated with these interactions. Second, I consider studies of spatial competition among cryptic metazoans. These have provided substantial support for the importance of overgrowth competition and they fostered the notion of competitive networks (*e.g.*, Jackson and Buss 1975), a potential mechanism for promoting coexistence in natural assemblages. Third, territorial competition among fishes is examined in some detail, especially those studies contributing to the notion that the acquisition of territories by juvenile fishes is a stochastic processes likened to a lottery (*e.g.*, Sale 1974). Other forms of interspecific competition occurring in coral communities are considered below only in terms of how they relate to these three examples.

5.2.1 INTERSPECIFIC COMPETITION AMONG CORALS

The competitive mechanisms involving scleractinian corals have been reviewed previously by several authors (*e.g.*, Sheppard 1982, Huston 1985a, Lang and Chornesky 1990). These corals engage in pre-emptive, overgrowth, and chemical competition as they acquire and defend space or are overgrown. In addition, Sheppard (1982) and others have noted that as corals use their extracoelenteric mesenterial filaments to interact with their neighbours, they may be acquiring nutritional resources. Thus these corals may be engaging in a form of consumptive competition. Huston (1985a) and many other authors have noted that rapid upwards growth of some corals, especially pocilloporids and acroporids, results in overtopping of other corals (a form of overgrowth competition which does not involve actual contact between competitors). Although overtopping clearly reduces light levels reaching overtopped colonies and thus has the potential to reduce the growth and survivorship of shaded corals, this is not always the case (Sheppard 1981, Stimson 1985). Lang and Chornesky

(1990) identified eight different competitive mechanisms employed by scleractinian corals. They are as follows: mesenterial filaments, sweeper tentacles, sweeper polyps, mucous secretions, histoincompatibility, overgrowth, overtopping, and water-borne chemical competition. Furthermore, as one expands the number of taxa interacting with these hard corals, one finds an even more diverse set of mechanisms associated with interspecific competition. Below, I highlight some of the major contributions to this fascinating subject.

Early observations of interspecific competition among corals at Heron Island indicate that overgrowth competition between branching and encrusting corals favored the former growth form (Connell 1973). Connell also noted overgrowth of branching coral by the large fleshy alga *Hydroclathrus clathratus.* Furthermore, he provided an example of how superior capabilities in one form of competition could be offset by another mechanism. In exposed pools on the north reef crest at Heron Island, overtopping by the tabular coral *Acropora hyacinthus* is typically favored (Figure 4.2). However, when this coral occurs close (within 2 cm) to massive corals like *Leptoria* sp., its growth is inhibited by mesenterial filaments extruded from the coelenteric cavity. Contact with these filaments can kill coral tissue at the growing edge of encroaching colonies.

During 1963-1972, Connell (1976) recorded a total of eighty two, interspecific interactions among twelve species of corals occurring in a single square meter on the reef crest at Heron Island. Thirty two of these interactions involved overtopping (indirect overgrowth), thirteen involved direct contact and active overgrowth. In the remaining thirty seven interactions, there was contact but no growth by either interacting colony over the observation period. These latter interactions are referred to as standoffs in that they tend to delay the rate of competitive exclusion and can persist for extensive periods of time. Using both indirect and direct interactions, Connell (1976) determined that the species rankings for these mechanisms "are similar, with two striking exceptions" (Table 5.2). The massive coral *Leptoria phrygia* never was observed to overgrow or be overgrown by another coral, yet it was the most aggressive in using mesenterial filaments. Connell placed it at the top of a digestive hierarchy based on interactions with several *Acropora* spp. (Connell 1973, 1976). However, he also noted its slow rate of growth and the probable use of mesenterial filaments as a defensive mechanism rather than one to secure additional space at the expense of competing corals. The other exception was *Acropora hyacinthus* which was ranked high in terms of overtopping ability, yet low in terms of the digestive hierarchy.

Connell (1976) also noted that rankings were not good predictors of the outcome of competitive exclusion in this system. This is especially true in cases where there were "reciprocal interactions of species of equal rank and the ability of lower-ranked species to sometimes stand-off species of higher rank". The former was illustrated by the similar overtopping capabilities of *A. hyacinthus* and *A. digitifera* (Table 5.2). The latter was most common in direct interactions involving *A. digitifera* or *A. palifera.* Interactions such as these contribute to an overall high frequency of symmetrical competition in this coral assemblage (85%). This distinguishes coral assemblages from those in which asymmetrical competition was so common [*i.e.*, subtidal epibenthic assemblages dominated by clonal organisms (Connell and Keough 1985) and the variety of terrestrial and aquatic situations evaluated by Connell (1983) and Schoener (1983)].

TABLE 5.2 Number of direct overgrowth (D) and indirect overtopping (I) interactions between species at Heron Island. Left column indicates successful species in ranked order. Data for interactions involving *Leptoria phrygia* were excluded (after Connell 1976).

Winning species	D/I observations — Losing species										
	1	2	3	4	5	6	7	8	9	10	11
1		1/1						0/1		1/1	
2			1/1	1/0			0/1		0/1	1/0	0/1
3		0/1				0/1	0/1		0/1	0/1	0/1
4			1/0						0/1		
5											
6										0/1	
7									1/0		0/1
8											
9											1/1
10											0/1
11											

Species designations: 1 = *Acropora hebes*, 2 = *Acropora digitifera*, 3 = *Acropora hyacinthus*, 4 = *Acropora valida*, 5 = *Acropora squamosa*, 6 = *Acropora nasuta*, 7 = *Acropora humilis*, 8 = *Stylophora pistillata*, 9 = *Acropora palifera*, 10 = *Pocillopora damicornis*, 11 = *Porites annae*

Competitive rankings among corals and the notion of a digestive hierarchy originated with the doctoral dissertation research conducted by Judy Lang on Jamaican corals. By placing pairs of different species together in laboratory aquaria, she was able to rank each species on the basis of the number of subordinate species they attacked and killed using mesenterial filaments. Thus Lang (1973) reported a hierarchical ranking which distinctly characterized the degree of aggressiveness of each species (Table 5.3). The most aggressive mussid and meandrinid corals in this hierarchy are generally small corals which "are comparatively minor constituents of Jamaican coral communities". Lang (1973) noted that "they occur with maximum density and number of species between 6 and 60 meters in the mixed and buttress zones of the reef crest, and in the fore-reef terrace and upper fore-reef slope zones of the seaward slope, where rapidly expanding acroporid, agariciid, and poritid corals are also abundant (see section 2.2.6). It is therefore not surprising that natural instances of interspecific aggression are commonly seen in these densely populated reef zones". It is also apparent that the most common coral species typically found on these reefs (*e.g.*, *Acropora palmata*, *A. cervicornis* and *Montastrea annularis*) are not among the most aggressive in the digestive hierarchy. Likewise, those species known to rapidly colonize substrate by larval settlement (*e.g.*, agariciid and poritid corals) are among the least aggressive corals (Table 5.3). These observations would then appear to support the position that mesenterial filaments generally function as a defensive or pre-emptive competitive mechanism.

Logan (1984) evaluated the digestive hierarchy among corals in Bermuda to emphasize geographic variation in the proposed hierarchy attributed to Lang (1973).

Based on replicated laboratory experiments, the rankings of corals in Jamaica and Bermuda are similar yet there are some discrepancies (Tables 5.3 and 5.4). For example, *Meandrina meandrites* had the most subordinate species in Bermuda as it clearly dominated interactions in all replicates with most other species, whereas *M. meandrites* and *Isophyllia sinuosa* were of equal rank in Jamaica. Likewise, the relative ranks of several species (*e.g.*, *Montastrea annularis, M. cavernosa*, and *Diploria labyrinthiformis*) differed somewhat between locations. More importantly, Logan (1984) noted the prevalence of variable interspecific interactions. Standoffs, mutual damage, and damage to each coral in separate encounters were reported as multiple outcomes of interspecific interactions. In the laboratory, the most aggressive coral *M. meandrites* engaged in variable interactions with *Isophyllia sinuosa, Scolymia cubensis*, and *Dichocoenia stokesi.* In the field, all pairwise encounters between species with more than seven observations were also variable (there were 21 such pairs). Although Logan (1984) attributed some of this variation to uncontrolled environmental variation, competitive mechanisms other than mesenterial filaments were also implicated.

TABLE 5.3 Hierarchical ranking of species based on the aggressive use of mesenterial filaments by common Jamaican corals (after Lang 1973).

Rank	Species	Number of subordinate species for the most aggressive corals
1	1	48
1	2	48
2	3	46
2	4	46
2	5	46
3	6	45
4	7	42
5	8	41
5	9	41
6	10	40
6	11	40
7	12	
8	13	
9	14 & 15	
10	16,17 & 18	
11	19	
12	20&21	
13	22&23	
14	24	
15	25	
16	26 & 27	

Species designations: 1 = *Mussa angulosa*, 2 = *Scolymia lacera*, 3 = *Isophyllia sinuosa*, 4 = *Mycetophyllia ferox*, 5 = *Meandrina meandrites*, 6 = *Mycetophyllia reesi*, 7 = *M. aliciae*, 8 = *M. lamarckiana*, 9 = *M. danaana*, 10 = *Scolymia cubensis*, 11 = *Isophyllastrea rigida*, 12 = *Montastrea annularis*, 13 = *Diploria labyrinthiformis*, 14 = *D. strigosa*, 15 = *D. clivosa*, 16 = *Montastrea* cavernosa, 17 = *Colpophyllia natans*, 18 = *Manicina areolata*, 19 = *Acropora palmata*, 20 = *A. cervicornis*, 21 = *Eusmilia fastigiata*, 22 = *Agaricia* spp., 23 = *Leptoseris (Helioseris) cucullata*, 24 = *Siderastrea* spp., 25 = *Porites* spp., 26 = *Madracis* spp., 27 = *Stephanocoenia* spp.

TABLE 5.4 Ranking of seventeen coral species based on the aggressive use of mesenterial filaments in laboratory experiments in Bermuda (reprinted with permission after Table 1 in Logan 1984). © Springer-Verlag 1984.

Rank	Species	Number of subordinate species	Relative frequency of variable interspecific interactions
1	5	15	3/16
2	3	14	7/16
3	13	12	8/16
3	Ds	12	7/16
3	16	12	9/16
4	12	11	7/16
5	10	10	10/16
6	14	9	8/16
7	Pa	5	11/16
7	Ff	5	11/16
8	Md	3	8/16
8	Mm	3	9/16
8	O	3	10/16
8	Sm	3	9/16
9	Pp	2	10/16
9	24	2	10/16
10	Af	0	8/16

Numeric species designations as in Table 5.3; other codes are as follows: Pa = *Porites astreoides*, Pp = *Porites porites*, Ff = *Favia fragum*, Md = *Madracis decactis*, Mm = *M. mirabilis*, O = *Oculina* spp., Sm = *Stephanocoenia michelinii*, Ds = *Dichocoenia stokesi*, Af = *Agaricia fragilis*.

Among scleractinian corals, another competitive mechanism employed in direct interactions between adjacent colonies involves the use of elongated sweeper tentacles. The very existence of these structures appears not to have been recognized until 1973. den Hartog (1977) was the first to suggest their function in interspecific competitive interactions which was further characterized as a defensive form of spatial competition in *Montastrea cavernosa* by Richardson *et al.* (1979). Furthermore, Richardson *et al.* (1979) speculated that sweeper tentacles could fend off attacks from the mesenterial filaments of *M. annularis*, a more aggressive species in the digestive hierarchy (Table 5.3). A similar situation in the eastern Pacific was documented by Wellington (1980). Although *Pavona* spp. are dominant over *Pocillopora* spp. in terms of digestion by mesenterial filaments (Glynn 1974), field observations substantiated that the latter species generally are competitively superior and cause significant tissue damage to the former species. Using pairwise interaction experiments in the field, Wellington (1980) found initial damage to pocilloporids caused by mesenterial filaments within 2 days of contact. One to two months later, the development of sweeper tentacles by pocilloporids had been induced and they were causing "necrosis and sloughing of the *Pavona* tissue" in 25 out of 34 interactions. Thus the use of sweeper tentacles reversed the relative competitive abilities of these corals based on digestive dominance and facilitated the eventual overtopping of slow growing massive pavonids by the faster growing branching pocilloporids.

Bak *et al.* (1982) documented the reversal of the competitive dominance relationships between *Madracis mirabilis* and three other more highly ranked species (*Eusmilia fastigiata*, *Montastrea annularis*, and *Agaricia agaricites*, see Table 5.3) on reefs in the Netherlands Antilles. In transplant experiments in which *M. mirabilis* and *E. fastigiata* were placed together in the field (<1 cm apart, but not touching), high partial colony mortality in the former species during the first three weeks was followed by decreasing mortality during the next 12-17 weeks. There were substantial increases in mortality to the latter species during these last few weeks. Likewise, within 5 hours of contact with *M. annularis* and *A. agaricites*, transplanted colonies of *M. mirabilis* were attacked by mesenterial filaments and mortality of the tissues of this subordinate species was evident. However, sweeper tentacle development in *M. mirabilis* over 8 weeks was strongly associated with increased mortality in *M. annularis* and *A. agaricites* 1-16 weeks after initiation of these experiments. All colonies of *M. mirabilis* bearing sweeper tentacles caused mortality in these other two species. Contact involving colonies lacking sweeper tentacles had much more variable consequences. Adding to this variability in the outcome of interspecific interactions was the induction of sweeper tentacles in a few colonies of *A. agaricites*. "This always resulted in the death of the adjacent *M. mirabilis* tissue."

Chornesky (1983) further explored the factors stimulating development of sweeper tentacles in *Agaricia agaricites*. In natural interspecific encounters observed on a Jamaican fore-reef, sweeper tentacles were initially present in 47-57% of interactions with four other scleractinian species, the zoanthid *Palythoa caribaeorum*, and the encrusting gorgonian *Erythropodium caribaeorum* (Table 5.5). Over an additional 10-month period, sweeper tentacles were observed in 65-80% of the colonies. Colonies which did not have sweeper tentacles at the beginning of experimental encounters developed them in approximately 30 days. Some of these colonies were responding to direct contact with opposing species, but others were engaged in encounters with fixed, 1-20 mm gaps between colonies. By the end of these experiments after 50-80 days, sweeper tentacles had developed in 40-90% of the colonies (Table 5.5). In additional experiments designed to distinguish between the effects of tactile contact (using artificial tentacles), tissue damage from the mesenterial filaments of *M. annularis*, and lesions created with hydrochloric acid, sweeper tentacles developed only in colonies exposed to mesenterial filaments. These studies documented localized sweeper tentacle development to tissues quite close (within 5 mm) to contact zones with other aggressively interacting cnidarians.

In a more detailed photographic study of interactions between pairs of interacting colonies in Jamaica (*i.e.*, ten *Agaricia agaricites-Montastrea annularis* and fourteen *Agaricia agaricites-Porites astreoides* pairs), Chornesky (1989) closely monitored changes in the position of colony edges, gap size, and mid-points between live tissue borders for 20 months. In 23 of these 24 interactions, "there was no consistent pattern of advance of one coral at the expense of the other". Instead, there were repeated cycles of reciprocal injuries by mesenterial filaments and sweepers tentacles followed by increases in gap size, regression of sweeper tentacles, and colony growth. The spatial dynamics of these encounter zones where characterized by frequent attacks and counter attacks, "transient shifts of coral edges", and oscillations in mid-point positions and gap size. As a consequence of repeated reversals in the dominance relationships between interacting colonies, there was little or no exchange of spatial resources over the observation period. Thus these mechanisms delay competitive exclusion and have the potential to promote the coexistence of competitors in these coral assemblages. These consequences are predicted to be "particularly important (1) in habitats where overtopping is rare, (2) among assemblages of shade-tolerant corals which persist below the canopies of overtopping species, or (3) in interactions between overtopping species which compete directly with one another" (Chornesky 1989).

TABLE 5.5 Relative frequency of colonies of *Agaricia agaricites* bearing sweeper tentacles in natural and experimental interactions. Sample sizes are given in parentheses (after Chornesky 1983).

	Natural			Experimental			
Opposing species	Initial		10-months	Direct contact		1-20 mm apart	
Porites astreoides	57%	(28)	71%	50%	(6)	--	--
Madracis decactis	47%	(17)	65%	80%	(5)	83%	(18)
Montastrea annularis	47%	(15)	80%	64%	(11)	56%	(34)
Montastrea cavernosa	--	--	--	--	--	90%	(10)
Palythoa caribaeorum	56%	(14)	--	40%	(5)		
Erythropodium caribaeorum	57%	(9)	--	50%	(6)		

In another long term photographic study, Karlson (1980) monitored overgrowth and standoff interactions among corals, zoanthids, and encrusting gorgonians in a Jamaican back reef zone. Although overtopping by small colonies of *Acropora palmata* did occur in this habitat, most of the substrate was unconsolidated rubble on which low-relief encrusting growth forms were most common. Repeated observations of naturally occurring encounters over 16 months indicated that the numerically dominant *Zoanthus solanderi* primarily engaged in standoffs. There were no apparent growth changes throughout the study at colony margins in contact with spatial competitors. Since this zoanthid was also overgrown by aggressive species like *Erythropodium caribaeorum* and *Palythoa caribaeorum*, it was not considered to be competitively superior. Furthermore, there were no obvious signs of tissue damage or gaps at the sites of these interactions and structures like mesenterial filaments and sweeper tentacles are not known to occur in zoanthids. Direct contact between *Z. solanderi* and the coral *Porites astreoides* inhibited the normal growth of the coral. These interactions are thought to be mediated by growth inhibitors contributing to the delayed expression of competitive dominance. The prevalence of *Z. solanderi* (and its congener *Z. sociatus*) in this habitat has been linked to a number of adaptations associated with intermittent disturbance (*i.e.*, storms) including those promoting pre-emptive competition (Karlson 1980, 1983, 1988). Interactions between the corals *P. astreoides* or *A. agaricites* with the two aggressive species noted above resulted in the corals being overgrown or in standoffs (Karlson 1980). Although it is unknown if these alternative outcomes are related to variation in the occurrence of sweeper tentacles in corals, the situation is further complicated by the discovery that the encrusting gorgonian *E. caribaeorum* also employs these structures as it competes for space (Sebens and Miles 1988).

Chemically mediated competition among corals can involve allelochemicals which promote directional growth and the frequency of some competitive encounters (Wahle 1980). At the other extreme, allelochemicals cause tissue necrosis and growth inhibition (Sammarco and Coll 1992). Wahle (1980) described how the hydrocorals *Millepora alcicornis* and *M. complanata* direct the growth of attack branches towards the gorgonians *Plexaura homomalla* and *Briareum asbestinum* to form "handlike structures that eventually contact, abrade, encircle, and encrust" nearby gorgonians (within 30 cm). Overgrowth of captured gorgonians by these hydrocorals can eventually cover the entire colony. In a 10-month experiment using *Millepora-Plexaura* pairs, it was shown that 1) the growth of attack branches was detectable after 1.5 months, 2) these branches were produced in response to living but not dead gorgonians, and 3) the response was directed more at target colonies positioned up-current rather than cross-current relative to hydrocorals. Thus attack branches do not represent a random response to inanimate erect objects and they are most likely stimulated by some water-borne chemical released by living gorgonians. Although the hydrocorals are only facultative encrusters, it appears this specific response to gorgonians may benefit hydrocorals by eliminating the need for them to produce their own skeletal support as they exploit their competitive advantage over gorgonians.

Tissue necrosis and growth inhibition are among the common outcomes of competitive encounters involving soft corals (Sammarco and Coll 1992). In the first of a series of experimental studies conducted on the Great Barrier Reef, Sammarco *et al.* (1983) reported the outcome of direct contact and non-contact encounters between the scleractinian corals *Porites cylindrica* and *Pavona cactus* and the soft corals *Lobophytum pauciflorum*, *Sinularia pavida*, and *Xenia* sp. Small soft coral colonies were transplanted into monospecific stands of scleractinian coral and monitored after four and seven months. The results document partial colony mortality, growth inhibition, and bleaching of scleractinian corals in response to these encounters. The magnitude of these effects on scleractinian corals varied among the species of soft corals they encountered. However, no soft coral suffered any apparent detrimental effects. In general, contact encounters had more severe consequences than did non-contact encounters, yet water-borne allelopathic effects were especially strong in non-contact encounters between *L. pauciflorum* and *P. cylindrica*. Similar results have been reported for encounters lasting 35 weeks between these same two scleractinian species and three additional soft corals (*Sarcophyton ehrenbergi*, *Nephthea brassica*, and *Capnella lacertiliensis*) (Sammarco *et al.* 1985). In addition, some scleractinian corals were overgrown by soft corals and necrosis to soft coral tissues was also reported. Necrosis of scleractinian tissue increased over the duration of this experiment, while that of soft corals decreased as colonies responded to contacts by secreting a "protective saccharide layer" (Sammarco *et al.* 1985). These studies highlight the high degree of interspecific variability in chemically mediated competition for space between scleractinian and soft corals.

The effects of local environmental conditions on the outcome of pairwise interactions between the soft coral *Clavularia inflata*, the encrusting gorgonian *Briareum stechei*, and the scleractinian *Acropora longicyathus* were investigated by Alino *et al.* (1992). In the central region of the Great Barrier Reef, mid-shelf reefs (*e.g.*, Britomart) are dominated by scleractinian corals and many inshore reefs (*e.g.*, Pandora) often have higher abundances of soft-bodied cnidarians. On Pandora, Alino *et al.* (1992) reported that scleractinian corals were being overgrown by soft corals and gorgonians "on a large scale" from 1981-1987 (based on photographs by T. Done). In a controlled transplant experiment, colonies were collected from Pandora, pairwise encounters were established on Pandora and Britomart, and growth changes were monitored bi-monthly for one year. In spite of some significant effects due to handling and transplantation, the salient result of this experiment is the change in the competitive relationship between *C. inflata* and *A. longicyathus.* On Pandora, the former species clearly overgrew the latter in encounters involving direct contact, while on Britomart this overgrowth "dropped to nil levels". Although both species were negatively affected by the new environment and *C. inflata* was reported to induce tissue necrosis in *A. longicyathus*, significant encroachment by this branching coral over substrate previously occupied by *C. inflata* was noted (*i.e.*, overtopping). Thus local environmental factors influence the expression of competitive dominance and further contribute to the variability in the outcome of interspecific competition. The putative environmental factors responsible for these effects in this transplant experiment include predators (*e.g.*, butterflyfishes), water turbidity and light regimes, and nutrient levels.

Similar patterns of overgrowth and allelopathic interactions have been anticipated and, in some cases, documented among non-coral inhabitants of coral communities. In his review, Sheppard (1982) noted toxic allelochemicals, aggression, and feeding interference as potential competitive mechanisms in a diverse set of taxa including sponges, sea anemones, bryozoans, and ascidians. Sponges are known to overgrow a variety of corals and, as in soft corals, allelochemicals have been implicated as competitive mechanisms facilitating their successful exploitation of space on reefs (*e.g.*, Porter and Targett 1988, Rinkevich *et al.* 1992). There is even a report of more rapid growth of the sponge *Terpios* sp. over living coral than on non-living substrate (Bryan 1973) and sponges on some reefs appear to be displacing scleractinian corals (Aerts and van Soest 1997). Likewise, recent evidence substantiates large increases in the abundance of ascidians (*e.g.*, *Trididemnum solidum*) at the expense of scleractinian corals on reefs (Bak *et al.* 1996).

At this point, it should be quite clear that interspecific competitive interactions among corals and other sessile taxa in coral communities can be intensely aggressive and quite complex. Multiple mechanisms are involved and any single interaction can be extremely variable in space and time (Lang and Chornesky 1990). The consequences of this complexity on community structure are difficult to predict even in the absence of modifying influences of the physical environment, consumers, and disturbances. However, some general patterns appear to be evident under restricted conditions. Rapid, undisturbed growth of tabular and/or branching corals can result in their numerical dominance as they overtop other growth forms (asymmetrical competition). At the same time, they may favor shade-tolerant forms. Symmetrical interactions among overtopping species should result in the coexistence of competitors. Direct contacts in many competitive interactions promote overgrowth and competitive exclusion, while other aggressive responses and counter measures result in standoffs, reversals, and competitive coexistence. Shifts in community structure as a result of these latter interactions are likely to be quite slow or not to occur at all. Chemical competition can be highly species-specific as competitors evolve mechanisms to successfully overgrow, defend against being overgrown, or escape these intense encounters. Thus, competitive exclusion or coexistence can result from these interactions. These alternative consequences are predicted to influence community structure quite differently [note the large difference in species diversity predicted for communities under exclusion and coexistence competition in the Menge-Sutherland model (Figure 1.11)].

5.2.2 SPATIAL COMPETITION AMONG CRYPTIC METAZOANS

The diverse assemblage of encrusting organisms which occur on the undersurfaces of foliaceous corals, walls and overhangs, and in crevices and caves has been extensively studied by Jackson and his colleagues on coral reefs in the Caribbean Sea. The primary taxa in this assemblage include colonial animals like sponges, bryozoans, and ascidians, and solitary forms like serpulid annelids, brachiopods, bivalves, and ahermatypic corals (Jackson *et al.* 1971, Jackson 1977a). In a 26-month experiment in which settlement panels were placed into artificial cryptic environments at a depth of 40 m off Discovery Bay, Jamaica, Jackson clearly documented how these colonial organisms colonized and overgrew solitary forms to become the dominant organisms (Figure 5.1).

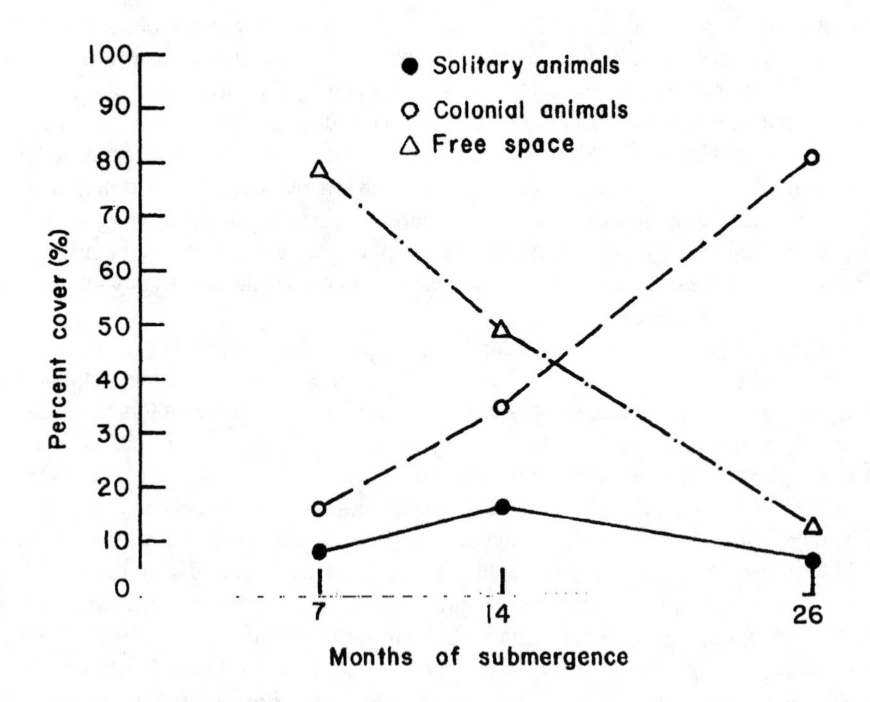

Figure 5.1 Relative abundance (percent cover) of solitary and colonial animals and free space (bare substrate) on panels placed in artificial cryptic environments at 40 m off Discovery Bay, Jamaica (redrawn after Jackson 1977a). © 1977 by The University of Chicago.

Free space was initially colonized by mostly solitary forms which occupied 80% of the substrate after 7 months. As colonial animals increased in abundance reaching 80% after 26 months, spatial competition by overgrowth became intense and the relative abundance of solitary animals decreased. Even some colonial species (*e.g.*, the bryozoan *Stomatopora* sp.) encountered severe overgrowth competition and became restricted to spatial refuges on bivalve shells (Buss 1979 and section 1.4.3). After 26 months, the relative abundance of colonial animals in these artificial cryptic environments matched those observed in natural cryptic environments (70.9-86.5%). Free space "is almost nonexistent" in these environments (Jackson 1977a). These results support the position that relative competitive abilities are major determinants of abundance patterns and species replacements in this assemblage.

The high abundance of colonial organisms observed after 26 months was strongly correlated with the depletion of suspended particulate food observed in these cryptic environments. Buss and Jackson (1981) reported that 89.2% of the bacteria and 79.4% of the naked cells (mostly unarmored flagellates) were retained by the suspension feeders in this assemblage (Table 5.6). Furthermore, they noted that most of the colonial organisms at the end of the experiment were sponges, a taxon well known for efficient suspension feeding (Reiswig 1971). Changes in the relative abundance of sponges and bryozoans over the course of this experiment (Table 5.6) appear to support the speculation that these two taxa actually compete for food as sponges "deplete the level of naked cells sufficiently to limit bryozoan populations". However, Buss and Jackson (1981) noted alternative explanations for the reduction in bryozoan abundance including the possibility that they are differentially eaten by predators and the substantiated fact that sponges generally overgrow bryozoans in this assemblage.

TABLE 5.6 Relative abundance (percent cover) of sponges and bryozoans on panels and retention estimates for suspended bacteria and naked cells in artificial cryptic environments (after Buss, L.W. and Jackson, J.B.C. 1981. Planktonic food availability and suspension-feeder abundance: evidence of *in situ* depletion. *Journal of Experimental Marine Biology and Ecology* 49, 151-161, with kind permission from Elsevier Science - NL, Sara Burgerhartstraat 25, 1055 KV Amsterdam, The Netherlands).

Time			Percent retained	
(Months)	Sponges	Bryozoans	Bacteria	Naked cells
7	5.4%	7.6%	5.73%	13.23%
14	28.1%	15.8%	64.60%	51.23%
26	74.2%	6.2%	89.20%	79.40%

Successful overgrowth by sponges may represent a form of chemical competition in that they are well known to contain a diverse range of bioactive compounds (Berquist 1978, Porter and Targett 1988). Jackson and Buss (1975) tested homogenates from nine species of sponges as well as two colonial ascidian species against three bryozoans (*Steganoporella* sp. nov., *Stylopoma spongites*, and *Reptadeonella violacea*), the brachiopod *Argyrotheca johnsoni*, the bivalve *Basilomya goreaui*, and two serpulid species. The bryozoan species were among the most common found under foliaceous corals on Jamaican reefs (Jackson 1979b), while *A. johnsoni* was "the most characteristic brachiopod" on these surfaces at depths of 30-67 m (Jackson *et al.* 1971). Sponge and ascidian tissue samples (2.5-20 ml) were scraped from foliaceous corals, ground and homogenized, and then diluted to a standard volume of 100 ml (Jackson and Buss 1975). The responses of the target species were assessed for 10 minutes after exposure to homogenates. All controls exhibited movement and feeding activities in this time period and all treatment effects were invariant among five replicates. Homogenates from four sponge and one colonial ascidian species killed at least one bryozoan species, yet none of the homogenates killed all bryozoan species nor any of the solitary forms tested. The sponge *Ectyoplasia ferox* caused the most death among bryozoans and the bryozoan *S. spongites* was most susceptible in that three of the eleven homogenates resulted in its death. These results suggest that the responses of the target species are species-specific. Furthermore, they "demonstrate the potential significance of allelochemicals in competition for space between ectoprocts (bryozoans), sponges, and ascidians". Since encounters among these colonial taxa are common in these cryptic environments, allelopathy may explain the dominance of sponges over bryozoans in the long-term experiments reported in Buss and Jackson (1981).

Three critical observations in their study led Jackson and Buss (1975) to formulate a novel hypothesis regarding the maintenance of species diversity in this cryptic assemblage and in other similar systems. First, they noted the extremely low levels of obvious predation and disturbance as well as the extreme lack of free space. The ability to acquire new space then requires superior overgrowth capabilities as species compete for this limited space. Second, they recognized the very high number of species (> 300) being maintained in the assemblage. This rivals the most diverse assemblages of metazoans known on earth. Lastly, they reported to have observed overgrowth relationships among three encrusting species in which no single species appeared to be the clear dominant. Rather than being hierarchical, these relationships formed "competitive networks". Jackson and Buss hypothesized that these networks maintain high diversity (promote coexistence) and reduce the likelihood that competitors are excluded from the assemblage. In the theoretical literature, these nonhierarchical relationships have been predicted to cause "quasicycles" of species replacements, so competitors tend to coexist with one another rather than to exclude inferior competitors from the assemblage (Yodzis 1978, 1989). If competitive networks are formed by random assembly processes, they are predicted to be quite common in species-rich assemblages (Yodzis 1978, 1989). On the other hand, if competitive relationships represent evolutionary responses to intense competition for space, they are more likely to occur in less speciose assemblages (Karlson 1985) such as that described by Buss (1986) (see section 1.4.3). The selective advantages associated with species-specific competitive mechanisms are clearly embedded in the network hypothesis (Jackson and Buss 1975, Buss 1976).

In order to determine whether competitive relationships among species in the cryptic reef assemblage actually form competitive networks or are hierarchical, two separate collections of foliaceous corals from depths of 11-30 m were examined. The first documented a very high frequency of overgrowth competition among a set of twenty common species. Buss and Jackson (1979) observed 152 of a possible 190 pairwise interactions to find that the colonial ascidian *Didemnum* sp. could overgrow all the species in this assemblage except for one sponge species (Table 5.7). Four sponge species were highly ranked in terms of overgrowth frequencies, while bryozoans were generally ranked much lower. In addition, the outcome of overgrowth interactions among bryozoans were much more variable than among sponges (Table 5.7). Regarding the network hypothesis, Buss and Jackson (1979) reported numerous instances in which lower ranked species overgrew more highly ranked species. The net effect of these overgrowth patterns was the formation of "highly complex nonhierarchical patterns of interference competitive abilities". For example, the most highly ranked ascidian was overgrown by the third-ranked sponge #4 which in turn was overgrown by the fourth-ranked sponge *?Toxemna* sp. This sponge was overgrown by the ascidian to form a 3-species network. Another includes the unnamed bryozoan #10 (ranked twelfth in this network!), the fifth-ranked sponge #5, and *?Toxemna* sp. (Table 5.7). In spite of the general advantage of sponges over bryozoans noted by Buss and Jackson (1981), eight of forty nine bryozoan-sponge interactions invariably favored the bryozoan species and four of these were cases in which lower ranked bryozoans overgrew more highly ranked sponges (Buss and Jackson 1979, Table 5.7).

TABLE 5.7 Observed overgrowth interactions among twenty common encrusting species in the Jamaican cryptic reef assemblage. Most interactions were observed at least five times (after Buss and Jackson 1979). © 1979 by The University of Chicago.

Taxa	Rank	Number of subordinate species	Relative frequency of variable interspecific interactions
Colonial ascidian			
Didemnum sp.	1	15	0/16
Sponges			
Sponge #1	11	2	1/10
?*Toxemna* sp.	4	11	2/16
?*Tenaciella* sp.	2	13	1/19
Sponge #4	3	12	2/15
Sponge #5	5	10	0/18
Sponge #6	10	4	2/19
Sponge #7	9	5	1/12
Ectyoplasia ferox	7	8	2/14
Bryozoa			
Steginoporella sp. nov.	7	8	8/19
Stylopoma spongites	12	1	5/18
Reptadeonella violacea	12	1	6/17
Trematooecia aviculifera	11	2	3/7
Disporella sp.	12	1	5/14
Smittipora levinseni	11	2	1/13
Bryozoa #9	6	9	2/15
Bryozoa #10	12	1	3/13
Foraminifera			
Gypsina sp.	6	9	6/18
Coralline alga #1	8	7	2/16
Coral			
Madracis sp.	9	5	1/15

Overgrowth observations between sponge species are more difficult to interpret due to their capacity to co-occur as epizoic forms rather than as clear competitors for space (Rützler 1970, Sarà 1970).

A second collection of foliaceous corals was conducted to more fully characterize the variable intraphyletic overgrowth interactions occurring among the most common bryozoans in the cryptic reef assemblage. Bryozoans dominated the undersurfaces of these corals occupying 49.9% of the substrate area (Jackson 1979b). Interspecific encounters resulted in overgrowth in 98% of 117 observed interactions and most overgrowth (91%) involved seven of fifteen bryozoan species. Rankings of successful overgrowth interactions indicate that *Parasmittina* sp. was the most aggressive. It overgrew other species in 79% of 19 encounters (Table 5.8). Overall, there were 25 cases in which a lower-ranked species overgrew a more highly ranked species. The third-ranked *Steginoporella* was overgrown most frequently by lower ranked species, but there were even three cases in which *Parasmittina* was overgrown. Thus the overgrowth relationships among these species form competitive networks rather than simple hierarchical sequences. Jackson (1979b) concluded that these complex interactions are influenced by a number of factors which, as a whole, contribute to variability in the outcome of interspecific competition (Table 5.8) and the formation of networks. These factors include 1) the capacity for frontal budding [an innovation which facilitates the elevated growth of colonies (see Lidgard and Jackson 1989)], 2) redirection of the growing edge of colonies (thus influencing the encounter angle of attacks and the probability of successful overgrowth), and 3) resistance to fouling of the colony surface (fouled surfaces were commonly overgrown in this study).

TABLE 5.8 Outcome of overgrowth interactions for seven bryozoan species in the Jamaican cryptic reef assemblage. All species were observed in at least ten intraphyletic interactions, but only 13 of 21 interspecific pairs were observed more than once (after Jackson 1979b).

Species	Rank	Wins	Losses	Ties	Frequency of overgrowth by lower-ranked species	Relative frequency of variable specific interactions
Parasmittina sp.	1	15	3	1	3	2/3
Stylopoma spongites	2	13	5	0	5	1/3
Steginoporella sp. nov.	3	35	25	0	12	5/5
Reptadeonella violacea	4	31	29	2	5	3/6
Microporella sp.	5	5	8	0	0	1/4
Smittipora levinseni	6	5	14	1	0	1/2
Cribrilaria	7	3	17	0	0	1/4

Subsequent studies conducted on this cryptic reef assemblage at depths of 10 and 20 m indicate that both sponges and bryozoans differ substantially in terms of their depth distributions and their location under foliaceous corals. Bryozoans were generally more common at 10 m and near the growing edges of corals, while sponges predominated in deeper water and farther from growing edges (Jackson and Winston 1982, Jackson 1984). Correlated with these differences were a number of potentially important factors which can influence the abundance of these competitors. These include predation (which is substantially higher in shallow water), the availability of food (which is probably higher at the edge of corals), and the availability of free space at growing edges where unimpeded colony growth and recruitment of juvenile colonies can occur. Among the common bryozoans, *Steganoporella* sp. nov. was found to be particularly abundant in edge assemblages at 10 m, while *Reptadeonella costata* (=*plagiopora*? =*violacea*?) was more common more than 20 cm from edges at this depth (Jackson 1984, Jackson and Hughes 1985). In contrast, *Stylopoma spongites* was more common at the edges of foliaceous corals at 20 m. Thus, the structure of this assemblage is spatially complex and discretely patchy. Futhermore it would appear that this structure is less controlled by purely competitive processes than originally conceived by Jackson and Buss (1975). The effects of competition are modified by the presence of spatial refuges and the heterogeneity created by consumers, disturbances, and patchy colonization processes (*e.g.*, Menge and Sutherland 1987, Figure 1.11). Nevertheless, there is considerable overlap in spatial distributions and the intensity of overgrowth competition in this assemblage, especially in deeper or more cryptic locations, cannot be denied. Competition must be considered an important component of the selective regime influencing these encrusting organisms.

5.2.3 TERRITORIAL COMPETITION AMONG FISHES

Instead of covering the extensive literature on all aspects of interspecific competition among fishes in coral communities (see reviews in Sale 1980, 1984, 1991a, Roughgarden 1986, Montgomery 1990), I focus here only on a few seminal contributions to our understanding of territorial competition. Because territorial behaviors result in the predictable association of relatively small fishes (*e.g.*, damselfishes) with specific sites within these communities, they lend themselves to manipulative experiments and have yielded significantly to our general understanding of how competition can influence community structure. As noted previously, Low (1971) provided some early experimental evidence for the occurrence of interspecific competition. This study was conducted on an outer reef flat at Heron Island where the herbivorous *Pomacentrus chrysurus* (=*flavicauda*) defended small (approximately 2 m^2) territories of coral rubble and sand. Within territories, Low (1971) documented aggressive interspecific interactions between this species and thirty eight species of primarily herbivorous or omnivorous fishes. Among sixteen species reported not to elicit this aggressive behavior, fourteen were carnivores. Eight captured *P. chrysurus* which were released outside territories also did not exhibit aggressive behavior. Following the removal of six *P. chrysurus* from their territories, there were large increases (as high as 600%) in the numbers of fishes entering these sites after only 2-5 days (Table 5.9). Within defended territories, Low (1981) estimated that 93% of the observed 4.2 aggressive encounters per 10-minute interval were devoted to defending algal resources from interspecific competitors. "This impressive expenditure of time and energy... emphasizes its importance" in the life of this damselfish. At the scale of individual territories, these aggressive behaviors effectively exclude competitors.

TABLE 5.9 Herbivorous (H) and omnivorous (OM) fishes observed in six damselfish territories before and after the removal of *Pomacentrus chrysurus* (=*flavicauda*). Observation periods were 10 minutes per territory per day (after Low 1971).

Species observed	Before removal			After removal	
	Jan.10[a]	Jan.18	Jan.20[a]	Jan.20	Jan.22
P. chrysurus (H)	0	0	0	8	7
Other pomacentridae (H/OM)	1	0	1	4	9
Scaridae (H)	0	0	0	3	60+
Chaetodontidae (H/OM)	0	0	0	0	3
Other roving fishes	0	0	0	1	3
Gobiidae					
Fusigobius spp. (OM?)	5	8	7	7	7
Blenniidae					
Salarias fasciatus (H)	6	3	7	12	15
Escenius mandibularis (H)	6	2	4	2	12
Other benthic fishes	2	4	1	5	15
Total fishes	20	17	20	42	131

[a]These are supportive data from six additional unmanipulated territories.

In a second study noted for the documentation of interspecific competition, Robertson *et al.* (1976) examined schooling in the facultatively territorial, striped parrotfish *Scarus iserti* (=*croicensis*). Along the Caribbean coast of Panama, this parrotfish forms three types of social units: nonterritorial stationary groups, territorial groups generally composed of the largest individuals, and roving schools which move widely over the reef to feed on algae (Shapiro 1991). Robertson *et al.* (1976) documented how parrotfish schooling exposes these fish to lower rates of attack from territorial individuals and allows them to feed on algae at higher rates than non-schoolers (*i.e.*, "loners" in groups of fewer than five individuals). In experiments in which the pugnacious damselfish *Stegastes* (=*Eupomacentrus*) *planifrons* were removed from their territories, feeding rates by territorial, schooling, and lone non-territorial parrotfish all increased dramatically above control levels. In addition, the feeding rates of four other benthic-feeding fishes associated with parrotfish schools (*Acanthurus coeruleus, Chaetodon capistratus, C. ocellatus*, and *C. striatus*) also increased significantly in response to damselfish removals. Thus not only did Robertson *et al.* (1976) substantiate territorial competition over algal resources between damselfishes and parrotfishes, but they also documented associational benefits to multiple species (including predators) which join parrotfish schools.

In Jamaica, Williams (1978) also documented interspecific competition by repeatedly removing damselfish from territories for a period of 7 weeks. The study site was located in a leeward reef environment at Discovery Bay which was dominated by the staghorn coral *Acropora cervicornis*, corals heads of *Montastrea annularis* and *Diploria* spp., dead coral rubble, and extensive sand corridors. Prior to removals, all but one of 18 experimental territories on *A. cervicornis* were occupied by adults of the threespot damselfish *Stegastes planifrons*. The exception was occupied by the dusky damselfish *S. dorsopunicans*. Most dusky damselfishes at this site were restricted to peripheral territories on coral rubble. Initially, vacant territories created by damselfish removal were recolonized by progressively smaller threespot damselfishes. These individuals were most likely non-territorial damselfish prior to the removals in that almost no turnover was observed among occupants of control territories. However, with the continued removal of damselfishes, turnover in controls increased as did colonization of vacant territories by immature threespot and dusky damselfishes. By the end of the experiment, dusky damselfishes represented 10% of all the removed damselfishes and it occupied 33% of the experimental territories. Interspecific competition by threespot damselfishes normally excludes these dusky damselfishes from their territories.

Also at this site, Williams (1981) used manipulative field experiments to test for the existence of interspecific competition among three common herbivores: the threespot damselfish *Stegastes planifrons* and two sea urchins, *Diadema antillarum* and *Echinometra viridis* (see preliminary studies noted in section 3.3.2). Within 27 patches of the staghorn coral *Acropora cervicornis*, all nine possible combinations of three treatments per urchin species were maintained for two weeks. Urchin densities were either doubled, reduced to zero, or not altered in controls. In another experiment, territorial damselfish were repeatedly removed from half of 34 coral patches over a 10-day observation period. Densities of these putative competitors during February of 1976 were 4.4 damselfish, 5.8 *Diadema*, and 10.1 *Echinometra* per m^3. In general, the removal of one sea urchin species resulted in increased densities in the other. Urchin additions "tended to inhibit population increases of the other (urchin) species, except under the influence of threespot damselfish intervention". Damselfishes tended to interfere more with *Diadema* than *Echinometra.* Densities of the former urchin responded rapidly to damselfish removals by increasing 57% within one day. Densities of *Echinometra* increased by 49% three days following the removal of damselfish. Williams (1981) concluded that "interference competition by the threespot damselfish appears to mediate competition between the two echinoids" thus allowing them to coexist in this system. Thus this example of territorial competition documents direct and indirect effects of interspecific interactions involving damselfishes.

In contrast with the studies noted above, Sale (1974) examined territorial competition at a larger spatial scale so as to include the spatial configuration and occupants of multiple territories. At Heron Island, *Pomacentrus chrysurus* is one of several similar species which exploit algal food resources and maintain territories on coral rubble. Sale (1974) referred to this group as a guild (Root 1967), although one might also refer to it as an ensemble as suggested by Fauth *et al.* (1996). This group includes at least eight pomacentrid species which co-occur within reef crest and flat environments (3-5 species per site). Over observation periods of 4-8 months, Sale (1974) mapped the spatial distribution of territories within six, 200-m^2 rubble patches. In spite of significant recruitment and mortality to damselfishes over this period, the total number of individuals of each species remained remarkably stable. Sale reported that territory boundaries persisted "over long periods of time" as long as fish were not disturbed. When damselfish territories were vacated because of death or emigration, recruitment occurred by newly settled juveniles or by adults previously on the periphery of the study area. Experimental removal of fourteen adult fishes [five *Stegastes* (*Pomacentrus*) *apicalis* and nine *Pomacentrus wardi*) from one mapped area resulted in a large shift in the proportion of territories occupied by these two species. After one day, most of the area was occupied by adults which had extended their territories from the periphery. Much of the space which had previously been occupied by *P. wardi* at this site was occupied by *S. apicalis* following the removal procedure. After 13 months, the territorial boundaries bore little resemblance to those before the experimental treatment.

The facts that undisturbed territories were stable and that multiple species in the guild co-occurred for extended periods of time indicates that territorial competition does not lead to competitive exclusion at scales including multiple territories. Species within this guild share the spatial resource. Furthermore, the evidence indicates that space on these reefs was in limited supply and rapidly utilized once made available. In fact, Sale (1974) hypothesized that the primary mechanism promoting the coexistence of these fish species depends upon the nature of competition for space as fish settle from the plankton and begin to occupy space in the benthos. He likened this process to a lottery in which the first larva to encounter available space would win that space. Once settled, juveniles can easily defend this space from further colonization. This probabilistic process is largely unpredictable in that it requires vacant territories on the reef as residents die or emigrate and passive larval dispersal from natal habitats by currents. Thus this mechanism has the potential to explain species coexistence and the high diversity known for fishes in coral communities. Rather than invoking the conventional niche-based notions of resource partitioning and limiting similarity as a mechanism for promoting species diversity, Sale (1974) used a nonequilibrial explanation. This was a harbinger of a major shift in ecological paradigms which has come to dominate current trends in community ecology (section 3.4, Chesson and Case 1986, Ricklefs and Schluter 1993).

Sale (1977) further developed this lottery hypothesis for the maintenance of the high diversity of coral reef fishes by arguing that there is considerable overlap among co-occurring species in their utilization of both spatial and food resources. In addition, he noted that the availability of space on coral reefs is spatially and temporally unpredictable. It is generated by predation and disturbance events on scales of "meters rather than kilometers" and "months rather than decades". Thus a shifting mosaic of very limited available space may have favored the evolution of large numbers of widely dispersed offspring in these fishes. Once these offspring settle and begin defending their own territories, they become quite sedentary. In general, they do not engage in energetically costly searches for new territories nor do they "oust prior residents from suitable space". If this lottery hypothesis is correct, Sale (1977) predicted that the species composition at any particular site is unstable in that it will not recover to its prior state following disturbances. Instead, the recovery process will be unpredictable being governed by stochastic recruitment events. Secondly, Sale (1977) predicted the diversity of coral reef fishes within habitats should be directly correlated with the rate of small-scale disturbances. More disturbed habitats (*e.g.*, those with greater exposure to storm damage or higher densities of predators) will have more available space over longer periods of time for the recruitment of a wider variety of larval fishes.

It is competition for space among these fishes as they enter the benthic environment that will dictate the structure of these fish guilds.

The suggestion that lottery competition controls the species composition of reef fish guilds has contributed greatly to vigorous debate among ecologists over the relative influences of larval dispersal, recruitment, and interspecific competition on the structure of ecological communities. Furthermore, it has stimulated empirical studies particularly on larval dispersal and recruitment processes in coral communities. In an extensive review of the literature mostly published after 1979, Doherty and Williams (1988) evaluated three alternative general models explaining how populations of coral reef fishes are replenished by the migration of juveniles and recruitment by larvae. The first is a classical niche-based equilibrium model which invokes density-dependent regulation of population sizes. The second model invokes partitioning of a limited resource among multiple species within a guild (*i.e.*, the lottery hypothesis) and predicts that density-dependent processes will result in an equilibrial density of the guild, yet species composition and the density of each species will be unstable. The third is purely nonequilibrial with unstable guild structure and population densities as populations fluctuate due to extrinsic disturbances and other density-independent processes. Interspecific competition is unimportant under this third model.

Based on their evaluation, the lottery hypothesis "can be eliminated with most confidence" in that they found only limited empirical support for this model. A similar degree of skepticism over this hypothesis appears in the review by Montgomery (1990). In addition, Doherty and Williams (1988) found little evidence for resource limitation on the densities of individual species or guilds of coral reef fishes. In contrast, there was substantial support for nonequilibrial explanations invoking recruitment limitation. Many natural populations of coral reef fishes appear to exist well below the resource-based carrying capacity of the environments in which they live (*e.g.*, Robertson *et al.* 1981). Sale (1991b) apparently agreed with some of this assessment in stating that "I see no way in which organisms with the type of life history typical of reef fishes could engage in interspecific competition according to the tenets of conventional theory (*e.g.*, MacArthur 1972, Diamond 1978)." He, like many others, noted that these fishes are iteroparous, highly fecund, their larvae are widely dispersed, and they experience very high rates of early mortality. Thus it should not be surprising that they often occur "below 'carrying capacity.' Yet when their numbers rise and competition takes place, they are predestined to enter into lotteries" (Sale 1991b).

These conclusions are reminiscent of the decades-old debate over the importance of density-dependent regulation of population sizes vs. the extrinsic control of population fluctuations by stochastic, density-independent processes [see the seminal contributions of Nicholson (1933), Andrewartha and Birch (1954), and den Boer (1968)]. Such debates have persisted in the literature (*e.g.*, Hassell *et al.* 1989, den Boer 1991) often because the authors continue to use different criteria to test these alternative models, studies are conducted on different scales, and compromise positions are thus almost impossible to reach. However, most authors would appear to agree that coral reef fish assemblages should receive more empirical attention especially directed at integrating 1) the biology of fish larvae and juveniles with that of adult fishes, 2) dispersal and metapopulation dynamics, and 3) the influence of scale on the assessment of community structure.

In so doing, Roughgarden (1986) suggested that we might view "competition communities" on the basis of the population structure of most component species (open vs. closed) and the degree to which their resources can be partitioned. Thus, assemblages of coral reef fishes might be considered competition communities in which local populations are largely open (thus decoupling resource availability from immigration and population "births") and resources are partitionable. Although competition may have important effects on the dynamics of these communities, they are fundamentally different from assemblages dominated by organisms with largely closed populations (*e.g.*, island birds and lizards, lacustrine fishes, pond snails, etc.) or those in which resources are not partitionable (*e.g.*, intertidal and subtidal hard substrate assemblages). He also noted the increasing volume of evidence supporting the notion that coral reef fishes do compete for food and that partitioning of food resources is more common among top carnivores than among fishes of lower trophic status. There is even an example of habitat partitioning and strong interference competition for food within a guild of herbivorous, territorial surgeonfishes at Aldabra (Robertson and Gaines 1986). In similar fashion, Montgomery (1990) also noted evidence for resource partitioning and size-dependent competition between adult fishes in coral reef communities.

Thus I conclude this section with the comment that these issues are still subject to debate and much more empirical work is needed. In fact, Robertson (1996) recently called for a re-evaluation of the importance of interspecific competition among territorial coral reef fishes as a consequence of his long-term study of Caribbean damselfishes. On the coast of Panama, there are six congeners of the genus *Stegastes* which co-occur. Adults of *S. planifrons*, *S. dorsopunicans* and *S. diencaeus* aggressively defend territories on coral reefs there, while adults of *S. leucostictus*, *S. partitus* and *S. variabilis* are much less aggressive. The former species are strongly reliant on algae for food and they typically maintain "single-owner" territories. The latter species have a wider variety of feeding habits and two of these species are known to share territories. During a 4-6 year experiment, Robertson (1996) removed aggressive damselfishes from sixteen isolated patch reefs (>94% of these were *S. planifrons* and the remainder were *S. dorsopunicans*), removed the less aggressive *S. partitus* from eight others, and maintained sixteen patch reefs as controls. Prior to the experiment, *S. planifrons* accounted for 56.3%, *S. partitus* 31.5%, *S. variabilis* 10.4%, and *S. dorsopunicans* 5.8% of 3003 damselfishes on these patch reefs. In response to the removal of *S. planifrons*, populations of *S. partitus* and *S. variabilis* doubled in size and the spatial distribution of the former species shifted from primarily peripheral to more central locations on patch reefs. Removal of *S. partitus* had no effects on the abundance of congeners. Hence, Robertson (1996) concluded that the abundance and distribution of *S. planifrons* and *S. partitus* are governed by asymmetric competition "and not by a competitive 'lottery' among species that have equal abilities to hold space, and among which space is allocated on a first-come, first served basis".

Furthermore, Robertson (1996) countered the notion that the observed patterns were recruitment-driven by noting that the increase in the abundance of *S. partitus* in response to the removal of *S. planifrons* required four years to reach a maximum (this is approximately equivalent to four generations). This increase "was not associated with a similar trend in recruitment". Nevertheless, active interspecific competition is not universally important in determining observed distribution and abundance patterns. Robertson (1996) recognized the general role of habitat partitioning and refuges in reducing the importance of interspecific competition among many coral reef fishes and even among some of the congeners examined in this study. Ultimately, he suggested that "generalizations about the relative influence of inter- and intraspecific competition on the abundances of such fishes will only become possible through experimental studies that examine effects of competition in relation to long-term population and recruitment dynamics". At present, such studies are few and far between.

5.3 Overview

Interspecific competition is complex, spatiotemporally variable, yet pervasive in coral communities. This complexity is enhanced by a rich variety of competitive mechanisms (see Table 5.1) and by interactions between competition and other factors (*e.g.*, predation, disturbance, and environmental conditions). Competitive exclusion, pre-emption, symmetrical and asymmetrical competition, competitive reversals, induction of competitive mechanisms, niche partitioning among competitors, and "coexistence competition" have all been documented in coral communities. Species turnover as a result of interspecific competition may be rapid, exceedingly slow, or absent. The prevalence of pre-emption in some assemblages of corals (and their allies) greatly diminishes the potential influence of interspecific competition on distribution and abundance patterns. On the other hand, rapid overgrowth and overtopping by branching corals in other assemblages is thought to be the most important determinant of these patterns. Some competitive mechanisms are specific to particular combinations of species and are likely to represent evolved responses favoring exploitation of competitive advantages or reduction of competitive disadvantages. In cryptic environments, encrusting metazoans engage in intense competitive interactions which strongly influence abundance patterns and the rate of species turnover. Among territorial fishes, distribution and abundance patterns can be strongly influenced by interspecific competition, yet other factors can also be important (*e.g.*, habitat partitioning, refuges, recruitment dynamics).

6 CONSUMER-RESOURCE INTERACTIONS

6.1 Coupled populations, food webs, and interactive community processes

The exploitation of trophic resources by consumers in ecological communities has a number of potential consequences on community structure. As noted previously, consumers can influence and even stabilize the spatial distribution, density, species composition, and diversity of consumed species (chapters 1 and 3). They can also influence the rate and direction of community succession (chapter 4). Some of these effects are direct consequences of consumption, while others are indirect. Consumers can mediate the outcome of interspecific competition or influence trophic levels below that of the species they consume. These effects of predators on lower trophic levels oppose the influences of resources on higher trophic levels. In some trophic interactions, either top-down or bottom-up effects may predominate. In others, both effects may occur simultaneously [e.g, the classic 10-year oscillations in the density of snowshoe hares is now thought to be controlled by fluctuations in the abundance of both plants and predators (Krebs *et al.* 1995, Stenseth 1995)].

Thus the study of any particular consumer-resource interaction should be conducted with the understanding that it is embedded within a web of interactions, which may include important interaction chains and modifications (see section 1.4.1). Abrams (1987) cautioned field ecologists working on specific consumer-resource interactions to "be aware of the large range of possible indirect effects that may occur". It is also important to recognize that strong consumer-resource interactions can selectively favor the evolution of morphological, chemical, and behavioral adaptations which reduce the risks of mortality among resource species or increase the probability of successful resource exploitation by consumers. Being subject to natural selection, these interacting species have the capacity to undergo evolutionary changes which can occur over relatively short periods of time [*e.g.*, some life history attributes of freshwater fishes respond to shifts in predation pressure in as little as four years (Reznick *et al.* 1997)].

In considering the theoretical basis for studying consumer-resource interactions, one can begin with the fundamental notions of functional and numerical responses (*e.g.*, Yodzis 1989). The functional response occurs as individual consumers in a population alter their rate of consumption with changes in resource density. Typically this rate increases with resource density to reach some asymptote representing consumer satiation, tradeoffs associated with the capture and handling of the resource, or changes in consumer behavior. The numerical response describes how the consumer population is influenced by changes in resource density in terms of the per capita growth rate (Yodzis 1989) or population size (Ricklefs 1990). This population response incorporates changes in the rate of consumption per consumer (the functional response), but is given in terms of the number of consumers produced as a consequence of resource consumption.

Following the characterization of functional and numerical responses, consumer-resource interactions can then be represented mathematically with a pair of differential equations describing the dynamics of the interaction as a closed system (Yodzis 1989). A complete two-dimensional depiction of this system is created by plotting consumer density as a function of resource density in what is called phase space. One can graphically illustrate the direction and magnitude of changes in the location of any point in this space (*i.e.*, the temporal trajectory of any density combination) along with consumer-resource isoclines which characterize equilibrium conditions. In section 3.1, it was noted that one or multiple equilibria may characterize a given dynamical system. Thus the trajectory of changing consumer-resource densities may converge towards point equilibria (or oscillating limit cycles). However, phase space may also include unstable regions where trajectories diverge and even reach the axes where one or both species are predicted to go extinct.

Classic models of consumer-resource dynamics include those by Lotka (1925) and Volterra (1926) which represented these interactions as oscillatory limit cycles. Further embellishments on these models as well as the introduction of alternative models over the years have resulted in a range of predicted patterns. These include the stable coexistence of each species at point equilibria, damped and undamped oscillatory fluctuations, local extinctions, persistence of interactions due to the presence of environmental heterogeneity or refuges for resource species, phase shifts between multiple equilibria due to external perturbations or internal threshold phenomena, consumer switching behaviors involving multiple resource species, catastrophic population outbreaks of consumers, and even population fluctuations due to deterministic chaos (see Yodzis 1989, Ricklefs 1990). Thus empirical studies of consumers and resource species have focused on a commensurate range of topics spanning the gamut from behavioral to community ecology. At the most fundamental level, these studies address the dynamics of consumption as the primary determinants of the abundances of these interacting species. This is a logical consequence of viewing these interactions as a closed system.

A broader perspective regarding consumption in ecological communities is attained by focusing on the food web in which specific trophic interactions are embedded (see sections 1.1 and 1.4.1). Food webs have structure in terms of the number of interspecific connections between resource species and consumers, number of trophic levels (food chain length), and complexity due to such phenomena as omnivory, cannibalism, and intraguild predation [*i.e.*, predator-prey interactions between competitors within a guild (Holt and Polis 1997)]. The dynamical properties of food webs are generally evaluated on the basis of stability criteria or in terms of the constraints on the process of community assembly (Yodzis 1989). For example, energetic constraints, body size and design constraints, environmental perturbations, and evolved feeding adaptations all are thought to limit the assembly process (Pimm 1984, Lawton 1988).

The study of food-web properties has resulted in some very broad generalizations and some degree of controversy. Lawton (1988) summarized the empirical evidence and theoretical explanations for ten of "the most interesting" patterns thought to characterize food webs in natural communities. His review included recognition of the early contributions to this subject by Shelford (1913) and Elton (1927), the seminal books published by Cohen (1978) and Pimm (1982), and a review of 113 published food webs (Briand and Cohen 1987). Cohen (1989a) succinctly presented five, food-web patterns as laws: food "cycles are rare, chains are short, and there is scale-invariance in the proportions of different kinds of species (top, intermediate, and basal species), in the proportions of different kinds of links (top-intermediate, top-basal, and intermediate-basal links), and in the ratio of links per species" (Cohen 1989a). Scale-invariance means that a particular food-web attribute does not change with the number of species in the web. This assessment was based on the evaluation of 23 terrestrial, 32 freshwater, and 58 marine food webs [only one of these marine webs was based on a coral reef community (see section 6.2)].

Sugihara *et al.* (1989) analyzed an additional 60 food webs collected from invertebrate-dominated communities in freshwater and terrestrial environments. They generally confirmed the existence of scale-invariant properties of food webs. Only the proportion of basal species varied with species number as it decreased slightly in larger food webs. In contrast, the proportions of top and intermediate species were scale invariant. A more thorough re-evaluation of these same 60 food webs failed to confirm most of these putative scale-invariant properties (Martinez 1994). Large food webs tended to have more links per species and a higher proportion of intermediate species than webs with fewer species. The proportions of top and basal species and links between them decreased with the size of food webs. Only the proportions of top-intermediate and intermediate-basal links appeared to be scale-invariant.

With this controversy, it becomes difficult to summarize the properties of real food webs in simple quantitative terms without some qualifications regarding scale-dependent variation. Sugihara *et al.* (1989) considered three food-web properties:

• Food-chain length is typically restricted to five or fewer trophic levels (*i.e.*, ≤ 4 links). However, exceptionally long food chains have been reported for planktonic assemblages in the open ocean where there may be as many as 10 links (Briand and Cohen 1987). Cohen (1994) highlighted this general pattern of longer food chains in the open ocean by contrasting them with shorter chains in terrestrial, freshwater, and coastal habitats.

• The proportion of observed links in a food web relative to the maximum possible (network connectance) decreases with increasing numbers of species in a web (also see Yodzis 1989). The product of species number and network connectance is thought by some to be scale-invariant. Sugihara *et al.* (1989) reported a mean value of 2.1 (range 0.5-3.9) for this product.

• The ratio of predator to prey species in a food web generally varies from 1-3 although Sugihara *et al.* (1989) reported one extreme example with a ratio of 14. Using eight published studies with 16.7-93 species, Martinez (1994) reported scale-dependent values for the proportions of different kinds of species, different kinds of links, and the ratio of links per species (Table 6.1). These values match the patterns established by his analysis of the 60 food webs.

TABLE 6.1 Food-web properties reported in the literature in terms of number and proportion of species, links per species, and proportion of links among top (T), intermediate (I), and basal (B) species (after Martinez 1994). © 1994 by The University of Chicago.

Number of species	Proportion of species			Proportion of links				Links per species
	T	I	B	T-I	T-B	I-I	I-B	
16.7	29%	53%	19%	35%	8%	27%	30%	1.9[1]
17.2	13%	80%	7%	30%	2%	51%	18%	4.0[2]
26.4	10%	85%	5%	20%	1%	70%	9%	6.0[3]
92	28%	68%	4%	30%	0.5%	60%	10%	4.4[4]
93	1%	86%	13%	0%	0%	91%	9%	11.0[5]
44	18%	68%	14%	23%	3.2%	55%	18%	5.0[6]
75	11%	73%	16%	--	--	--	--	6.3[7]
30	0%	90%	10%	--	--	--	--	9.6[0]

Sources: [1] Cohen *et al.* (1990), [2] Warren (1989) Open water community, [3] Warren (1989) Margin community, [4] Hall and Raffaelli (1991), [5] Martinez (1991), [6] Goldwasser and Roughgarden (1993), [7] Winemiller (1990), [8] Polis (1991).

In another recent re-evaluation of food webs, Goldwasser and Roughgarden (1997) found that "almost all known food-web properties" are seriously biased due to the incompleteness of the data. Using their own food web constructed for a terrestrial community on the island of St. Martin (Goldwasser and Roughgarden 1993) as well as a published data base of food webs (Cohen 1989b), their analysis revealed a strong effect of sampling intensity on most food-web properties. Thus it would appear that most food webs represented in the data base have been too poorly sampled to generate an unbiased perspective of food-web properties. Only the estimated number of species in a food web and the standard deviation associated with food-chain length were insensitive to sampling effort. The former was artificially limited to the total number given in the data base and thus not a "truly robust" attribute of food webs in the field. The latter was only a minor property Goldwasser and Roughgarden had used in their initial description of the St. Martin food web. This assessment clearly indicates the premature nature of any generalizations about food-web properties. As noted over a decade ago by Lawton (1988) and Paine (1988), more thorough studies are needed.

There is now accumulating evidence to support the notion that the biological interactions depicted by food webs (or more generally by interaction webs) have important effects on community structure and ecosystem processes. This is occurring despite the enormous complexity of real food webs (see Polis and Strong 1996) and the apparent difficulties associated with characterizing their fundamental properties. Strong experimental evidence from terrestrial, freshwater, and marine communities indicate that trophic interactions have significant direct and indirect effects. Some of the best marine examples have already been highlighted in chapters 1 and 3. Three non-marine studies recently featured in *Ecology* offer evidence that predators and herbivores in lakes can even influence nutrient recycling through consumer-mediated, top-down effects (Vanni and Layne 1997, Vanni *et al.* 1997). Both top-down and bottom-up effects can predictably influence primary producers in old-fields (Schmitz 1997). This same result was noted earlier for the herbivorous snowshoe hare. These examples illustrate how "the issue of using field experiments to quantify the dynamical couplings and the form and strengths of species interactions is a current focal point of food-web ecology" (Schmitz 1997).

Two final general points regarding trophic interactions can be made before proceeding to consider coral communities. First, it should be recognized that food-web diagrams typically oversimplify the complexity of the real world. The very identification of a single trophic level for many species is inappropriate given the ubiquity of omnivorous feeding habits. Polis and Strong (1996) highlighted omnivory (including detritivory and saprophagy) and intraguild predation as major components of food-web complexity. In addition, they emphasized interactions such as symbioses (including those between zooxanthellae and many coral reef invertebrates) and disease-host interactions as important components of "highly reticulate" trophic interactions. If we are to understand how natural communities really function, ecologists need "to rethink, quantify, model, and test the interplay" between this complexity and community dynamics (Polis and Strong 1996). After all, there is considerable support for the view that general patterns of trophic structure and dynamics actually do characterize broad categories of organisms (Hairston and Hairston 1997). In other words, the notions of trophic levels and food webs are useful tools for conceptualizing some features of ecological communities.

A second related point is that the relative magnitude of the direct and indirect effects of consumer-resource interactions varies with the influence of other factors. Hixon and Menge (1991) noted how habitat complexity and the relative availability of prey refuges in a habitat interact with predation and interspecific competition to influence community structure. At a slightly larger spatial scale, one can consider "spatial subsidies" provided to local habitats as resources or consumers move among habitats (Polis and Strong 1996). Interactions between species commonly found in different habitats can be important. For example, Hay (1981) provided evidence for the exclusion of competitively superior algae from coral reef habitats in the Caribbean by reef-associated herbivores. These algae were relegated to the limited hard substrates found on adjacent sand plains. Herbivores clearly limited the membership of algal assemblages in one habitat, but not the other. Thus the influence of consumer-resource interactions may be evaluated across larger spatial scales encompassing multiple habitats or even regional gradients in environmental conditions and recruitment. For example, consider the variation in fish trophic guilds across the central Great Barrier Reef (section 2.3) or recruitment gradient noted for intertidal organisms from the Gulf of California to the Bay of Panama (section 1.4.1).

6.2 Food-web interactions of coral reef fishes: an example

Hiatt and Strasburg (1960) conducted an extensive evaluation of the food and feeding habits of 233 species of coral reef fishes at Arno, Bikini, and Eniwetok Atolls in the

Marshall Islands. Based on this study, they generated a partial food web with five trophic levels (Figure 6.1). This web was among the 113 webs used in the analysis of food-web properties by Briand and Cohen (1987). It is a partial web because it does not include invertebrates other than zooplankton and corals. Furthermore, it represents the trophic relationships of slightly less than 40% of the fish species known for this region. In addition, one should note that interactions are represented in this web to occur between broad categories of species rather than individual species. Only 27 links are depicted in Figure 6.1, whereas many of the interactions they characterize are much more species-specific. Thus important trophic interactions such as those involving individual keystone species (section 3.3), specialized consumers, or particularly well defended resource species cannot be addressed using this web.

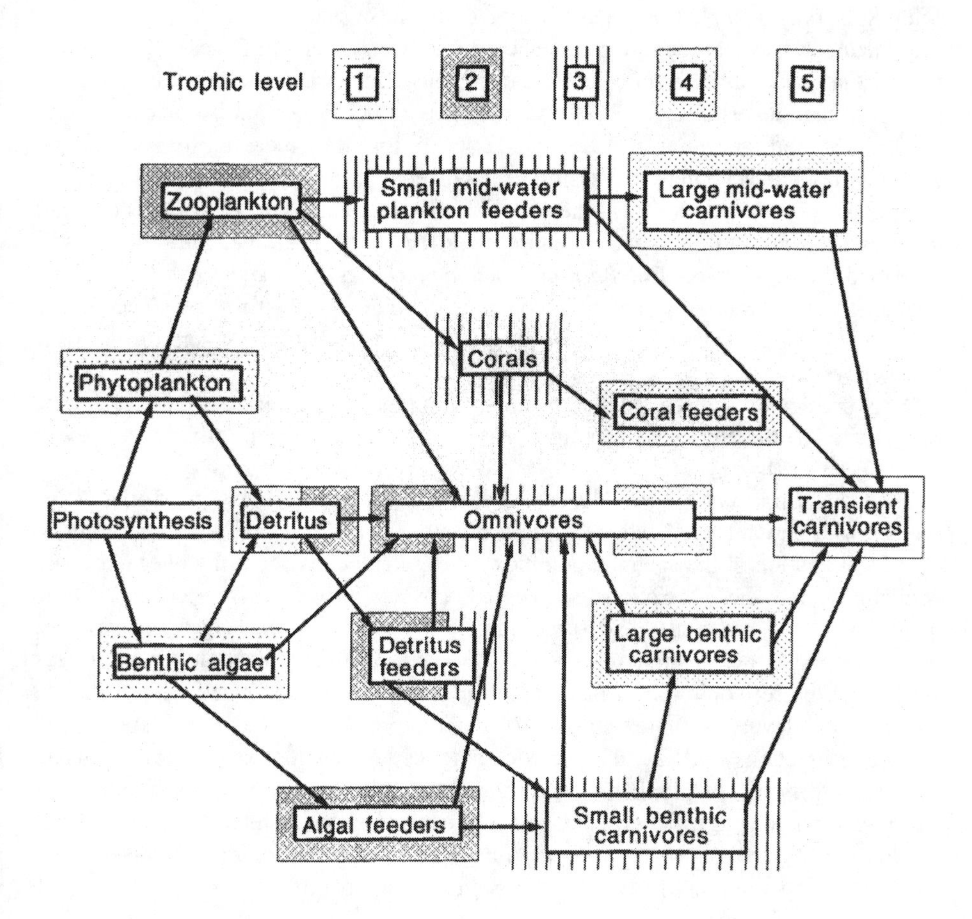

Figure 6.1 A food web for coral reef fishes at Arno, Bikini, and Eniwetok Atolls in the Marshall Islands (redrawn after Hiatt and Strasburg 1960).

Nevertheless, Hiatt and Strasburg (1960) noted several general patterns which emerged from their detailed study:

- Herbivory was more prevalent among evolutionarily advanced taxa than among primitive, "highly carnivorous" forms. The direct exploitation of plant food resources by fishes evolved with the radiation of specialized herbivores [most notably the surgeonfishes (Acanthuridae), damselfishes (Pomacentridae), and rabbitfishes (Siganidae)]. Overall, herbivory in this coral community included the consumption of phytoplankton by zooplankton and benthic algae by a variety of fishes representing sixteen families.
- Only eight species of detritivorous fishes representing three families were noted. Thus most detritus apparently enters the food web of this community through invertebrate consumers.
- Similarly, only a single scavenger was noted, the nurse shark.
- Zooplankton was fed upon by corals and only thirteen species of fishes.
- A very large and diverse set of other carnivorous fishes was represented by the three highest trophic levels in the web. These included bottom feeders, mid-water predators, corallivores, and large, transient, top predators. In that top predators in many marine communities are becoming increasingly endangered [*e.g.*, see Dayton *et al.* (1998) regarding "missing animals" and the lost megafauna of California kelp communities], these carnivores in the Marshall Islands should be noted. Hiatt and Strasburg (1960) identified the gray sharks *Carcharhinus melanopterus* and *C. menisorrch*, the smooth dogfish *Triaenodon obesus*, the barracuda *Sphyraena genie*, and the tunas *Gymnosarda nuda*, *Katsuwonus pelamis*, and *Euthynnus affinis yaito* as top predators on the fifth trophic level.
- Thirty species of omnivorous fishes representing eleven families and 13% of all the fish species in this study were assigned as a group to the second, third, and fourth trophic levels (Figure 6.1).

Hiatt and Strasburg (1960) also emphasized the strong association of many fish species with particular habitats on these coral atolls. Thus the food web represents an oversimplification of the spatially complex set of interactions occurring over the entire community. Specific associations were noted between individual fish species and tidal pools, reef flats, surf zones and surge channels, sandy habitats, ledges and caverns, mid-water and surface communities, and coral heads. As a generalization, top carnivores were much more mobile covering a wide range of habitats compared with fishes on lower trophic levels. As noted above, many herbivorous fishes have quite specialized feeding adaptations and are thus restricted to certain habitats and types of algae. For example, some surgeonfishes were noted to feed exclusively on the filamentous algae growing on compacted sand. In addition, restricted habitat associations and limited vagility are reinforced by many of the behavioral adaptations associated with territorial competition in herbivorous fishes (see section 5.1.3).

6.3 Plant-herbivore interactions in coral communities

As the role of herbivorous damselfishes and sea urchins as keystone species in coral communities has already been developed (section 3.3), I use a broader perspective here to emphasize a range of topics included in the study of plant-herbivore interactions. At a 1990 symposium devoted to the study of these interactions in all marine benthic communities, several contributions specifically focused on coral communities or included them in more general reviews. For example, Polunin and Klumpp (1992) presented a study of carbon flux on Davies Reef in Australia linking algal primary productivity to projected yields in fisheries directed at macrograzers, planktivores, invertebrate-feeders, and piscivores. Brawley (1992) reviewed the role of small invertebrate grazers in a variety of marine communities including several examples from coral reefs. Hay and Fenical (1992) considered the role of algal chemistry in mediating interactions with herbivores. Horn (1992) focused on the evolutionary convergence of feeding and digestive mechanisms in herbivorous fishes. Morse (1992) briefly summarized research on the influence of algae on larval settlement and metamorphosis in a variety of invertebrates including some herbivorous species (see section 4.2.3). Branch *et al.* (1992) considered the effects of algal "gardening" by herbivores (most notably the territorial damselfishes featured in sections 3.3.2 and 5.1.3). Spencer (1992) reviewed the importance of invertebrate grazers and other taxa on the erosion of substrates (coral reef environments provide some of the best information on "biological abrasion rates"). Lastly, Davies (1992) reviewed the nutritional significance of endosymbiotic associations between zooxanthellae and such anthozoans as corals, anemones, and zoanthids. Thus the symposium ranged through topics in nutritional physiology and anatomy, chemical ecology, larval biology, community ecology, fisheries, evolution, and geology. Below, I highlight three of these topics as particularly important aspects of plant-animal interactions which have not been developed in earlier chapters of this book.

6.3.1 DINOFLAGELLATE-CORAL INTERACTIONS

The endosymbiotic associations between zooxanthellae (*i.e.*, dinoflagellates in the genus *Symbiodinium*) and anthozoans is perhaps one of the most important examples of symbiosis in any marine benthic community. These species interact in a mutually beneficial manner to provide a nutritional basis for coral growth and thus food for coral consumers and physical structure to the habitat. Since coral communities often flourish in waters characterized by low nutrient levels (see review by Sorokin 1993), the symbiotic association with dinoflagellates is likely to influence the distribution and abundance of corals since zooxanthellae promote nutrient retention and recycling within the coral host (Muscatine and Porter 1977). In particular, zooxanthellae are known to take up and retain excretory ammonium from the host (Muscatine and D'Elia 1978). They recycle it back to the host as alanine (Lewis and Smith 1971, Trench 1971) and possibly other amino acids (Davies 1992). Using the widely distributed coral *Stylophora pistillata*, Rahav *et al.* (1989) estimated that up to 90% of the nitrogen utilized for protein synthesis in zooxanthellae is recycled within the coral host. Although this high level of recycling probably represents an overestimate because nitrogen lost to the environment was assumed to be negligible, it nevertheless emphasizes the potential for this symbiotic association to facilitate the occurrence of corals in nutrient-poor environments.

Photosynthesis by zooxanthellae generates oxygen and fixes carbon at variable rates depending on the species and several environmental factors (*e.g.*, light, temperature, depth, water clarity, etc.). Within the coral host, the symbionts generally occur at high density in gastrodermal cells. Photosynthetically fixed carbon may be utilized by these symbionts or translocated to the host chiefly in the form of glycerol or fatty acids (Davies 1992, Figure 6.2). Again using *Stylophora pistillata*, it has been determined that more than 95% of the fixed carbon can be translocated to the coral (Muscatine *et al.* 1984). This organic carbon is either metabolized by the host, stored as lipids, or released to the environment as mucous-lipid and utilized as food by a variety of heterotrophic consumers (Davies 1992). Other consumers feed directly on coral polyps, on crushed coral skeleton and the associated biota, or on the released gametes and larvae of corals.

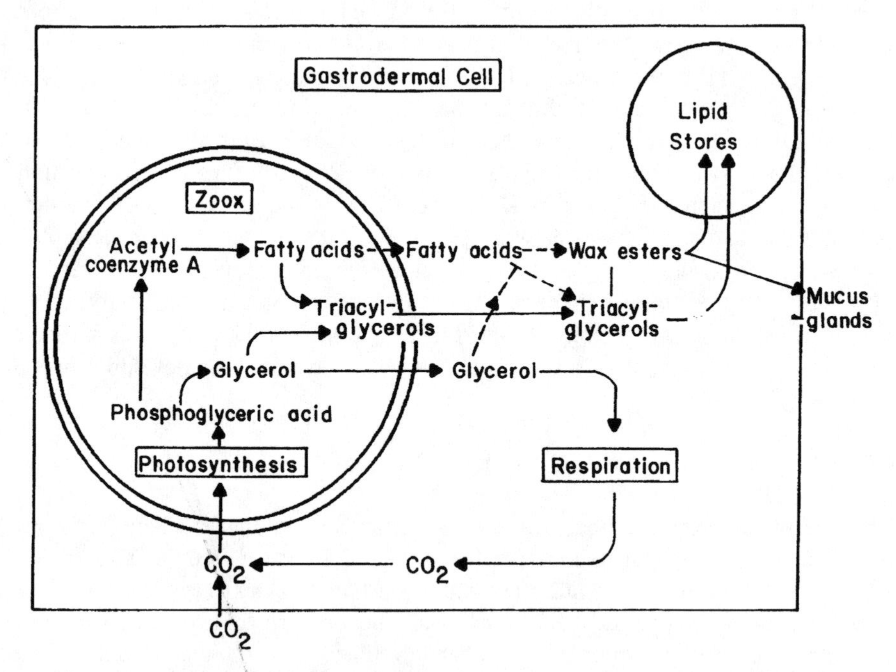

Figure 6.2 Major probable routes of carbon metabolism in an anthozoan gastrodermal cell with zooxanthellae and exposed to light. Broken lines indicate routes for which evidence is uncertain (redrawn after Davies 1992 in *Plant-Animal Interactions in the Marine Benthos* by permission of Oxford University Press).

In order to determine how much photosynthetically fixed carbon is produced and potentially available to consumers, one can use energy budgets which have been generated for a number of coral species. These permit the evaluation of primary production and heterotrophy as sources of energy along with expenditures for maintenance, growth, and reproduction. Davies (1992) identified a study by Edmunds and Davies (1986) as a particularly detailed analysis of the fate of photosynthetically fixed carbon in the coral *Porites porites*. They used small (approximately 300-mg dry tissue) coral branch tips to estimate that coral fragments produce 9.1 ml O_2 d^{-1} (equivalent to 191 joules d^{-1}) under ideal sunny conditions. They estimated that much of this energy was consumed by the respiration and growth of the symbiont (22%) or the coral host (33%), while 45% was lost to the environment "probably as mucous-lipid" (Davies 1991, 1992). These estimates indicate translocation of 78% of the fixed carbon. On sunny days, a substantial excess of fixed carbon was produced, but deficits on cloudy days requires corals to utilize lipid stores or alternative methods of energy acquisition. Although Edmunds and Davies found little evidence to suggest that heterotrophic consumption of zooplankton was important in this coral species, zooplanktivory is thought to be much more important in other species as a source of energy [or phosphorous (Johannes *et al.* 1970)]. For example, Porter (1974, 1976) estimated that up to 10-20% of the energy requirements of the "voracious carnivore" *Montastrea cavernosa* could be met by this form of heterotrophy.

Davies (1991) determined the energy budgets for three species of corals in Hawaii (*Pocillopora damicornis*, *Montipora verrucosa*, and *Porites lobata*) under three different light conditions. As in Edmunds and Davies (1986), only photosynthetic sources of energy from zooxanthellae were considered. Under "ideal" conditions with no clouds, the measured integrated irradiance at a depth of 3 m was 14.4 E m^{-2} d^{-1}. Integrated irradiance under "normal" conditions with sporadic clouds was 11.9 E m^{-2} d^{-1} and on "overcast" days it was 6.1 E m^{-2} d^{-1}. Davies (1991) measured the rates of respiration and photosynthesis using standard 10 g pieces of coral and then estimated daily energy budgets for these corals under these ambient light conditions. A net excess of energy representing 19.3-32.4% of the total production (355-551 joules d^{-1}) was estimated for coral fragments of these three species under ideal light conditions. Under normal conditions, somewhat smaller excesses were estimated. Overcast conditions were projected to result in net energetic deficits of 26.4% in *P. damicornis* (59.4 joules d^{-1}) and 7.8% in *P. lobata* (23.9 joules d^{-1}). In that the lipid stores in these coral fragments were estimated at 2.57-7.43 kJ, Davies (1991) postulated that the three species could tolerate extended periods of overcast conditions on the order of 28-114 days. Thus the association of zooxanthellae with corals represents an important trophic relationship which can potentially promote the growth of corals even during occasional long periods of suboptimal environmental conditions.

6.3.2 ALGAE, ENERGY FLOW, AND FISHERIES PRODUCTION

Recognition of the importance of photosynthesis by benthic algae and zooxanthellae in coral communities dates back at least to the now classic study by Odum and Odum (1955) on a reef at Eniwetok Atoll. They estimated a very high rate of primary productivity in this community and determined that benthic plants (including zooxanthellae) accounted for 92% of the total production. Only 8% was attributed to planktonic sources. Primary productivity on coral reefs rivals the most productive levels found anywhere on earth (Ryther 1959, Whittaker and Likens 1973). Sorokin (1993) characterized the general level of productivity on coral reefs at 5-10 g C m^{-2} d^{-1} although there are specific local examples of rates as high as 60 g C m^{-2} d^{-1} (Odum and Odum 1955). Odum and Odum (1955) also reported very low levels of available nitrogen and phosphorous in the ambient waters around Eniwetok thus recognizing the potential importance of nutrient cycling in the "algal-coelenterate complex". However, their quantitative assessment indicated that zooxanthellae represented only 6% of the plant biomass. Filamentous green algae embedded in coral skeleton were the major primary producers in this community. Thus studies of energy flow through coral communities cannot be restricted to the carbon fixed by zooxanthellae. Overall, Odum and Odum (1955) found that primary productivity and respiration in the coral community were approximately balanced, so this coral community was not a net sink or source of energy. A recent extensive review of the ratio of production to respiration in coral communities appears to confirm this point. In spite of high levels of productivity, P/R ratios "are usually well balanced" at 0.7-1.5 (Sorokin 1993).

In an assessment of primary productivity, consumption, and fish production across a reef flat at Davies Reef in the central Great Barrier Reef, Polunin and Klumpp (1992) estimated gross primary productivity of benthic algae (mostly algal turf and crustose coralline algae) at 3.0 g C m^{-2} d^{-1}. Photosynthesis by zooxanthellae was considered relatively unimportant in that the coral-symbiont unit was assumed to metabolize most of the fixed carbon. The proportions of fixed carbon passed on to macrograzers (*e.g.*, large molluscs, parrotfishes, damselfishes, blennies, and surgeonfishes), mesograzers (amphipods, copepods, other crustaceans, small molluscs, and polychaetes), or left as detritus were estimated at 14%, 16%, and 14%, respectively (Figure 6.3). Polunin and Klumpp further partitioned these three carbon pathways on the reef into five algal-based food chains. These included various forms of piscivory, invertebrate feeding, microbial decomposition, and particulate feeding processes. These pathways (along with three additional planktivorous food chains based on relatively small extrinsic sources of carbon) were modelled to predict the potential yield of fisheries under various fishing scenarios. When fisheries targeted only piscivorous fishes, the model predicted the lowest yields for Davies Reef at 0-2 t km^{-2} yr^{-1}. Slightly higher yields were predicted for fisheries targeting only herbivorous fishes (4 t km^{-2} yr^{-1}). When fisheries targeted herbivorous fishes, invertebrate feeders, and planktivores, the highest yield estimates were generated at 14-23 t km^{-2} yr^{-1}. This represents 0.5-0.8% of the gross primary productivity of the reef (Polunin and Klumpp 1992) and approaches some of the highest fishing yields reported for coral reef fisheries [21-40 t km^{-2} yr^{-1} (Russ 1984c, Munro and Williams 1985, Galzin 1987)].

"Fishing is no doubt the most important human activity on coral reefs" with significant effects on fish populations and the entire community (Russ 1991). The direct population consequences of fishing include the following:

• The total catch for many targeted species in places with high fishing effort [*e.g.*, the Philippines and Jamaica (noted in section 3.3.1)] has been decreasing even with large increases in fishing effort.

• Evidence supports the conclusion that high fishing effort substantially increases the total mortality rates (the sum of fishing and natural mortality).

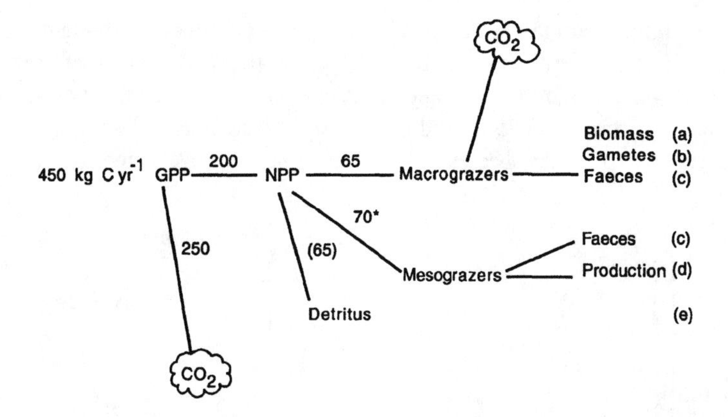

Figure 6.3 Estimated flux of organic carbon derived from benthic algal photosynthesis on a 410-m² reef-flat transect at Davies Reef. Gross (GPP) and net (NPP) primary productivity are indicated. Macrograzer food chains pass organic carbon to piscivores (a) or to picking carnivores and piscivores (b). Mesograzer food chains pass organic carbon to mesocarnivores, picking carnivores, and piscivores (d) or to microbial decomposers, microbivores, particle feeders, carnivores, and piscivores (e). Fecal material is also passed to microbial decomposers (c) (redrawn after Polunin and Klumpp 1992 in *Plant-Animal Interactions in the Marine Benthos* by permission of Oxford University Press).

• Comparisons between unexploited and heavily fished populations clearly demonstrate large shifts towards younger and smaller individuals in the latter.
• The density and biomass of larger piscivorous fishes are greatly reduced on heavily fished reefs.
• The influence of this reduction on recruitment by larval fishes may be important in some fisheries, but the evidence is equivocal at the present time.
• Among the many families of sequentially hermaphroditic fishes, heavy fishing may influence the sex ratio, social structure of their populations, and reproductive success.
• Intense fishing may result in local extinctions, but there is no evidence that "fishing has modified patterns of distribution of coral reef fishes on scales of tens of kilometers or greater".
• Although not well documented, evidence suggests that behavioral responses to intense fishing (especially spearfishing) can influence zonal distributions and the location of sleeping sites of targeted fishes.

Russ (1991) also noted some indirect effects on fish populations as fishing activities modify the habitat. Most destructive is the use of explosives, but fishing techniques employing traps and stone weights on "scarelines" also can result in local damage. In particular, he noted reductions in cover by live coral, the structural heterogeneity of the reef habitat, and availability of shelter for reef fishes.

In reviewing the effects of fishing at the community level, Russ (1991) particularly emphasized the role of interspecific interactions in mediating the responses among trophic groups and competitors. Due to their large size, aggressive habits, and good taste, predatory fishes are thought to be more vulnerable to intense fishing than other trophic groups (Munro and Williams 1985). At Apo Island in the central Philippines where fishing intensity is high, the species richness of large predators (mostly serranids, lutjanids, and lethrinids) has been estimated at 0.1-0.4 species per 750 m^2 of reef (Figure 6.4). Following the establishment of a protected reserve in 1981, predator species richness increased steadily from 1983-1988 within the reserve reaching 2.7 species per 750 m^2. Outside the reserve, predator species richness remained relatively unchanged.

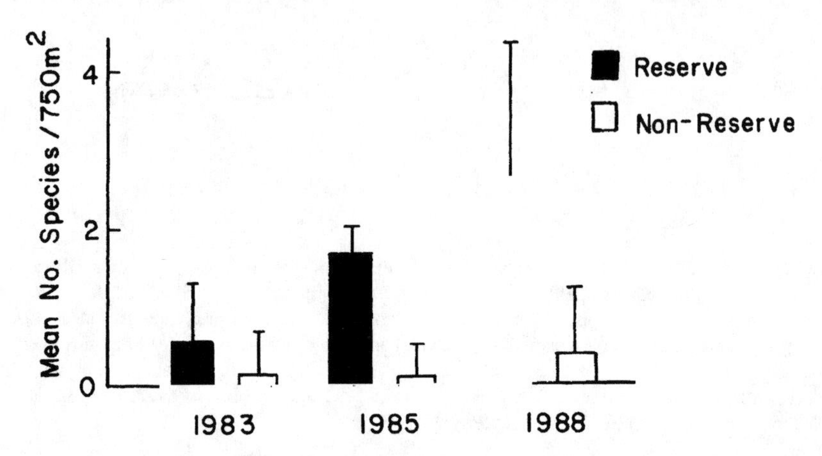

Figure 6.4 Mean species richness of predatory fishes (± 1 SE) at Apo Island inside and outside of a protected reserve established in 1981 (redrawn after Russ 1991 and personal communication).

However, in spite of their general vulnerability to fishing and the large response of predators to protective measures, the evidence for corresponding responses among their prey species appears to be equivocal. A similar lack of evidence for responses among predatory fishes to intense fishing targeted at lower trophic levels (*e.g.*, herbivorous fishes in Jamaica or planktivorous fishes in the central Philippines) was also noted. Russ (1991) suggested two attributes of coral reef communities which may be responsible for these less than dramatic responses observed among coral reef fishes.

• First, there are many species of carnivorous fishes in these communities which are opportunistic and quite general in their feeding habits. "Any increase in abundance of a species of prey following the removal of one of its predators is likely to lead to functional responses of other predators which may switch their attention to the abundant prey item" (Russ 1991).

• Second, because of highly variable recruitment in reef fishes, any response to reduced predation (or elevated prey densities) "may be very difficult to detect" (see sections 1.4 and 5.1.3 on the relative importance of biological interactions and recruitment). Given the large scale of most fisheries and the non-experimental nature of much of the quantitative data used to assess them, this is likely to continue to be a problem until experimental approaches become more common.

As a final point regarding coral reef fisheries, I note the enormous difference in the trophic status of caught fishes across differentially fished communities. The intensely fished locations noted by Russ (1991) include Sumilon Island in the central Philippines. It has been estimated that more than half the fishery there is comprised of planktivorous fishes. This supports the view that such targeted coral reef fisheries can achieve high yields in excess of 20 t km^{-2} yr^{-1} [this is greater than suggested by Polunin and Klumpp (1992) who estimated a potential yield from planktivores of 3 t km^{2} yr^{-1} for Davies Reef]. It also illustrates a major shift away from the traditional view that benthic algal productivity is responsible for most of the yield in coral reef fisheries.

Russ (1991) and others have noted the proximity of some communities to large land masses where inputs of extrinsic organic matter and nutrients can be substantial. Thus sustainable fisheries strategies for these communities are likely to be quite different from those on offshore reefs or on oceanic atolls. In moderately fished coral communities, the yields of fishes of variable trophic status are more likely to be sustainable than in heavily fished communities. Based on energetic arguments, fisheries targeting predominantly piscivorous species at the highest trophic levels are only sustainable under relatively low fishing intensity. However, "even reefs that are 'lightly fished' have failed to sustain their promised yield" (Birkeland 1997; also see Jennings and Polunin 1996, Polunin and Roberts 1996). These generalizations regarding sustainability are based on fisheries models of productivity, energy flow, and harvesting. They refer only to the flux of matter through the community and imply nothing about the stability of coral communities under intense fishing pressure (see chapter 3).

6.3.3 ALGAL CHEMICAL DEFENSES AND HERBIVORY

Given the importance of benthic algae and herbivory in many coral communities, considerable recent attention has been focused on evolutionary responses to these plant-animal interactions. Herbivores consume much of the algae produced in these communities and levels of herbivory are reportedly among the highest known for any ecological community (Carpenter 1986, Hay and Fenical 1988, 1992, Hay 1991). Consequently, herbivory should represent a major selective force on the evolution of algal defenses. According to classical plant defense theory, plants evolve secondary metabolic compounds in response to herbivory while herbivores respond by evolving mechanisms to circumvent these defenses. Thus herbivores become increasingly more specialized to particular plant species (Howe and Westley 1988).

The distribution of these secondary metabolites among and within plants has been explained in terms of plant apparency (Feeney 1975, 1976, Rhoades and Cates 1976) or resource availability theory (Coley *et al.* 1985). Apparent plants are described as large, long-lived organisms which are hypothesized to employ generalized "quantitative" defenses against all herbivores (*e.g.*, high concentrations of relatively immobile and indigestible compounds like cellulose, hemicellulose, lignins, tannins, and silica). Unapparent plants are small and ephemeral, so "qualitative" defenses suffice (*e.g.*, low concentrations of such simple toxins as alkaloids, cyanogens, glucosinolates, and terpenes). According to plant apparency theory, specialized herbivory evolves only among consumers of unapparent plants. "While apparency theory implies that herbivore foraging efficiency strongly influences plant defense, resource availability theory assumes that inherent growth physiology, photosynthetic capability, and nutrient availability determine the amounts and kinds of defenses that plants use" (Howe and Westley 1988). Furthermore, plants and parts of plants more costly to replace are predicted to be more heavily defended and greater herbivory is predicted to occur on less-valuable plant parts (Coley *et al.* 1976).

In coral reef communities, many common apparent plants (*e.g.*, the algal genera *Halimeda*, *Laurencia*, and *Dictyota*) produce small organic compounds (*e.g.*, terpenes) rather than the quantitative defenses predicted from theory (Hay and Fenical 1992, Paul 1992a). Terpenoids from tropical genera are the most prevalent of all secondary compounds to have been isolated from marine algae (Hay 1991, Paul 1992a). This is in spite of the facts that quantitative defenses (*e.g.*, polyphenolic compounds like phlorotannin) are known to deter tropical herbivores (Van Alstyne and Paul 1990) and algae on coral reefs are exposed to a diverse array of generalist herbivores (Hay 1991, Hay and Fenical 1992). In general, polyphenolic compounds were thought to occur mostly in temperate brown algae and to be almost completely absent in closely related tropical species (Steinberg 1989, 1992, Van Alstyne and Paul 1990). Hay and Fenical (1992) also noted that resource availability theory predicts that the algae found in the high light and low nutrient environments on coral reefs should have higher polyphenolic concentrations than those found in less nutrient-limited environments. Thus the reported distributional patterns were opposite to those predicted from both theories. Furthermore, there was apparently no evidence supporting the assumptions that quantitative defenses are more costly to employ in marine algae than are qualitative defenses or that herbivores of algae evolve tolerance more easily to qualitative defenses than to quantitative defenses. Thus Hay and Fenical (1992) suggested that the traditional conceptual basis for studying plant defenses (which was largely developed from knowledge of terrestrial plant-insect interactions) is not appropriate for the study of marine algal-herbivore interactions.

With a wider range of sampling of algal species particularly in the western Atlantic, we now know that polyphenolic compounds are more common in the tropics than previously supposed. Targett *et al.* (1992) countered the notion of a strong contrast in the levels of polyphenolics found in temperate and tropical algae. They found high concentrations of these compounds in six of eight algal species sampled in Belize. Steinberg (1992) responded by suggesting that this evidence might apply only to this location and thus not be representative of the tropical, western Atlantic. Targett *et al.* (1995) confirmed very high concentrations of polyphenolics throughout this region, thus shifting the debate to consider the inter-oceanic contrast. Possible explanations may include unique coevolutionary features of the algal-herbivore interactions in each ocean or other evolutionary phenomena unrelated to these interactions. The high levels of polyphenolics found in western Atlantic algae may very well be consistent with plant apparency theory. However, why these quantitative defenses are not employed in response to herbivory on tropical Indo-Pacific algae remains unclear.

After rejecting classical plant defense theory, Hay and Fenical (1992) suggested five "new directions" for research in this area. They are worth mentioning, but I would add that further tests of classical theory are also appropriate. The evidence noted above by Targett *et al.* (1995) supports this view as does that by Bolser and Hay (1996). This latter study provided evidence for generally stronger chemical defenses among tropical algae relative to those in closely related temperate species. The suggested directions from Hay and Fenical (1992) are as follows:

• Given that chemical defenses can strongly deter feeding by both visual and non-visual herbivores, many alga may have evolved to mimic well defended species or to exploit the benefits of being associated with such species. Mimicry has been largely unstudied in marine algae, but associational benefits have been detected (*e.g.*, Littler *et al.* 1986, Hay 1986). Littler *et al.* (1986) were the first to experimentally document associational benefits among marine plants. At Carrie Bow Cay in Belize, they found significant associations between several palatable algal species and the unpalatable, numerically dominant *Stypopodium zonale.* This alga is known to be chemically well defended, toxic to fishes, and only lightly grazed by herbivorous crabs and sea urchins. The experimental removal of *S. zonale* was followed by rapid losses of palatable algae due to grazing surgeonfishes and parrotfishes, while controls experienced minimal or no losses at all. Thus the interactions between palatable algae and herbivorous fishes were strongly modified by the presence of the unpalatable species.

• Given that algae recruit to these benthic communities as quite small plants, they are thought to be extremely vulnerable to herbivory. High, size-dependent mortality among algae can result in size refuges for large plants and may select for refuge exploitation by small plants, size-dependent resource allocations for plant defenses, and avoidance behaviors among herbivores based on protective, aposematic coloration. As evidence, Hay and Fenical (1992) noted "the iridescent coloration of several chemically rich" brown algae, "the bright red tips on some species of chemically defended *Laurencia*" (both being particularly noticeable in small plants), and "extremely high concentrations of the most deterrent metabolites" in young plants or in newly produced portions of *Halimeda* (Hay *et al.* 1988, Paul and Van Alstyne 1988). Thus there may be important ontogenetic shifts in plant defenses.

• The study of secondary metabolites in marine algae has apparently been so focused on their role as plant defenses that other potential functions have been largely ignored. Hay and Fenical (1992) mentioned the inhibition of diseases or settlement by invertebrate larvae as alternative ecological foci and suggested that the functional role of these compounds should be evaluated by a closer inspection of where and how they are stored within plants, release mechanisms, and the ambient concentrations of metabolites near or on plant surfaces.

• Hay and Fenical (1992) suggested that the allocation of compounds within and among individuals is an important, yet unstudied source of natural variation in populations of marine plants. Understanding such variation is fundamental if we are to appreciate how natural selection acts on evolving plant-herbivore interactions. Thompson (1988, 1994) has extended this perspective to include variation in the outcome of interspecific interactions across geographic mosaics. One of the main features of this approach is that it considers coevolution as a dynamic process operating at a geographic scale above that of the local population yet below that of the species. To my knowledge, this approach has yet to be applied in any study of coevolution in the sea.

• Another largely neglected area is the study of inducible defenses in marine algae. This is surprising in that it is a well known phenomenon in terrestrial plants and aquatic invertebrates (especially in clonal forms which typically incur only partial mortality when attacked by consumers). "Inducible defenses appear favored over constitutive ones when the probability of contacts by biological agents is high but unpredictable, the cues associated with contact are reliable and not fatal, and the fitness costs offset some of the benefits of the defense, favoring intermittent deployment" (Harvell 1990). Steinberg (1994) reported that the well known phlorotannins which deter herbivory in temperate brown algae are constitutive rather than inducible defenses in *Ecklonia radiata* and *Sargassum vestitum.* He suggested that intense herbivory was much too predictable to favor inducible defenses in these temperate forms. Given that herbivory is thought to be even more intense in the tropics, inducible defenses may not be an important feature of plant-herbivore interactions in coral communities. In fact, the presence of high concentrations of secondary metabolites in small tropical algae (see above) also suggests that herbivory is too intense and predictable to favor inducible defenses. However, there is little evidence on which to base generalizations at this time.

6.4 Predator-prey interactions in coral communities

The interactions between predators and their prey in coral communities have received extensive attention over the years. This is especially so because of the emphasis on keystone species in community ecology (section 3.3) and the importance of coral reef fisheries throughout the tropics (section 6.3.2). In addition to these important ecological contexts, predation (like herbivory) can also be considered from an evolutionary perspective. Predation can strongly influence the evolution of chemical, morphological, behavioral, and life history responses among prey species. Such evolved attributes limit the extent to which active predation can influence the observed distribution and abundance of prey.

Among the notable contributions regarding predation in coral communities, are three general treatments of this subject by Glynn (1973, 1988a, 1990a). Glynn (1973) described the general ecology of coral reefs in the western Atlantic including what was known about predation specifically on corals. Corallivory appears to have been largely unstudied at the time, but was becoming more appreciated as a common phenomenon in coral communities. For example, the extensive treatment of the feeding habits of West Indian fishes by Randall (1967) included only four corallivorous species. Glynn (1973) recognized the significant discovery of only five invertebrate corallivores during the preceding decade. Beyond the identification of these predators, Glynn (1973) noted the capacity of some like the polychaete *Hermodice carunculata* to cause extensive damage to corals.

At the sixth International Coral Reef Symposium held in Townsville, Australia, Glynn (1988a) provided an overview of predation on coral reefs. In particular, he emphasized how predators control community structure through a variety of direct and indirect effects. Examples of devastating direct effects of corallivory included those due to the asteroid *Acanthaster planci* throughout much of the Indo-Pacific (section 3.3.3), the gastropod *Drupella* spp. at a number of sites also in the Indo-Pacific (Moyer *et al.* 1982, Turner 1994), and the echinoid *Eucidaris thouarsii* in the Galápagos Islands (Glynn *et al.* 1979). Glynn (1988a) also highlighted three types of indirect effects observed in coral reef communities. Trophic linkage [an interaction chain *sensu* Wootton 1993 (section 1.4.1)] occurs when a consumer species reduces the abundance of a prey species which in turn causes an increase in the abundance of a competitor. One of two examples noted by Glynn (1988a) was a putative positive response by *Porites compressa* to selective corallivory by *A. planci* on *Montipora verrucosa* in Hawaii (Branham *et al.* 1971). However, this study was strictly observational and the indirect effect was not actually demonstrated. A better experimental example indicating this form of indirect effect was conducted by Cox (1986) also in Hawaii. The butterflyfish *Chaetodon unimaculatus* selectively preyed on *M. verrucosa,* reduced this coral's growth rate, restricted its distribution on the reef, and appeared to reverse the competitive relationship between *M. verrucosa* (the competitive dominant) and *P. compressa.*

Glynn (1988a) noted three indirect behavioral effects of predation in coral communities. The territorial behaviors of damselfishes can strongly influence corallivores and herbivores (see Glynn and Wellington 1983 and sections 3.3.2 and 5.1.3)]. The defensive behaviors by decapod symbionts can limit corallivory on host corals (see Glynn 1983 and section 3.3.3). Fish foraging behaviors can alter the availability of prey to other consumers (see Robertson *et al.* 1976, Karplus 1978, Ogden and Lobel 1978, Abrams *et al.* 1983). Lastly, Glynn (1988a) identified chemically mediated indirect effects whereby predators reduce their own susceptibility to being consumed by using the chemical defenses of their prey (see Gerhart 1986, Paul 1992b).

Many of the general points made in Glynn (1988a) and elsewhere also appear in his extensive review of the effects of large consumers on coral community structure (Glynn 1990a). In addition, Glynn (1990a) gave considerably more attention to the impact of consumers on the spatial distribution patterns of coral reef organisms. Within a reef community, spatial heterogeneity is enhanced by the presence of sheltered sites in which prey can hide. Thus intense predation on herbivores can result in large differences in the abundance of plants (a trophic cascade and interaction chain *sensu* Wootton 1993) depending on the distribution of such shelters. For example, Randall (1965) attributed the conspicuous zones of bare sand (devoid of sea grasses) around reefs in the Virgin Islands to intense grazing primarily by parrotfishes and surgeonfishes. He associated this pattern with the proximity of sheltered sites on the reefs. Later, Ogden *et al.* (1973) associated "halos" around patch reefs also in the Virgin Islands with intense nocturnal grazing by the echinoid *Diadema antillarum.* During daylight hours, these herbivores remain in sheltered sites on the reefs rather than in the nearby grass beds. Some coral species also exhibit discontinuous distributions on coral reefs which are associated with the intensity of corallivory. For example, on Indo-West Pacific reefs such as those around Guam, *Pocillopora damicornis* is restricted to reef flat, margin, and lagoonal habitats (Neudecker 1979). When Neudecker transplanted 118 colonies to the fore-reef habitat on three fringing reefs, all of them experienced heavy corallivory (primarily by butterflyfishes and triggerfishes) though none were completely killed.

In contrast with corals, many other coral reef invertebrates are toxic to fish. This is especially true of soft-bodied species which live in habitats exposed to predatory fishes (Vermeij 1978, Bakus 1981, Paul 1992b). For example, Gerhart (1984) found high concentrations of prostaglandin A_2 in the West Indian gorgonian *Plexaura homomalla.* He documented the negative effects of this chemical on fishes (but see Pawlik and Fenical 1989). He also noted that this gorgonian "is readily consumed by *Cyphoma gibbosum*", the predatory gastropod which specializes on several species of gorgonians and employs their chemical defenses as its own (Gerhart 1986). One of the chemically best-defended gorgonians in the West Indies is *Briareum asbestinum.* This species employs 5-15 diterpenoid compounds which Harvell *et al.* (1993) have found to vary significantly among habitats on reefs and among sites in the West Indies. Although these defensive compounds appear to be "genetically fixed" rather than inducible in this gorgonian, it is not known how differences among populations are maintained. Terpenoids are also common among gorgonians and soft corals in the Indo-Pacific where the toxicity of these organisms has been widely reported (*e.g.*, Coll *et al.* 1982, La Barre *et al.* 1986). However, the adaptive significance of these compounds remains uncertain as they have been associated with multiple processes including predator defense, competition for space (section 5.1.1), inhibition of growth and settlement, and even reproduction (Sammarco and Coll 1992).

One of the most toxic compounds found in any marine invertebrate is palytoxin which occurs in several species of zoanthids (Moore and Scheuer 1971, Attaway and Ciereszko 1974). These soft-bodied forms are especially successful in shallow subtidal and intertidal habitats throughout the tropics where they are preyed upon by only a small number of species. In the Caribbean, the polychaete *Hermodice carunculata* and damselfishes are significant predators (Sebens 1982b, Karlson 1983) as are butterfly-fishes (Randall 1967). Sebens (1982b) conducted predator exclusion experiments to link the observed intertidal distribution patterns of zoanthids with the intensity of predation. In contrast, Karlson (1983) experimentally excluded *Diadema antillarum* in a shallow subtidal habitat to illustrate how *Zoanthus sociatus* is one of the few species able to tolerate extreme herbivory/grazing by this echinoid. In the Indo-Pacific, the filefish *Alutera scripta* is known to feed on the highly toxic *Palythoa tuberculosa* (Hashimoto *et al.* 1969). In fact, *A. scripta* is itself highly toxic and thus may use the defenses of its prey as its own defense against predators.

In attempting to document the most toxic region of the zoanthid body, Kimura *et al.* (1972) concluded that the toxicity levels found in *P. tuberculosa* are "essentially defined by the presence of eggs". Thus extreme toxicity was linked to reproductive success as gravid individuals or spawned eggs and larvae were thought to experience reduced levels of predation. A recent evaluation using Caribbean zoanthids detected palytoxin in several species. However, the location of palytoxin was highly variable within and among colonies and was "not correlated with their reproductive cycle" (Gleibs *et al.* 1995). Here again the adaptive significance of toxic substances in soft-bodied invertebrates remains uncertain. As with algal toxins, there are likely to be significant functional differences among species within a region and between faunal provinces.

The problem of establishing the adaptive significance of toxic compounds in prey species can also be extended to question the evidence for predation being a major selective force in coral communities. Theoretically in benign environments where densities are not recruitment-limited, intense predation is predicted to control species diversity and strongly favor the exploitation of spatiotemporal refuges (Menge and Sutherland 1987; see section 1.4.1). Jones *et al.* (1991) specifically reviewed the evidence regarding the importance of predation by coral reef fishes on the benthic community of corals and other invertebrates. This focus was justified because benthic invertebrate feeders are generally among the most common trophic groups represented among coral reef fishes (Table 6.2). Within this group, predators on mobile invertebrates (especially crustaceans) appear to be more prevalent than corallivores or predators on sessile invertebrates.

TABLE 6.2 The relative frequency of trophic groups among coral reef fishes in seven selected studies (after Jones *et al.* 1991).

	Study						
Trophic group	1	2	3	4	5	6	7
Herbivores	26	13	22	7	15	18	20
Planktivores	4	12	15	18	20	15	38
Benthic invertebrate feeders	49	44	27	56	53	41	33
Coral feeders	6	1	-	9	5	9	-
Sessile animal feeders	8	6	-	13	3	-	-
Mobile animal feeders	35	37	-	34	45	-	-
Omnivores	13	7	-	10	4	19	-
Piscivores	10	25	38	7	8	4	8
Others (*e.g.*, cleaners)	-	-	-	2	-	2	1

1= Hiatt and Strasburg (1960), 2 = Randall (1967), 3 = Goldman and Talbot (1976), 4 = Hobson (1974), 5 = Williams and Hatcher (1983), 6 = Sano *et al.* (1984), 7 = Thresher and Colin (1986).

As a protocol, Jones *et al.* (1991) outlined some necessary steps in demonstrating the importance of predation:

• The predators must be identified through an analysis of diet and feeding selectivity. Several experimental studies were noted to have been conducted "without this fundamental information". Jones *et al.* (1991) also noted that the extreme flexibility in the feeding habits of many fishes is often not fully appreciated. For example, three species of predatory fishes in the Caribbean, which had specialized on *Diadema antillarum* as their primary prey prior to the mass mortality of this echinoid (section 3.3.1), switched their feeding habits in very different ways. The triggerfish *Balistes vetula* switched to feed primarily on crabs and chitons (Reinthal *et al.* 1984). The toadfish *Amphichthys cryptocentrus* switched to feed on a wide range of mobile invertebrates, while another toadfish *Sanopus barbatus* switched to become primarily a piscivore (Robertson 1987)]. Such diversified feeding capabilities greatly extends the range of possible interactions involving these predators.

• Spatiotemporal variation in the abundance of predators and their foraging behaviors must be documented. From this information, one can then infer how predation pressure might vary in space and time.

• Experimental manipulations are required to establish the causal effects of predators on the distribution, abundance, and structure of prey populations. Their evaluation of the experimental techniques employed in such studies led Jones *et al.* (1991) to conclude that "all methods have their problems and all results when viewed in isolation are equivocal".

• Nevertheless, they encouraged the use of multiple observational and experimental techniques and recommended that we be more critical in assessing the limitations of these methods. In so doing, the importance of predation in these communities can be evaluated relative to other processes like recruitment limitation, interspecific competition, and disturbance (Jones *et al.* 1991). However, we should be careful to avoid making overly simplistic generalizations (Fisher 1994, see section 1.3).

Three recent reviews on predation and grazing by fishes and invertebrates in coral communities (Carpenter 1997, Hixon 1997, Pennings 1997) essentially reiterate the message that these feeding activities can have a variety of strong direct and indirect effects. Hixon (1997) focused primarily on corallivorous and herbivorous fishes in emphasizing the effects of consumption, schooling behaviors, and territoriality (see section 3.3.2). Pennings (1997) presented the general conceptual framework for studying indirect effects [as in section 1.4.1, Miller and Kerfoot (1987), and Glynn (1988a)]. He emphasized that we have much to learn about these effects in coral communities. Carpenter (1997) contrasted invertebrate corallivores and herbivores based on those generating "major effects" (essentially consumers as keystone species) with the large majority of species which cause only "minor effects" as consumers in these communities.

Basing this distinction on consumption rather than on a broader view of stabilizing interactions (see section 3.3) led to the suggestion that predation on live corals by most corallivorous species "does not have a dramatic influence on coral-reef community structure". However, distinctions based only on consumption ignore the potential for important indirect effects of the association between corals and corallivores. For example, Carpenter (1997) recognized the important interaction between the coral symbionts *Trapezia* spp. and corallivores like *Acanthaster planci* (see above and section 3.3.3), yet categorized corallivory by these crustaceans to have only minor effects. The major effect in this example involves all three members of the interaction (the coral, the corallivorous symbiont, and the corallivorous asteroid). One could quite easily categorize the symbiont as another keystone species based on the experimental evidence [see Glynn (1976) and section 3.3.3]. In similar fashion, other consumers noted to generate only minor effects due to consumption may be involved in important indirect effects in these communities. Consequently, they may merit either keystone species status or recognition as one of several species involved in strong, but diffuse interactions of a consumer guild. Such effects can potentially contribute to compensatory responses to perturbations, trophic flexibility, the complexity of interaction webs, and the stability of coral communities. The keystone status of individual consumer species such as *Diadema antillarum* or *Acanthaster planci* may merely reflect the unstable nature of taxonomically impoverished conditions due to overfishing or other disturbances to these communities. Thus the emergence of new keystone species from among those consumers now thought to generate only minor effects in coral communities may be a consequence of ongoing or future perturbations.

6.5 Overview

Direct and indirect effects of consumer-resource interactions are pervasive in coral communities. These interactions influence such fundamental processes as energy flow, fisheries production, larval settlement and metamorphosis, bioerosion, and the evolution of chemical defenses. As emphasized in chapter 3, the very stability of many of these communities is strongly controlled by consumers acting as keystone species. Thus I would emphasize this aspect of consumer-resource interactions as perhaps the most important focus for future study. Since keystone species status is spatiotemporally variable and difficult to predict in complex communities, more effort should be expended to determine who they are and what mechanisms control their abundances. In general, consumer-resource interactions in coral communities are influenced by a variety of complicating factors. These include: 1) the large number of species and potential interactions, 2) multiple energy pathways, 3) variation in species-specific feeding habits and trophic modes, 4) habitat complexity, 5) scale-dependent spatial variation in the occurrence of particular interactions, and 6) overfishing. Nevertheless, the importance of understanding how consumers influence community stability cannot be emphasized enough.

7 DISTURBANCE

7.1 Disturbance and species coexistence

As noted throughout previous chapters, disturbance is an integral part of the dynamics of ecological communities. The two major notions I have emphasized are: 1) disturbances represent perturbations to several biological processes (*e.g.*, community assembly, succession, recovery, and competition), and 2) disturbances can occur predictably enough to represent an important part of the selective regime. Thus the immediate consequences of a particular disturbance may be primarily destructive, but may also differentially favor species with adaptations promoting good colonizing abilities or resistance to the disturbance (see section 3.2). In that evolutionary responses to disturbance can potentially influence a number of life history attributes (*e.g.*, longevity, growth, clonal replication by fragmentation, recruitment by larvae), they contribute to the variance in the collective response of the entire community. Life history patterns are known to be constrained by species-specific factors (Begon and Mortimer 1986, Stearns 1992), so each species in a community should respond uniquely. The simple dichotomy between disturbed and undisturbed communities is unlikely to explain the full extent of this variation. However, the theoretical basis for studying the influence of disturbance on community structure incorporates such contrasts (as well as exposure gradients) and provides a starting place for examining this phenomenon in real communities. In what follows, I consider this general body of theory. This is followed with examples of empirical studies evaluating the effects of disturbances on coral communities.

From a theoretical perspective, the most important influence of disturbances on ecological communities is their role in promoting species coexistence. Certainly the patch dynamics of our best-studied epibenthic communities are strongly affected by biological and physical disturbances. Likewise, the notion that disturbance enhances species diversity across the local landscape has been a central theme in the marine literature. Notable early contributions include the thought provoking contribution by Connell (1978) and several other influential studies already noted in section 1.4 [also see Dayton and Hessler (1972) and Sousa (1979)]. The ideas developed in these contributions continue to influence general theory today (*e.g.*, see Huston 1979, 1985a, 1994, Hastings 1980, Menge and Sutherland 1987, Petraitis *et al.* 1989, Caswell and Cohen 1991, 1993).

Connell (1978) is often cited for his presentation of the "intermediate disturbance" hypothesis as an explanation for the high species diversity observed for trees in tropical rain forests and corals on tropical reefs. The general idea that catastrophic disturbances represent perturbations to communities which "set back, deflect, or slow the process of return to equilibrium" was well established at the time [*e.g.*, see Gleason (1917) and

Andrewartha and Birch (1954)] and there were several prior contributions which specifically noted a peak in local species diversity at intermediate levels of disturbance. In spite of clearly defining his focus on the tropics, Connell (1978) was criticized for neglecting to mention the large body of work from temperate forests and marine epibenthic communities which illustrate this relationship between species diversity and disturbance (see Fox 1979). One example of this earlier work was presented in a contribution by Grime (1973). He described how environmental stress and disturbances offset competitive processes among herbaceous plants and generate high species diversity ("species density") at intermediate levels of these perturbations. Another example was presented by Horn (1975) in a model of temperate forest succession. He noted that "intermediate disturbances produce higher diversity than either very high or very low levels".

In emphasizing the role of disturbances on the diversity of two major tropical assemblages, Connell (1978) contrasted the view that these communities represent ordered states with a nonequilibrial perspective. The former assumes that assemblages are generally at or near equilibrial conditions. The latter assumes that disturbances are so prevalent that these "assemblages seldom reach an ordered state". As these opposing views continue to be debated today, a brief review of the arguments is appropriate here. Connell (1978) specifically evaluated six models which he categorized as equilibrial or nonequilibrial explanations for high diversity.

- The "intermediate disturbance" hypothesis was the first of three invoking nonequilibrial conditions. According to this hypothesis, diversity in a community fluctuates in response to the opposing influences of disturbance and interspecific competition. Inferior competitors are eliminated during periods of low disturbance or between disturbances, while only species resistant to disturbance or capable of rapid colonization persist under highly disturbed conditions.
- The "equal chance" hypothesis posits that diversity fluctuates with the opposing influences of colonization from the available species pool and local extinction events. There are no significant differences among ecologically similar species in terms of colonization ability, invasion resistance, or environmental tolerances. Thus the community trajectory through phase space represents an unpredictable stochastic process. It is similar to that described by Hubbell (1979) and Hubbell and Foster (1986) in the community drift model (see section 3.4).
- The "gradual change" hypothesis invokes perpetual changes in the environment such that no equilibrium is ever reached. This was suggested years ago by Hutchinson (1961) and supported more recently by the growing body of information documenting the influence of climate change on ecological communities (see section 3.4).

Connell (1978) offered three equilibrial explanations for the high diversity of tropical assemblages.

- The "niche differentiation" hypothesis invokes evolved differences in adaptations to the local environment and supposes that "each species is competitively superior in exploiting a particular subdivision of the habitat". Thus the total diversity of an assemblage depends on the degree of habitat specialization and the range of habitats (see section 2.2 for discussion of within- and between-habitat diversity).

• The "circular networks" hypothesis invokes nonhierarchical competitive abilities (*i.e.*, networks) and the notion of "coexistence competition" used later by Menge and Sutherland (1987) (see sections 1.4.1, 1.4.3, and 5.1.2). Under this hypothesis, there are no competitive dominants in the assemblage as no species outcompetes all others. As originally envisioned, species replacements are predicted to involve three or more species using multiple competitive mechanisms to interfere with one another.
• The "compensatory mortality" hypothesis suggests that competitively dominant species experience disproportionately high mortality (due to factors unrelated to competition) thus offsetting their competitive advantage. If species are competitively similar, mortality may be highest among the most common species (*i.e.*, frequency-dependent mortality). Thus diversity is maintained because competition does not eliminate inferior species.

Connell (1978) concluded that the high diversity of tropical assemblages is largely maintained in a nonequilibrial state by "frequent disturbances or as a result of more gradual climatic changes". This is in spite of the fact that the six hypothetical explanations are not mutually exclusive and there is evidence supporting all of them. Connell based his conclusion on evidence from his work on the coral reefs at Heron Island (see sections 1.4.3, 4.2, and 5.2.1) and the rain forests of Queensland, Australia, as well as on other published work at the time. As a consequence of this evaluation, two of the three nonequilibrial explanations for high diversity were supported as major paradigms explaining tropical diversity patterns [see Chesson and Case (1986) and section 3.4 for further discussion of these issues]. Both processes maintain higher diversity than would be achieved if they were to reach equilibrium because inferior competitors are not eliminated from local assemblages. Species coexistence is promoted by the spatiotemporal heterogeneity created by disturbances within habitats in the former case, and by much larger-scale shifts in environmental conditions in the latter. As noted by Chesson and Case (1986), the former explanation predicts a stable coexistence of species which differ in terms of their colonizing and competitive abilities. The latter predicts no such stability especially when applied to assemblages of long-lived organisms like forest trees or corals. A second difference deals with the importance of historical phenomena. All classical equilibrium theories and some nonequilibrial notions like the intermediate disturbance hypothesis make predictions which are totally independent of history. In contrast, the "gradual change" hypothesis makes predictions in which aspects of past history can be very important (Chesson and Case 1986).

A somewhat different approach to the role of disturbance in ecological communities was hypothesized by Huston (1979) in a "dynamic equilibrium model of species diversity"; Connell (1978) referred to it as the "rate of competitive displacement" hypothesis. Instead of using models predicting competitive equilibria or the maintenance of high diversity by single factors like predation (Paine 1966) or disturbance (Connell 1978), Huston (1979) argued that "diversity may be maintained by periodic population reductions or a low rate of competitive displacement" and that it is the dynamic equilibrium between these two processes which generates observed diversity patterns. This hypothesis "assumes that most communities exist in a state of nonequilibrium where competitive equilibrium is prevented by periodic population reductions and environmental fluctuations". High species diversity and prolonged periods of coexistence are predicted at low to intermediate rates of competitive displacement and infrequent population reductions. "Diversity may be reduced either by competitive displacement (and eventually exclusion) or by a high frequency of population reduction which does not allow some competitors to recover between disturbances" (Huston 1979). In that this hypothesis allows for extended periods of coexistence with only intermittent intense competition, species are eliminated from communities at very slow rates. Thus community structure is more likely to be influenced by stochastic phenomena and unique historical events (Chesson and Case 1986).

Huston (1985a) used this dynamic equilibrium model to explain the patterns of coral species diversity observed on coral reefs. On most reefs, diversity is low in shallow water and reaches its highest levels at depths of 15-30 m (section 2.2). This pattern is consistent across a range of disturbance regimes such that "no depth gradient of either biotic or abiotic disturbance can explain the change in diversity with depth". By noting that coral and algal growth rates vary strongly with the availability of light, he suggested that the rate of competitive displacement should vary with depth on a reef. Competition should be intense in shallow water, while reduced levels of competitive displacement in deep water should prevent the elimination of inferior competitors. Hence, "the effect of disturbance on reef community structure is interpretable only in the context of the rate of competitive displacement, which is equivalent to the rate of regrowth following disturbance". Huston emphasized the importance of a significant nonequilibrial interaction between coral growth rate and the frequency of disturbance as illustrated in Figure 7.1. At any particular disturbance frequency, a range of diversity predictions depend on the intensity of competition and thus should vary with depth. "Both shallow turbid sites and deep clearwater sites may have the appropriate combinations of growth rates and disturbance frequency to allow high levels of diversity". The predicted patterns of variation in species diversity depend solely on the interaction between competition and disturbance which, in turn, are controlled by the local physical conditions characterizing a site.

Huston (1994) expanded the notion of a dynamic equilibrium between disturbance and competition in considering global diversity patterns of not only corals, but most major groups of plants and animals [see book review by Lawton (1995)]. Once again his central tenet was that observed spatiotemporal patterns of species diversity across environmental gradients at multiple scales (*e.g.*, latitudinal, altitudinal, productivity, temperature, depth, etc.) are interpretable in terms of population growth rates. These rates are directly influenced by the rate of competitive displacement and the frequency (and intensity) at which disturbances reduce population sizes. A chapter dealing specifically with marine ecosystems covers coral reefs [essentially as in Huston (1979, 1985a)], pelagic systems in central oceanic gyres, the deep sea benthos, and the rocky intertidal zone. Relative to these other systems, corals reefs are characterized by high diversity across a wide range of disturbance frequencies (and intensities) and rates of competitive displacement (Figure 7.2). Rocky intertidal zones are characterized by high rates of competitive displacement (*i.e.*, they have potentially high population growth rates which may be offset by population reductions due to predation and/or disturbances as emphasized in section 1.4.1). The deep sea benthos is characterized by low disturbance frequencies (but see Dayton and Hessler 1972) and very low productivity resulting in low rates of competitive displacement (Figure 7.2).

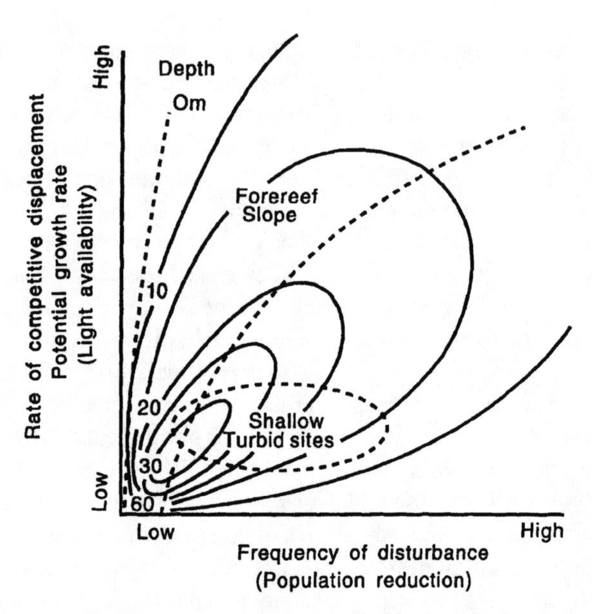

Figure 7.1 Predicted species diversity of corals based on light availability and the frequency of disturbance. High light availability varies with depth (given in meters) and is associated with high rates of coral growth and competitive displacement. The closed concentric curves represent contours of species diversity with the highest diversity predicted for the smallest closed curve in the lower left corner. The pair of dashed lines encompass the range of physical conditions found on fore-reef slopes. The dashed ellipse represents shallow turbid sites often found in back reef or lagoonal habitats (redrawn after Huston 1979, 1985a). © 1979 by The University of Chicago.

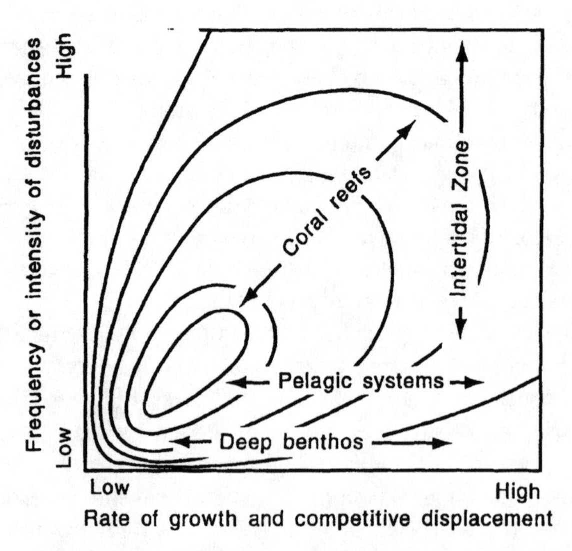

Figure 7.2 Range of productivity and disturbance conditions and the relative species diversity of four marine systems according to the dynamic equilibrium model. Contours are as in figure 7.1 (redrawn after Huston 1994).

This paradigm invokes local environmental conditions and population phenomena as primary determinants of species diversity patterns in these marine systems. The influence of larger-scale transport processes, other geographical phenomena, geologic history, and even the somewhat predictable periodic cycles associated with El Niño events and glaciation on diversity at local scales are mediated through these population processes. As a generalization, Huston (1994) stated that "the effect of disturbances can be best understood in the context of the environmental conditions that influence the growth and recovery of populations, the rate of competitive displacement, and the rate of dispersal or reimmigration into disturbed areas". If this perspective is correct, local community structure is primarily influenced through the interactive effects of disturbance and interspecific competition on population growth. Consequently, these effects must be considered across the spatiotemporal scales at which disturbances, competition, and population growth operate.

Another possibility is that ecological communities are largely non-interactive entities which are buffeted about by environmental disturbances and fluctuations. Their structure may have nothing to do with biological interactions like competition. Caswell (1976) explored the consequences of such a "neutral" model of community development on community structure. In this model, the colonization of species into a community and the growth in the number of descendants of each successful colonist are described as stochastic processes. These processes operate independently of the number and identity of prior colonists and their descendants. Furthermore, this model generates reasonable dominance-diversity patterns and species-area relationships which are in accordance with both theoretical arguments and empirical evidence (Caswell 1976). In order to estimate the impact of biological interactions in nature, Caswell contrasted predictions of the model with observed patterns in real communities. In many of the examples he evaluated (*e.g.*, assemblages of birds, fishes, insects, and trees), the neutral model predicted higher species diversity than was actually observed. The neutral model did not adequately predict community structure in a wide variety of cases. Thus one can infer that biological interactions in the real world generate lower diversity than predicted by the neutral model especially in the absence of abiotic disturbances. This result is consistent with the general notion that interspecific competition eliminates species from natural assemblages and that this process is offset by disturbances [*e.g.*, by intermediate levels of disturbance (Connell 1978, Menge and Sutherland 1987, Caswell and Cohen 1991, 1993) or by population reductions due to disturbances (Huston 1979)].

If disturbance generally promotes the coexistence of competitively inferior species with superior competitors, "then the pattern of disturbance and the life history patterns of the species involved must be well matched" (Armstrong 1976). That is to say that this mechanism for species coexistence only works if species live long enough and are dispersed over spatiotemporal scales large enough to take advantage of disturbed patches of habitat. Connell and Keough (1986) concluded their review of disturbance and patch dynamics in subtidal, epibenthic communities on hard substrate with a similar statement. The responses of natural communities to disturbances depend on disturbance "size, intensity, and frequency and also on the characteristics and life histories of the resident species". Given that many types of disturbance represent an important part of the selective regime, there is the potential for evolutionary responses promoting some degree of partitioning of the environment along disturbance gradients. This is analogous to the notion of partitioning of the regeneration niche in terrestrial plant communities as developed by Grubb (1977). Life history attributes like fecundity and the longevity of dispersed offspring (*i.e.*, larvae, seeds, resting stages, etc.) clearly can influence dispersal distances and patch colonization rates. Likewise, factors promoting adult longevity can be linked to the ability to tolerate aspects of the local disturbance regime.

The theoretical responses of species richness to the opposing influences of interspecific competition and disturbance are a consequence of two assumptions. Competitive abilities are assumed to be hierarchically ranked among a set of interacting species and disturbances are assumed to have a negative effect on population size. If one relaxes these assumptions, then the number of potentially coexisting species becomes a more complicated function of disturbance intensity/frequency than is typically predicted by the models noted above. Suppose that the outcome of interspecific competition is highly variable as in some of the examples of competitive overgrowth, reversals, and networks discussed in chapter 5. Suppose that the species in an assemblage are differentially susceptible to disturbance and that some may even be favored by disturbance. Sebens (1987) provided an example of this possibility using a simple extension of an earlier model by Armstrong (1976). He considered the coexistence of multiple species with variable (indeterminate) competitive ability across a disturbance gradient. His analysis of a three-species system revealed two interesting results.

- There were regions along the disturbance gradient which promoted the coexistence of two and even all three species.
- Single-species monopolies were predicted not only at very low and high disturbance rates (by the best competitor and best colonizer, respectively), but also at intermediate disturbance rates. Intermediate disturbance levels favored a species which could colonize space faster than the superior competitor, yet compete for this space better than the superior colonizing species.

Field examples which may conform with these ideas include the dominance of *Postelsia palmaeformis* in rocky intertidal habitats, *Alcyonium siderium* on subtidal rock walls, and *Zoanthus sociatus* and *Z. solanderi* on coral reefs (see sections 1.4.1 and 1.4.3). Prior to the mass mortality of *Diadema antillarum* in the Caribbean (see section 3.3.1), these zoanthids were noted to interact with a diverse set of algal and cnidarian species in the back reef of Discovery Bay, Jamaica. In spite of being poor competitors for space, they often formed nearly monospecific aggregations (Karlson 1980, 1983, 1988). Thus patterns of species coexistence across disturbance gradients may be much more variable than is often predicted by theory. Competitive exclusion occurs in these examples "because it is the best competitor that can withstand the prevailing scale of disturbance" which is predicted to form monopolies (Sebens 1987). The success of a particular species depends on a combination of traits influencing colonization, competition, and susceptibility to disturbance.

In review, disturbance theory predicts that species coexistence can occur at several spatiotemporal scales. Species may coexist within a local patch of habitat, within a set of patches in a homogeneous habitat, or in a more heterogeneous environment with a range of disturbance levels. Furthermore, species coexistence may occur as a result of 1) nonequilibrial disturbances perturbing a local assemblage away from competitive equilibrium, 2) nonequilibrial dynamics described by the opposing effects of slow competitive displacement and population reductions, and 3) more equilibrial dynamics invoking partitioning of the environment along disturbance gradients. As one considers the issue of scale, differential responses of species to disturbance and this mix of equilibrial and nonequilibrial dynamics can confound the interpretation of observed patterns of species coexistence. As noted by Fox (1979), many nonequilibrial explanations for species coexistence can be "understood in the context of macroscopic equilibrium". Thus careful specification of the spatiotemporal scale and species to be considered is appropriate when evaluating the effects of disturbance in real communities. Lastly, I repeat the message from chapter 3 that the community response to disturbance may be evaluated in terms of its resilience (recovery) and/or its resistance to disturbance. The former is an equilibrial notion, while the latter attribute may be associated with either equilibrial and nonequilibrial dynamics. Real ecological communities are likely to include elements of both.

7.2 Disturbances to coral communities: a potpourri

Because coral communities flourish in shallow tropical seas, they are exposed to a physical environment characterized by intense solar radiation, occasional major storms (cyclones/hurricanes), and sea level changes associated with fluxes in the global climate, catastrophic geologic events, and extreme tidal anomalies. Although many of the most destructive phenomena occur only rarely [*e.g.*, earthquakes (Stoddart 1972), volcanic eruptions (Tomascik *et al.* 1996), and extreme cooling events (Porter *et al.* 1982)], there are also a number of processes which disturb these communities more frequently but on much smaller spatiotemporal scales. Jackson (1991) noted that currents, wave forces, and sedimentation are the most common forms of physical disturbance on coral reefs in the Caribbean (particularly during storms). Smaller-scale events associated with predation, territorial behavior of damselfishes (algal gardening), and bioerosion are among the most important forms of biological disturbance. So coral communities can be disturbed across a wide range of spatiotemporal scales. Jackson (1991) characterized this enormous range in terms of the spatial extent, duration, and frequency of these disturbance processes on Caribbean reefs (Table 7.1). They almost completely encompass the full range of conceivably relevant macroscopic scales! In comparing Caribbean with Indo-Pacific reefs, Jackson also noted the major disruptive effects of population outbreaks by *Acanthaster planci* (section 3.3.3) and stated that "Indo-Pacific reefs are almost certainly even more disturbed" than the "highly disturbed" reefs of the Caribbean.

Given the staggering range of spatiotemporal scales indicated in Table 7.1, one might ask if some disturbances are more important than others. Thus one might focus attention on the corresponding scales at which such important disturbances operate. A ranking of importance would first require one to select appropriate criteria before culling through the various types of disturbance phenomena. Certainly disturbances which destabilize populations of keystone species (section 3.3) should be considered, especially with the potential for phase shifts to alternative community states. Other disturbances affecting guilds of consumers may disrupt the stabilizing influence of diffuse herbivory or predation. And catastrophic disturbances can obliterate the entire community thus shifting attention to important recovery processes at local and regional spatial scales (*e.g.*, the relative importance of colonization by coral fragments or larvae). In what follows, I present some notable examples of disturbances to coral communities. These include several natural catastrophic disturbances as well as some caused by humans. This is followed by a section addressing the question as to whether disturbances actually promote species coexistence in these communities as predicted by theory.

TABLE 7.1 Spatial extent, duration, and frequency of disturbances on Caribbean coral reefs (after Jackson 1991, Volume 41, Page 478). © 1991 American Institute of Biological Sciences.

Process	Spatial extent	Duration	Frequency
Predation	1-10 cm	Minutes-days	Weeks-months
Damselfish gardening	1 m	Days-weeks	Months-years
Coral collapse (bioerosion)	1 m	Days-weeks	Months-years
Bleaching or disease of individual corals	1 m	Days-weeks	Months-years
Storms	1-100 km	Days	Weeks-Years
Hurricanes	10-1000 km	Days	Months-decades
Mass bleaching	10-1000 km	Weeks-months	Years-decades
Epidemic disease	10-1000 km	Years	Decades-centuries
Sea-level or temperature change	Global	10,000-100,000 Years	10,000-100,000 Years

7.2.1 GENERAL REVIEW

In an early review, Stoddart (1969) described the processes contributing to the structure of coral communities and those which destroy it. The formation of coral reefs was interpreted in terms of several geological processes which control the relative height of the growing coral community and sea level. These include continental glaciation, migration of the earth's crust, and the formation and subsidence of oceanic volcanoes. Several physical attributes of shallow reef environments were highlighted as major local controls on coral growth and the geographic distribution of corals (*i.e.*, light, temperature, salinity, intertidal exposure, wave action and turbulence, and sedimentation). Although the above processes and environmental factors can contribute to both the growth and demise of corals, Stoddart (1969) stated that "the major cause of catastrophic coral mortality on reefs is mechanical destruction during tropical storms". This destruction differentially influences corals. Fragile branching corals are much less resistant than are massive corals. Likewise, corals on exposed seaward slopes generally experience more mortality than those in more protected lagoonal or back reef habitats. Other natural causes of catastrophic mortality in coral communities noted by Stoddart (1969) include diseases, siltation, and excessive freshwater from coastal rivers and rainfall during cyclones. Differential responses among corals to these processes include a wide range of tolerances to exposures to freshwater and high sediment loads. Tolerance to several human activities which cause catastrophic mortality are generally poor. As examples, Stoddart (1969) specifically mentioned weapons testing, the construction of harbors, military bases, airfields, and roads, and the exploitation of corals for industrial purposes.

In that several reviews of both natural and human-caused disturbances have appeared in the literature in recent years, I present in tabular form an overview of the wide range of disturbance phenomena they cover (Table 7.2). As in Stoddart (1969), major tropical storms are emphasized in most reviews. For example, a special issue of *Coral Reefs* devoted to the effects of disturbance on coral reefs dynamics (Hughes 1993) also emphasized such storms. This was the primary subject of most of the contributions. As indicated in Table 7.2, not all disturbances can be categorized simply as either natural or caused by man. Several notable phenomena including population outbreaks of *Acanthaster planci*, the spread of diseases, mass mortality events, and coral bleaching events may be caused primarily by man, by natural processes, or by a combination of the two. In most cases, documenting the causal factors is extremely difficult and thus there is much speculation and a wide range of opinions regarding actual causes of these phenomena.

TABLE 7.2 Natural and human-caused disturbances to coral communities. Other disturbances with unknown causes are noted separately.

Natural disturbances	Human-caused disturbances
Major storms including cyclones/hurricanes[1,2,4,5,6,7]	Chronic pollution including sewage and eutrophication[1,3,4,5,6,7]
Vulcanism[1,4,5,6]	Dredging[1,3,5,7]
Extreme chilling[1,4,5,6]	Blasting including nuclear explosions, dynamite fishing and construction[1,3,4,5]
Red tides[1,6]	Increased runoff and siltation due to urban development, forestry, and agriculture[1,5,6,7]
Extreme low tides sometimes associated with heavy rainfall[1,2,4,5,6,7]	Extreme pollution events including oil spills and use of chemical detergents[3,4,5,6,7]
El Niño warming events[5,6,7]	Thermal stress from power plants[3,4,5]
Earthquakes[5]	Heavy metals and radioactivity[3]
Chronic wave action and currents[2]	Ciguatera[3]
Sedimentation[2]	Species introductions[3]
Herbivory and predation[2,7]	Sedimentation[4,5,6,7]
	Anchor damage[4]
	Mining[4,5,6]
	Shell and coral collecting[3,4,5]
	Overfishing including the use of poison[3,4,5,7]
	Tourism and recreation[3,5,6]
	Industrial development[6]
	Ship grounding[7]

Other disturbances
Population outbreaks by *Acanthaster planci*[1,2,3,5,6,7]
Mass mortality of *Diadema antillarum*[2,4,6,7]
Episodic coral bleaching events[4,5,6,7]
Coral diseases[5,6,7]

Sources: [1]Pearson (1981), [2]Huston (1985), [3]Salvat (1987), [4]Grigg and Dollar (1990), [5]Sorokin (1993), [6]Brown (1997), [7]Connell (1997)

Connell (1997) surveyed the literature in search of quantitative evidence regarding the effects of disturbance and the incidence of recovery from disturbance in assemblages of corals. He found 23 studies with at least 4 years of data on the percent cover of corals at 62 different sites on 24 different reefs throughout the world. The most significant result of this analysis was the large difference found for the effects of acute (short-term) and chronic (frequent) disturbances. The abundance of corals recovered to at least 33% of pre-disturbance or control levels in only 8 of 30 examples of chronic disturbances [and these all occurred after the disturbance (sewage in Kaneohe Bay, Hawaii) had been discontinued], but in 11 of 19 examples of acute disturbances. Failure to recover from chronic disturbances included examples of natural (*e.g.*, major storms) and human-caused (*e.g.*, oil spills, coastal development, overfishing) perturbations to coral assemblages as well as some due to unknown causes (*e.g.*, mass mortality of *Diadema antillarum*). Thus there was no significant difference in the proportion of recovered coral assemblages based on this criterion. One troubling result was the contrast between western Atlantic and Indo-Pacific coral assemblages. There were no recoveries from disturbances causing significant declines in coral cover in the western Atlantic. This is due to the high frequency of chronic disturbances which have destabilized coral community structure especially on the well-studied reefs of Jamaica and the Virgin Islands (see section 3.3.1).

Given that there are more than two dozen major causes of disturbance to coral communities (Table 7.2) and these can occur in a variety of combinations at any particular site (*e.g.*, overfishing, Hurricane Allen, and the mass mortality of *Diadema antillarum* along the north coast of Jamaica), I present below a more detailed description of only a few selected examples. This selection is restricted to some of the best studied coral communities and to natural disturbances. Readers interested primarily in man's impact on coral communities should refer to the reviews noted in Table 7.2 and to the symposium volume by Ginsburg (1993). In that major storms have been persistently emphasized as one of the most important forms of disturbance, I present two examples. First, the effects of Hurricane Allen on the coral community at Discovery Bay, Jamaica, are covered. This community was well studied prior to the storm in 1980 and the effects were examined on a community-wide basis by a variety of investigators with diverse expertise. Second, the effects of multiple cyclones as well as other disturbances on the well-studied reefs at Heron Island are considered. In keeping with the scale-dependent theme developed in chapter 1, large-scale disturbance to coral communities by El Niño warming events is presented as the last example.

7.2.2 HURRICANE ALLEN AT DISCOVERY BAY

Following a particularly long, hurricane-free period of 36 years (Woodley 1992), Hurricane Allen passed Discovery Bay on August 6, 1980. As it moved along the coast at a minimum distance of 45 km, it generated high winds, enormous waves (some 12 m high), and extensive damage to the coral community. This major storm was the second strongest of the century. Hurricane Gilbert in 1988 was stronger, but did less damage to the branching corals *Acropora palmata* and *A. cervicornis*. These corals had not yet recovered to previous abundance levels after the devastation from Hurricane Allen

(Woodley 1992). As an indication of the noteworthiness of Hurricane Allen, two correspondences regarding the destruction it caused were published in *Nature* (Woodley 1980, Lewis 1980) during October. Three more quantitative assessments were published in *Science* or *Nature* the following year (Woodley *et al.* 1981, Porter *et al.* 1981, Knowlton *et al.* 1981). This storm whose winds approached 300 kph did considerable damage to coastal homes, forests in the Blue Mountains, and banana plantations in the lowlands (Woodley 1980). The immediate damage to the coral community extended to depths below 50 m as branching corals were smashed and reduced to rubble, massive corals were toppled or split, and foliaceous corals were fractured or torn off the reef slopes. Woodley (1980) also noted the death or injury of other organisms including gorgonians, sponges, echinoids, and fishes as well as the removal of tons of sediment from reef slopes and terraces. Extensive abrasion by suspended sediment had particularly destructive effects on the community.

Quantitative verification of these effects first appeared in Woodley *et al.* (1981), a joint effort representing the collaboration of twenty authors. In general, the damage was shown to vary "with reef location, depth, and topography". Furthermore, significant differences in the extent of damage among taxa was attributed to their location, growth form, and type of construction. The spatial variability in damage occurred at several scales. For example, the northeast coast of Jamaica experienced more damage than other coastlines. In addition, the east fore-reef at Discovery Bay was more damaged than the west fore-reef. In terms of habitats, the shallow fore-reef and crest were more damaged than the back reef and deep fore-reef. Thus the spatial extent of disturbance by hurricanes can occur at smaller scales than indicated in Table 7.1. For example, death and damage to gorgonians on the fore-reef was much more extensive than that on the nearby back reef. Likewise, damage to corals and the echinoid *Diadema antillarum* was more extensive on the shallow fore-reef than in deeper water. Sloping and level reef zones experienced greater damage than did vertical walls at comparable depths, while the broad terrace on the east fore-reef afforded some protection to *Acropora palmata* at a depth of only 3 m. This dominant branching coral was completely leveled on the narrow west fore-reef. Woodley *et al.* (1981) also noted multiple examples of smaller-scale spatial variation in damage to massive coral heads of *Montastrea annularis* on the west fore-reef. In sand channels, only 35-64% of the heads survived, while 93% survived on reef lobes. This variability was accentuated as "surviving large, massive corals provided shelter for more fragile organisms in their lee, but those that toppled left paths of damage". Spatial variation due to sand abrasion, falling sand, and skeletal debris was associated with coral positions relative to patches of sand, sand channels, and chutes down vertical walls.

The extent of damage from Hurricane Allen varied significantly among taxa, growth forms, and with the mechanical characteristics of the skeleton of sessile species (Woodley *et al.* 1981). For example, at a depth of 6 m on the west fore-reef, there was a 94% reduction in the abundance (cm^2) of branching *Acropora* spp., a 23% reduction in the foliaceous and encrusting *Agaricia agaricites*, and only a 9% reduction in the massive *Montastrea annularis*. Fragile branching and foliaceous corals experienced more breakage and fragmentation than did thicker coral plates or stouter branching corals. Over all coral species, there was generally a large reduction in their percent cover especially on the shallow fore-reef. Woodley *et al.* (1981) estimated the magnitude of this reduction at 39% at 4 m, but only 0-1% at 10-33 m. Hughes (1994a) reported a reduction in percent cover of approximately 38% at 10 m, 16% at 15-20 m, and 10% at 35 m (Figure 3.3). Gorgonian colonies experienced 51% mortality at 7 m on the fore-reef and 98% of the surviving colonies were injured (Woodley *et al.* 1981). Among the sponges, ropelike forms were heavily damaged at 15 m with nearly half of them being broken off at the base and 67% of those remaining were injured. Other sponge growth forms experienced less damage in that only 31% had lost more than one third of their tissue. The influence of the type of skeleton on the extent of damage was evident in that brittle branching corals were severely damaged. In contrast, the flexible skeleton of gorgonians "proved more resilient". Likewise, tough sponges like *Ircinia* spp. "often appeared unaffected, whereas almost all soft, crumbly forms, such as *Neofibularia nolitangere*, were destroyed".

Among motile organisms, there were significant differences in the abundances of such keystone species as the echinoid *Diadema antillarum* and the damselfish *Stegastes planifrons* before and after Hurricane Allen. On the shallow fore-reef at two sites, densities of *D. antillarum* were reduced to 1% and 46% of pre-hurricane densities, while those in deeper water were less dramatically affected (Woodley *et al.* 1981). Surprisingly, densities of *S. planifrons* were not reduced by the hurricane. At several sites they remained approximately the same, but they actually increased 83% at one 18-m site on the west fore-reef. Woodley *et al.* (1981) and Kaufman (1983) noted some unusual behavioral changes in motile species immediately following Hurricane Allen including the following: *S. planifrons* became less aggressive, typical schooling behavior in the striped parrotfish *Scarus croicensis* was disrupted, several normally cryptic species were more active in the open, planktivorous fishes were seen feeding nearer reefs than was normal, and sharp increases were noted in the abundance of large predatory fishes (*i.e.*, snappers, groupers, and grunts). Some behaviors returned to normal in as little as 2-10 days after Hurricane Allen (*e.g.*, territoriality in damselfishes). However, some abnormal behaviors persisted for several weeks (particularly the more wide ranging foraging behaviors noted above).

In addition to the immediate damage caused by Hurricane Allen, there were delayed consequences resulting in death several months and even years later. Woodley *et al.* (1981) reported that most fragments of *Acropora cervicornis* and *A. palmata*, which had been alive 1-2 weeks following the storm, were dead after 16-22 weeks. Only four of 254 tagged fragments of the former species and 15 of 54 in the latter survived this length of time. Knowlton *et al.* (1981) estimated that this secondary mortality in *A. cervicornis* "was over an order of magnitude more severe than that caused by the immediate effects of the storm". The fragments of this coral experienced nearly continuous mortality over a 5-month observation period and "in some areas, tissue mortality still exceeded new growth... nearly 1 yr after the hurricane". The causes of this secondary mortality during the first few months were attributed to disease and the stress associated with fragmentation during the storm. However, five months after Hurricane Allen, most mortality was attributed to the activities of *Diadema antillarum*, *Stegastes planifrons*, and the corallivorous gastropod *Coralliophila abbreviata.*

Knowlton *et al.* (1981) further emphasized the potential importance of this form of secondary mortality by noting that the extensive immediate damage of the storm affected corals much more than corallivores. Thus "predation pressure per colony of surviving *A. cervicornis*" was substantially increased. They suggested that Hurricane Allen had destabilized predator-prey interactions so as to increase the mortality experienced by coral fragments several months after the storm. This is ironic in that fragmentation and asexual propagation are thought to be important life history attributes in *A. cervicornis.* The extreme severity of Hurricane Allen appears to have far exceeded the normal adaptive benefits of fragmentation in this species.

Knowlton *et al.* (1990) documented the collapse of populations of *A. cervicornis* along the north coast of Jamaica in 1982-1987. Furthermore, they supported the speculation that Hurricane Allen had destabilized predator-prey interactions. Three primary fore-reef sites at 10-13 m were selected to represent varying degrees of storm damage to "otherwise comparable staghorn reefs". These included a very exposed site between St. Ann's Bay and Priory, an intermediate site at Pear Tree Bottom, and a least exposed site near Montego Bay. At each site, coral cover was assessed multiple times and 28-80 apparently healthy, attached colonies were tagged and monitored for growth, survivorship, and specific evidence of death due to predators. Initially colonies with damage from predators or disease were avoided as were those within damselfish territories. These sites had all been dominated by *A. cervicornis* prior to Hurricane Allen. In 1982, total staghorn cover was 35-71% (0.4-21.4% live cover plus at least 34% dead rubble). By 1986, mortality among the tagged colonies was 92.3-100% (only 6 of the original 160 colonies survived). By 1987, estimates of live staghorn cover at all three sites had dropped to 0%.

The three most conspicuous causes of mortality were *S. planifrons*, *C. abbreviata* and the polychaete *Hermodice carunculata.* All three of these consumers remained abundant at these primary sites throughout the study. Knowlton *et al.* (1990) again presented the notion that predators were preventing the recovery of staghorn populations [due to a phase shift when predation rates increased to levels exceeding a predicted threshold]. However, also contributing to the demise of this coral, was the extensive growth of algae following the mass mortality of *D. antillarum* in 1983 (section 3.3.1). Knowlton *et al.* (1990) estimated fleshy algal cover in 1987 to have reached 82-88% of the substrate at the three primary sites. Thus multiple disturbances contributed to the destabilization of coral populations and community structure on these reefs. Furthermore, there was significant spatial variability in the relative importance of different disturbances. For example, Knowlton *et al.* (1990) noted that over the interval 1982-1986, there was an increase in the percent cover of *A. cervicornis* from 1% to 3% on Dancing Lady Reef off Discovery Bay. This increase corresponded with sharp declines in predator densities. This pattern was distinctly different from that described at the three primary sites and may have indicated some degree of recovery (rather than a phase shift) towards pre-hurricane staghorn cover on this particular reef.

7.2.3 MULTIPLE DISTURBANCES AT HERON ISLAND

Connell *et al.* (1997) recently evaluated the long-term dynamics of the coral assemblages at Heron Island to emphasize the effects of the type and scale of disturbances, recruitment, and pre-emptive competition. As noted previously, three shallow sites monitored at this location from 1962-1989 experienced considerable disturbance-related mortality (Tanner *et al.* 1994 and section 4.2.1). In this latest analysis, Connell *et al.* (1997) considered six sites around Heron Island. Three sites located on the north reef were exposed to extreme wave forces during cyclones (*i.e.*, on the intertidal crest, in shallow subtidal pools, and on the deeper slope at 1-14 m). Three sites to the southwest of the island were protected from these wave forces, but exposed to other disturbances (*i.e.*, to strong tidal currents on the slope at depths of 1-10 m, to intertidal exposure on the reef crest, and to multiple natural and human-caused disturbances on the inner reef flat). Among all disturbances, cyclones generating winds of up to 163 kph were the most severe. All major declines in coral abundance in excess of 40% cover occurred in years in which cyclones passed within 200 km of Heron Island. There were 35 such cyclones reported in 1946-1992.

During 1962-1992, the abundance of corals in permanent quadrats sharply declined at some sites in each of four cyclone-years (1967, 1972, 1980, and 1992), but not in six other years with cyclones in the area. The most significant declines occurred on the exposed crest in 1972 and in the exposed pools in 1967 when nearly all of the corals died. Much less damage occurred in deeper water and at more protected sites in these years. Connell *et al.* (1997) also documented a gradual decline in coral abundance on the protected crest. At this site, coral cover decreased from 67% to 7% (1969-1992) and the density of coral genets decreased from 90 per m^2 to 11 per m^2 (1982-1992). Although this site was significantly damaged by a cyclone in 1980, the long term declines were attributed to increased aerial exposure due to upward coral growth and greater drainage of this site caused by a ship grounding and channel erosion. At all three of these damaged sites, there was a significant reduction in the mean percent cover of corals in five cyclone-years, but not in other years (1976 was added as a cyclone-year because of Cyclones Beth, David, and Dawn and observed damage to trees at Heron Island). In contrast, there was no significant reduction in mean coral cover on the inner reef flat during cyclone-years. At this site, Connell *et al.* (1997) documented gradual declines from 1974-1989 resulting in the loss of almost all corals. They identified aerial exposure just like that at the protected crest, low rates of coral recruitment, and moderate damage caused by cyclones as probable causes of the decline in coral abundance.

In their evaluation of coral recovery from disturbances at Heron Island, Connell *et al.* (1997) stressed three aspects of the process. First, the distinctions between acute and chronic disturbances were emphasized [as discussed in Connell (1997) and section 7.2.1]. Second, disturbances which alter the environment directly were contrasted with those which do so only indirectly as a consequence of coral mortality. Third, severe disturbances followed by slow recovery due to the settlement of coral larvae were distinguished from more benign disturbances followed by faster recovery due to the regeneration and growth of coral fragments. Gradual changes in the environment unrelated to disturbances also influenced the recovery process as did variation in recruitment due to harsh physical conditions and pre-emptive competition. Chronic disturbance from frequent cyclone damage in the exposed pools repeatedly interrupted the recovery process, yet some degree of rapid recovery did occur at this site after the cyclones in 1972, 1976, and 1980. Coral cover in the exposed pools recovered to pre-1967 levels in approximately 10 years, while coral density had not fully recovered by 1992. Likewise, coral cover and density on the exposed crest did not recover from the acute disturbance caused by cyclones in 1972. This was attributed to major physical changes in the environment which increased the aerial exposure of corals at this site. At protected sites at Heron Island, "recovery from the limited cyclone damage ranged from moderate to nil" largely due to gradual physical changes in the environment at these sites.

In that recovery following acute disturbances could occur by regenerating corals *in situ*, by transport of coral fragments into the permanent quadrats, or by larval settlement, Connell *et al.* (1997) assessed the rate of recruitment of new corals at the four shallowest sites. As discussed earlier in section 4.2.2, the 30-year mean recruitment rates were 1.7-12.7 recruits per m^2 per yr. These rates were generally high on the protected crest, low on the exposed crest, and highly variable in the exposed pools and on the inner reef flat. In fact, annual recruitment rates were so variable that Connell *et al.* (1997) detected no consistent pairwise differences among sites. However, analysis of recruitment rates across all sites indicated that available free space significantly enhanced recruitment thus explaining some of the annual variability. This effect was strongest at the two exposed sites where high coral cover was associated with low recruitment rates and cyclones were responsible for generating much of the free space. At the more protected sites, high macroalgal cover was associated with low recruitment rates on the inner reef flat, while free space was never in limited supply on the protected crest. Thus, Connell *et al.* (1997) inferred that pre-emptive competition by adult corals and macroalgae significantly reduced coral recruitment. In addition, they noted that the physical modifications of the environment due to disturbances at sites on the exposed and protected crests subjected potential coral recruits to quite harsh conditions thus significantly reducing coral recruitment.

The assessment of variation in coral abundance and recruitment in response to disturbances by Connell *et al.* (1997) employed multiple sampling procedures (*i.e.*, quadrats, line transects, and belt transects) over several spatiotemporal scales. Thus they could evaluate the scales at which disturbances and recovery processes influence the dynamics of the assemblage. They identified five spatial scales ranging from replicate, 1-m^2 quadrats 1-5 m apart within a site to large-scale contrasts between exposed and protected sites up to 1850 m apart. Seven temporal scales using 1.6-18.6 yr intervals were identified. They found that declines in coral abundances due to cyclones and aerial exposure occasionally occurred across large expanses of shallow habitat and there was considerable variation among habitats in terms of cyclone damage. Nevertheless, these sources of coral mortality "were probably operating within each habitat at relatively small spatial scales" [generally at 10-100s of meters (Connell personal communication)]. Temporal scales of variation were short at shallow 0.5-5 m depths (approximately 10-yr periods between peaks caused by relatively rapid recovery and troughs by cyclone damage). These scales were much longer at the intertidal sites and in deeper water (≥20 yr). Spatial variation in coral recruitment rates were similar to those for coral abundances, but temporal patterns were quite different. There was no evidence for predictable fluctuations in the highly variable, annual recruitment rates at any site. By examining the effects of disturbances and recovery across a variety of spatiotemporal scales, Connell *et al.* (1997) highlighted the considerable variability in the effects of these processes. Furthermore, they emphasized the need to study these processes "at as many scales as possible, if we are to understand the mechanisms underlying these variations".

7.2.4 EL NIÑO EVENTS IN THE TROPICAL PACIFIC

In that the destructive effects of El Niño events are known for temperate epibenthic communities (*e.g.*, see section 1.4.2), it should be no surprise that these thermal anomalies also have devastating effects on coral communities in the tropics. In the eastern equatorial Pacific, sea level and surface sea temperatures rise, nutrient levels, productivity and the thermocline drop, rainfall in the region increases, and coastal areas particularly in Peru and Ecuador experience unusual flooding events (Glynn 1988b). The extremely severe El Niño of 1982-1983 resulted in a major bleaching episode in which "all species of reef-building corals" lost their zooxanthellae. Furthermore, 50-98% coral mortality was reported for sites from Ecuador to Costa Rica and two species of the hydrocoral *Millepora* were reported to have possibly undergone an unusual extinction event. This latter report was modified by Glynn and De Weerdt (1991) to indicate that extinction was likely only in an undescribed endemic species of the Panamic Pacific Province. The more widely distributed *M. platyphila* probably experienced a major range reduction by 1983 and had not yet recolonized reefs in this province by 1990. In the Galápagos Islands, higher sea level coupled with spring tides and unusual swell conditions resulted in the destruction of large patches of coral reef. In the western Pacific, El Niño events have also disturbed coral communities. For example, Yamaguchi (1975) reported the mass mortality of a variety of inner reef flat organisms on Guam as a result of the El Niño of 1972-1973. Likewise, Brown and Suharsono (1990) reported elevated water temperatures, coral bleaching, and 80-90% mortality among shallow reef flat corals in Indonesia as a result of the 1982-1983 El Niño.

A community-wide assessment of the 1982-1983 event has indicated a large range of effects throughout multiple trophic levels in coral and other marine communities of the eastern Pacific (Glynn 1988b, 1990b). The altered oceanographic conditions had negative effects on primary producers in that nutrient levels and salinity were lower than normal. Benthic populations were exposed to unusual sea level and wave conditions. Reduced primary and secondary productivity resulted in food limitation, reproductive failure, and mortality in a variety of consumer species. These included several species of seabirds throughout the region, the herbivorous marine iguanas in the Galápagos Islands, and four species of chiefly piscivorous pinnipeds. Coral species richness and diversity on reefs in the Galápagos Islands and in Panama were significantly reduced. Glynn (1990c) reported species losses of approximately 50-80% along transects sampled before and after the event, while "most measures of diversity declined to 0".

Other community level effects included 1) large changes in the relative abundances of important spatial competitors and consumers, 2) loss of coral framework and habitat for associated species, 3) invasions of new species as a result of the unusual oceanographic conditions, and 4) indirect disturbance effects occurring substantially after the 1982-1983 event. For example, Glynn (1988b) noted the red tide which occurred along the coasts of Costa Rica and Panama killing many corals down to depths of 10 m after the El Niño event was over. In two other examples, delayed mortality in corals was attributed to keystone species: 1) Non-pocilloporid corals (*e.g.*, *Gardineroseris planulata* and *Pavona* spp.) suffered extensive mortality from *Acanthaster planci* following the El Niño event (Glynn 1985, 1988b, 1990a). These corals had previously been protected from this predator as a consequence of the crustacean symbionts living with surrounding pocilloporid corals (see sections 2.3 and 3.3.3). 2) Some massive corals, which had experienced partial mortality as a direct consequence of the El Niño, suffered secondary mortality in 1985 due to *Stegastes acapulcoensis* as this damselfish established territories on dead coral substrate.

Given the extensive devastation due to the 1982-1983 El Niño, Glynn (1990c) concluded that this disturbance in the eastern Pacific was extraordinary in it severity and scope. During this century, only the 1925-1926 warming event matched the 1982-1983 event in terms of the magnitude of the thermal anomaly (it reached 8 °C above normal). However, these extreme warm temperatures did not persist as long as they did during the 1982-1983 event. Furthermore, based on the estimated age of living corals (*e.g.*, one large colony of *Gardineroseris planulata* on Uva Reef in Panama was estimated to be 192 years old) and on the thickness of the coral reef framework, Glynn (1988b, 1990c) inferred that no El Niño event as severe as that in 1982-1983 had disturbed these reefs in the region for at least 200 years. Glynn (1990c) suggested that perhaps it had been as long as 400-500 years since such an event had occurred in the Galápagos Islands. These severe events are thought to be natural, yet highly infrequent climatic anomalies associated with the Southern Oscillation (as mentioned in section 1.4.1). Colgan (1990) estimated that only 18-65 severe events have occurred during the last 6500 years. In that the recent frequency of all El Niño warming events is approximately once every four years, the vast majority of these are much less destructive than the 1982-1983 event. Yet severe events occur often enough to severely retard reef building in the eastern Pacific (Colgan 1990, Glynn and Colgan 1992).

Although these severe disturbances may preclude the relatively slow accretion of reefs, some species in the region appear to have been rapidly recovering from the 1982-1983 event. Glynn (1988b) noted evidence for the recovery of seabirds, marine iguanas, and fur seals with the resumption of normal oceanographic conditions (*e.g.*, upwelling, cool water, high primary and secondary productivity). However, recovery following some of the major changes in benthic communities (*e.g.*, loss of dominant species and species invasions) has occurred much more slowly or not at all. By 1986, the epibenthic community on subtidal rock walls in the Galápagos Islands was noted to be far from pre-1983 conditions. Coral communities were experiencing little recruitment of new corals by 1987 (Glynn 1988b) and only moderate levels of recruitment in a few species were reported by 1989 (Glynn 1990c). Furthermore, the coral communities were exposed to extremely high levels of bioerosion due to grazing sea urchins.

The vast expanses of dead coral substrate appears to have promoted recruitment of *Diadema mexicanum* to reefs in Panama and *Eucidaris thouarsii* in the Galápagos Islands. Densities of these sea urchins in 1989 were 5-10 times pre-1983 levels. This grazing activity was "reducing the availability of stable substrata for the settlement of coral larvae" and may also have been killing some larvae directly. Glynn (1990c) estimated the time required for the recovery of the least damaged coral reefs in the region to be a matter of decades. In contrast, the most severely damaged reefs with no living corals "may require centuries" to recover.

7.3 Do disturbances promote species coexistence?

Since all major theoretical explanations for how disturbances influence ecological communities include the enhancement of species coexistence, it is appropriate to examine this specific prediction using evidence from coral communities. In particular, I focus on how species composition and richness patterns are influenced by disturbance and how they change during the recovery process. Here I return to the evaluation by Connell (1997) to consider the eleven examples in which the community was judged to have recovered from acute disturbances (section 7.2.1, Table 7.3).

TABLE 7.3 Changes in the structure of coral assemblages in response to acute disturbances and at least four years of recovery (after Connell 1997).

Site	Disturbance	Disturbance promotes species coexistence	Species are differentially affected by disturbance
Heron Island			
Exposed crest[1,2]	1967 storm	No[2]/NP[1]	Yes[2]
Exposed pools[1]	1972 storm	NP	NP
Exposed pools[1]	1980 storm	NP	NP
Exposed slope[1]	1972 storm	NP	NP
Innisfail reefs[3,4]	*Acanthaster*	No	Yes
John Brewer Reef[4,5]	*Acanthaster*	NP	Yes
Pari reef flat[6]	1982-1983 El Niño	Yes	Yes
Tikus reef flat[6]	1982-1983 El Niño	Yes	Yes
Phuket Island			
Reef flat[7,8]	1986-1987 dredging	No[7]/NP[8]	Yes
Eilat[9]	1970 aerial exposure	Yes	Yes
Guam[10-12]	*Acanthaster*	Yes	Yes

Sources: [1]Connell *et al.* (1997), [2]Connell (1976), [3]Pearson (1974), [4]Pearson (1981), [5]Done (1985), [6]Brown and Suharsono (1990), [7]Brown *et al.* (1990), [8]Brown *et al.* (1993), [9]Loya (1976c, 1990), [10]Randall (1973), [11]Pearson (1981), [12]Colgan (1987).
NP = evidence not yet published

Connell (1997) used 33% of pre-disturbance coral percent cover as a criterion for recovery. I ask if the published record indicates how these communities changed sufficiently enough to permit an evaluation of the predictions that 1) disturbances promote species coexistence, and 2) species within these assemblages exhibit differential responses to disturbance. The former represents disturbance as a mechanism for diversification within the assemblage which may operate on ecological or evolutionary time scales. The latter is an essential element of the argument that disturbance represents an important component of the selective regime. Although the two predictions are not mutually exclusive, not all theories invoke both effects of disturbance.

• There were four examples from the long-term study at Heron Island which reported significant recovery from disturbances (Connell 1997). These occurred following the 1967 storm on the exposed crest, the 1972 storm in the exposed pools and on the exposed slope, and the 1980 storm in the exposed pools. None of these sites were severely devastated by these storms. Only 9-31% of the mean coral cover was lost, much more severe damage was noted for other disturbance events, and recovery was relatively rapid [Connell (1997), Connell *et al.* (1997), and section 7.2.3]. Very little of the data regarding changes in species composition and richness in response to these disturbances and subsequent recovery have been published, but Connell *et al.* (1997) indicated their intention to evaluate these changes in the near future.

Connell (1976) did report this data for one of the five, 1-m^2 quadrats on the exposed reef crest. The first censuses of corals in 1963 included eleven species in this quadrat. *Acropora digitifera*, *Porites annae*, and *Pocillopora damicornis* were most common representing 63% of 43 colonies. The storm of 1967 scattered coral fragments over the site, yet all eleven species (but only 29 colonies) were still alive in the quadrat two weeks afterwards (Connell 1976). By 1969, two scleractinian species (one colony each) were lost from the quadrat, while two colonies of a twelfth species (*Stylophora pistillata*) were gained. Both of these new colonies were absent in 1970 when the census resulted in 24 colonies and 9 species. Thus the number of species coexisting in this single quadrat remained nearly the same over the 1963-1970 interval in spite of the 1967 storm and the subsequent recovery of coral cover and density of coral genets.

Nevertheless, there were differences in longevity among species over this entire interval and in response to the 1967 storm. Colonies of some species lived through all of the seven years [*e.g.*, *Leptoria gracilis* (=*phrygia*) as noted in Connell (1973, 1976) and section 5.2.1]. Others suffered significant losses during the short 1965-1967 interval (*e.g.*, 31% of thirteen colonies of *Acropora digitifera*). Thus species differentially respond to disturbance and this long-held view is supported (see Stoddart 1969).

• A fifth example noted in Connell (1997) also comes from studies conducted on the Great Barrier Reef where population outbreaks of *Acanthaster planci* in the mid 1960s significantly disturbed the communities. On multiple reefs along the Queensland coast near Innisfail, Pearson (1974, 1981) described the damage and the recovery process. Coral cover and species richness were quite low in samples taken on damaged reefs (compared to undamaged reefs) in 1971. Using quadrat sampling along transects over a 7-27 m depth range on the seaward slope of each reef, mean coral cover in 1971 was estimated to be 4% on damaged reefs and 28% on undamaged reefs (Pearson 1974). Mean species richness was reported to be 5.7 and 8.1 species per m^2 on damaged ($n = 247$ quadrats) and undamaged ($n = 99$ quadrats) reefs, respectively. Pearson (1974) also noted that some particularly large colonies of *Diploastrea* and *Porites* had survived these outbreaks on damaged reefs.

Some level of recovery had already begun by 1971 as the density of small colonies was slightly higher on damaged reefs (7.1 versus 6.7 colonies per m^2). By 1972, recently recruited colonies were noted to represent 21% and 38% of all corals sampled on two damaged reefs (Feather and Ellison Reefs). By 1977, recovery was indicated by most measures of coral abundance and species richness along resurveyed transects. Pearson (1981) reported a mean of approximately 11.2 species per m^2 along 10 transects on eight damaged reefs and 12.2 species per m^2 along five transects on four undamaged reefs. Furthermore, he noted significant competition among corals on recovering reefs as early as 1975 when rapid growth and overtopping by acroporids was apparently eliminating smaller, slower growing species. However, in spite of lower coral abundance due to disturbance by *Acanthaster*, at no time during this study was species richness reported to be higher on damaged reefs than on undamaged reefs.

• South of Innisfail off Townsville, John Brewer Reef had been severely damaged not only by *Acanthaster* in 1969-1970, but also by Cyclone Althea in 1971 (Pearson 1981). Along permanently marked transects established at 3 m on the leeward slope and 6 and 12 m on the seaward slope, Pearson (1981) monitored the recovery of this reef between 1974 and 1978. Within five, 1-m^2 quadrats, hard coral cover increased from 5-7% to 20-45%, coral density from 26-53 colonies per m^2 to 42-88 colonies per m^2, and coral recruitment occurred at rates of 12-24 recruits per m^2 per yr.

Done (1985) sampled these same sites from 1980 through 1983 to document the continued recovery followed by the effects of another outbreak of *Acanthaster* in 1983-1984. Using 10 X 1 m phototransects, he reported the percent cover of hard corals to be 26-60%, 39-78%, 43-72%, and 2-5% in 1980, 1982, 1983, and 1984, respectively. During recovery, the rapid growth of six coral genera (*i.e.*, *Acropora*, *Mycedium*, *Echinopora*, *Merulina*, *Montipora*, and *Porites*) was noted. The 1983-1984 outbreak resulted in high mortality in all coral genera, but was most severe among the *Acropora*. All acroporid corals in the phototransects died as a result of this last major disturbance. In contrast, some surviving colonies not damaged at all by *Acanthaster* were noted among several genera (*i.e.*, *Porites*, *Galaxea*, *Goniopora*, *Tubastraea*, *Physogyra*, *Millepora*, *Pocillopora*, and *Stylophora*). Likewise, surviving remnants following partial mortality in these and other genera were noted. Although this reef had lost a great deal of coral cover, "a significant and diverse stock of surviving corals" remained.

Thus Done (1985) emphasized how these survivors contributed to the resilience of the community following this major disturbance. On the other hand, he also emphasized how the recovery of dominant acroporid corals on the reef would depend on successful larval settlement and recruitment. This genus is particularly diverse on reefs in eastern Australia [see Veron (1986) and section 2.2.2]. Therefore, it is possible that successful recruitment of acroporid corals could result in more species on this recovering reef than on undisturbed reefs where more competitive exclusion would be expected. However, it appears that data on the species composition and richness under these two conditions have not yet been published.

• Two more examples of disturbance and recovery were reported by Brown and Suharsono (1990) for coral assemblages exposed to the 1982-1983 El Niño warming event in Indonesia. Ten, 30-m line transects on each of the reef flats off Pari and Tikus Islands in the Palau Pari complex were established in 1981 and monitored at 1-2 year intervals through 1988. In 1981, the most common corals on these reef flats were *Acropora hyacinthus*, *A. formosa*, and *Heliopora coerulea* at Pari Island, and *H. coerulea* and *Montipora digitata* at Tikus Island. Between 1981 and 1983, the mean coral cover dropped from 22-26% to 2-3% "due to a decline in the branching coral genera, in particular *Acropora*, *Pocillopora*, and *Montipora*". Not one colony of twenty two species of *Acropora* survived this event. The total number of coral species dropped from 40-53 to 19-21 species per site.

By 1985, coral cover had recovered to 11-16%, total species richness was 33-43 species per site, and twenty one species of *Acropora* had re-colonized the reef flat on Pari Island (but only eight at Tikus Island). Further differences between sites were evident from 1985 through 1988. *M. digitata* dominated the reef flat at Tikus Island where it occupied 44% of the total coral cover in 1987. Interestingly, during the 1985-1987 interval at this site, percent cover of all corals dropped marginally to 15%, the abundance of *M. digitata* almost doubled as a result of "rapid growth", and coral species richness dropped again from 33 to 19 species. In contrast, the reef flat at Pari Island was dominated by *Porites nigrescens*, *P. cylindrica*, and *H. coerulea* in 1985-1988, yet species richness also dropped from 43 to 21 species. Fourteen of the twenty one species of *Acropora* which were present at this site in 1985 and 1987, had disappeared by 1988.

In both of these cases, the decline in species richness was associated with the increased dominance of at least one species which had been very abundant prior to the disturbance event. The rise and fall of species richness at these sites is consistent with the prediction that disturbance promotes species coexistence (Table 7.3) and is offset by the expression of competitive dominance.

• In 1979-1992, Brown *et al.* (1993) monitored the effects of multiple natural and anthropogenic disturbances at Phuket Island in Thailand. The coral assemblage on intertidal reef flats there are dominated by faviid and massive poritid corals (especially *Porites lutea*). This assemblage was exposed to a combination of acute and chronic disturbances including warming events, coral bleaching, aerial exposure, discharges from a tin-ore processing plant, and a dredging operation over this time period. At one disturbed and one control site one km away, multiple 10-m line transects were used to estimate coral cover, number of colonies, and species diversity. In response to a 1986-1987 dredging operation in which dredged material was impounded adjacent to the disturbed reef flat [thus exposing the corals to increased sedimentation of fine clay particles (Brown *et al.* 1990, 1993)], mean coral cover abruptly declined from 35% to 11% and then recovered to 30% within one year (Connell 1997). No such changes occurred at the control site.

Brown *et al.* (1993) noted that most of the damage associated with this disturbance resulted in partial colony mortality. This was followed by "a spectacular recovery" due to particularly "rapid regeneration" among massive poritid corals. These corals represented 74% of the total coral cover by 1990 ("compared to 40-55% in previous years"). Coral species diversity initially dropped in response to the disturbance, recovered to pre-disturbance levels within one year, and dropped again with the dominance of poritid corals. Nevertheless, the total species composition and richness reported for this site for 1983-1988 did not change (Brown *et al.* 1990). Ten of thirteen species declined in abundance in 1987, but none were excluded from the site. There were no additional species reported on the recovered reef by 1988. *Goniastrea favulus*, *Coeloseris mayeri*, and *Acropora aspera* did not decline in response to the dredging and increased sedimentation. Brown *et al.* (1990) specifically noted that this last coral did not appear to be adversely affected. Thus there were differential responses among species, but no published evidence for enhanced species coexistence (Table 7.3).

• In 1970, an extremely low tide occurred in the Gulf of Eilat exposing corals on reef flats to the air and very high temperatures for five days (Loya 1990). "The immediate consequence of the low tide was the death of approximately 80-85% of the hermatypic corals along the northern part of the Gulf (Loya 1975, 1976c)." Some of these extensively studied reefs (see section 2.2.3) were near two large oil terminals and therefore were exposed to chronic disturbances. A control reef further away was relatively free of oil pollution. On the control reef flat prior to the extreme low tide, coral cover was 39% and there were 35 colonies and 12 species per 10-m line transect (Loya 1990). After the disturbance event, coral cover plummeted to 9%, coral abundance to less than 10 colonies per transect, and species richness to less than 2 species per transect.

On the control reef, recovery from this extreme low tide was nearly complete by 1973. Coral abundance was approximately 48 colonies per transect and species richness was nearly 16 species per transect (Loya 1990). From 1973-1982, coral cover continued to increase from 30% to approximately 38%, but species richness declined slightly to less than 14 species per transect. This recovery was due to both the "massive recruitment of the most abundant species" (*e.g.*, *Cyphastrea microphthalma*, *Millepora dichotoma* and *Favia favus*) and "complete regeneration" of massive corals (Loya 1976c, 1990). Loya (1990) noted the "remarkable resilience" of this reef flat assemblage in that it had recovered so quickly. The chronically disturbed reefs exhibited no such recovery. The evidence indicates that species coexistence on the control reef flat was promoted by the disturbance. Ten new species were discovered in 1973 and the total number occurring along all transects increased from 33 to 39 species during 1969-1973 (Loya 1976c).

• The last example of recovery from an acute disturbance comes from Guam where the outbreak of *Acanthaster* devastated the coral assemblage in 1969-1970 (see Chesser 1969, Randall 1973, Pearson 1981, Colgan 1987). As noted in section 3.3.3, there were major losses in coral cover (especially in submarine terrace and seaward slope zones), species richness declined 29-44%, and preferred prey species among the *Acropora* and *Montipora* suffered much higher mortality than did non-prey species. Very rapid recovery in terms of species and generic diversity was evident within one year, while coral cover, density, and species richness required more time (Colgan 1987). By 1981, preferred prey species again dominated these coral assemblages. The estimates of total species richness at Tumon Bay prior to the outbreak in 1969 were 98, 73, and 57 species in the reef front, submarine terrace, and seaward slope zones, respectively (Randall 1973). By 1970, Pearson (1981) reported these values to be 70, 47, and 32 species for sites at Tumon Bay and nearby Tanguisson Point. Colgan (1987) (with Randall's assistance) re-analyzed the survey data in light of taxonomic revisions and reported species richness for Tanguisson Reef to be 60, 41, and 26 in 1970 and 66, 73, and 73 in 1981 (Table 3.1). Therefore, species coexistence appears to have been promoted by disturbance in the two deeper zones because species richness in 1981 exceeded pre-disturbance levels.

In summary, long-term studies of acute disturbances and the subsequent recovery process have indicated that the number of coexisting species within coral assemblages 1) abruptly decreases as a consequence of disturbance, 2) then recovers (and in some cases even exceeds) the pre-disturbance level as colonies regenerate and larvae settle, and 3) finally decreases again as dominant species begin to exclude inferior competitors. This pattern is consistent with the intermediate disturbance hypothesis (Connell 1978) and much of the rest of disturbance theory predicting that disturbances promote species coexistence (section 7.1).

Because the later stages of recovery often seem to be associated with intense competition and a predictable decline in species richness, stochastic, random walk models invoking diffuse and unpredictable competitive interactions do not appear to be applicable to many coral assemblages. The rates of competitive displacement in these assemblages (especially in shallow water habitats) appear to be high enough to generate predictable exclusion events. This is in accordance with the intermediate disturbance hypothesis (Connell 1978). Rogers (1993) reached a similar conclusion regarding the applicability of this hypothesis to assemblages on shallow coral reefs. The non-equilibrial competitive process described by Huston (1979, 1985a, 1994) in the dynamic equilibrium model may be more appropriate for coral assemblages occurring in deeper or more turbid water. Competitive exclusion in these habitats is likely to be very slow, less intense, and important only intermittently. Whether it is less predictable is open to question.

Some of the above examples did not support the prediction that disturbance promotes species coexistence. I suggest that the potential for a significant increase in species richness in these cases may have been precluded by the use of data from very small samples (*e.g.*, individual 1-m^2 quadrats). This result can also be precluded when studies are conducted in relatively impoverished habitats or regions where there is a low potential for species coexistence. Thus the local scale of measurement and the regional setting can influence the assessment of disturbance and recovery processes.

The widely held view that species differentially respond to disturbances is well supported by the examples noted above and elsewhere (*e.g.*, Stoddart 1969, Porter *et al.* 1981, Jackson 1991). Thus the influence of disturbance on the structure of coral communities is likely to include relatively complex shifts in species abundance patterns across disturbance gradients. Individual taxa are likely to differ substantially in how they are distributed over such a gradient. In most cases, species respond differentially to disturbance because of differences in their abilities to tolerate disturbed conditions, regenerate following disturbance, colonize space, and compete with one another. In this regard, simple contrasts between large categories of species (*e.g.*, fragile branching versus massive corals, preferred versus non-preferred prey species, species with rapid versus slow colonization rates, and those with superior versus inferior competitive abilities) contribute to predictable shifts in species composition as a consequence of disturbances and the recovery process. However, because of high species richness, the complexity of species replacement patterns (*e.g.*, Tanner *et al.* 1994 and section

4.2.1), the high frequency of disturbances, and the long time required to document the recovery process, it will take many years to fully understand species-specific responses to disturbances in most coral communities. Nevertheless, this represents an important objective because disturbances are such a major feature of the environmental setting in which these communities occur.

7.4 Overview

Coral communities are influenced by a wide variety of natural and human-caused disturbances operating at an enormous range of spatiotemporal scales. Ecological theory makes two simple predictions as to how communities respond to disturbances. First, disturbances should promote species coexistence. This may occur quite locally within patches or at larger spatial scales across multiple patches, disturbance gradients, or habitats. Second, species are predicted to differentially respond to disturbances over ecological and evolutionary time. This further promotes niche partitioning across disturbance gradients. It also increases the complexity of the response of the community beyond the simple predictions from theory.

Evidence from coral communities supports these predictions. Generally, species coexistence is promoted by disturbances because of tradeoffs associated with interspecific competition, colonization, regeneration, recruitment, and disturbance resistance. Furthermore, most studies support the prediction that species differentially respond to disturbances. Some severe disturbances or combination of disturbances can result in phase shifts in community structure (especially when disturbances are chronic or when they significantly alter the physical environment). Lastly, the response of coral communities to disturbances are not easily predicted. Although these responses may occur as a consequence of predictable processes, they are complicated by the large number of species in these communities, complex species replacement patterns, the high frequency of disturbances, and the long time often required for full recovery. Nevertheless, understanding these dynamic changes in coral communities should be given high priority.

8 LARGE-SCALE PERSPECTIVES

8.1 General considerations

Multiple processes operating across a wide range of spatiotemporal scales influence the structure of ecological communities. In aquatic communities, one can contrast small-scale shear forces with orbital forcing as vastly different yet important determinants of community attributes (section 1.3). Also operating at a wide range of scales are a myriad of disturbance phenomena which generate a great deal of heterogeneity in ecological communities (*e.g.*, see chapters 1 and 7). At the largest scales, some processes operate at nearly global dimensions. For example, oceanic currents spread the effects of El Niño warming events across the Pacific (sections 1.4.2 and 7.2.4), disperse the offspring of benthic species over long distances, and contribute to latitudinal gradients in recruitment (section 1.4.1). Lastly, the evolutionary responses of populations to prevailing features of selective regimes can modify and even reverse the effects of some physical factors on communities (sections 1.3 and 3.2). In order to understand the full range of these effects, we should examine the structure and dynamics of ecological communities at multiple spatiotemporal scales (Levin 1992, Giller *et al.* 1994).

One consequence of broadening the scale at which ecological communities are studied is the reunification of community ecology with the disciplines of evolutionary biology, systematics, biogeography, and paleontology (Ricklefs 1987, Ricklefs and Schluter 1993). Schluter and Ricklefs (1993a) explicitly described this reunion as a natural consequence of investigations into "diversity phenomena". With the attention of community ecologists having shifted to consider large-scale processes, the notion that "patterns of diversity are caused by a variety of ecological and evolutionary processes, historical events, and geographical circumstances" has begun to emerge. In order to convey the breadth of ideas contributing to this paradigm shift, Ricklefs and Schluter organized a major symposium through the Ecological Society of America. The subsequent symposium volume includes contributions from fifty one authors "who had active interests in species diversity and well-developed perspectives" (Ricklefs and Schluter 1993). In that I have already introduced several of the diverse ideas presented in this volume, my objective here is to focus specifically on the species richness of local assemblages as viewed from different spatiotemporal scales. The notions of saturation, convergence, limited membership, and regional enrichment are central to this presentation. I justify this focus on species richness based on its pervasive emphasis in the literature across multiple disciplines. Whenever possible, I attempt to expand this focus to consider more detailed information regarding species composition, relative abundance patterns, resource partitioning, and habitat associations.

In section 2.2, I introduced the general ideas that tropical habitats are generally saturated with species and the characteristic high diversity of many tropical taxa is due to habitat specialization (MacArthur 1965). Conventional theory predicts that biological interactions like competition and predation as well as physical attributes of the local environment will restrict the number of species occurring together in local assemblages. Thus local community structure is organized primarily by the principles of limiting similarity and niche differentiation. As a consequence, local diversity should be strongly correlated with such features of the environment as the diversity of resources (Schluter and Ricklefs 1993a). A second prediction generated by equilibrial models of community assembly, structure, and stability is that "independently assembled communities in similar habitats on different continents (or in different oceans) should (converge to) contain similar numbers of species" (Schluter and Ricklefs 1993a). A third prediction describes the quantitative relationship between local and regional species richness. "If competition is a strong force, diversity in local sites should be near saturation; that is diversity within any habitat type should reach a ceiling that is independent of the number of species in the regional pool" (Schluter and Ricklefs 1993a). Cornell and Lawton (1992) depicted this relationship as a saturation curve for a theoretical assemblage characterized by strong biological interactions (Figure 8.1). In the next two sections, I discuss the growing body of evidence regarding these last two predictions.

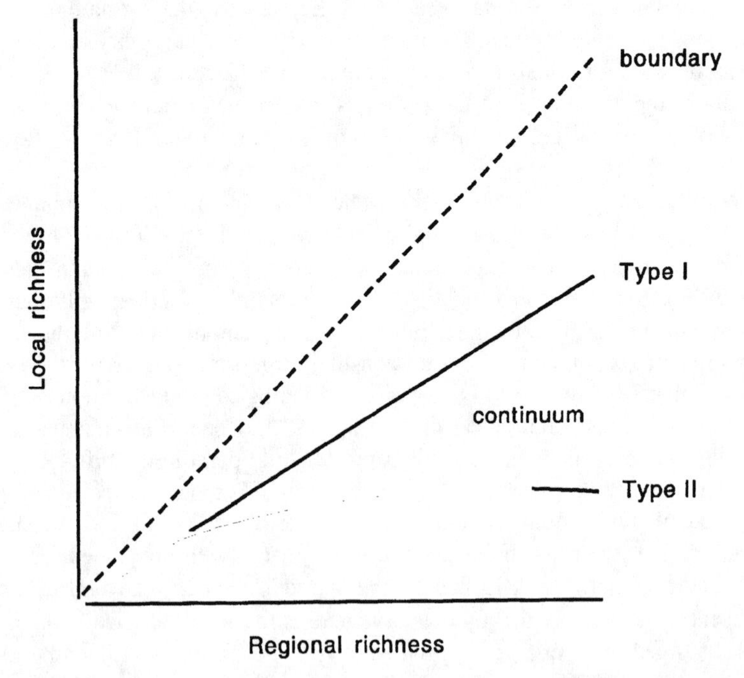

Figure 8.1 The theoretical relationship between local and regional species richness in ecological assemblages. The local richness of "Type I" assemblages is independent of biological interactions and increases as a linear proportion of regional richness (*i.e.*, regional enrichment). In "Type II" assemblages, local richness increases towards an upper limit as it becomes constrained by biological interactions and independent of regional richness. At the limit, the local assemblage is saturated with species. Patterns in real assemblages "probably fall on a continuum between these two extremes" (redrawn after Cornell and Lawton 1992).

8.1.1 CONVERGENCE OR DIVERSITY ANOMALIES AMONG REGIONS

This topic spans an enormous body of literature dealing with a variety of plants and animals from all over the world. Much of this work is reviewed in several of the contributions found in Ricklefs and Schluter (1993) and I shall only attempt a brief overview here. In general, there is mixed support for the prediction of convergence in the number of species within similar habitats from different places [see Samuels and Drake (1997) for a recent discussion of divergent, scale-dependent perspectives on this issue]. Because of the unique evolutionary histories of Australian plants and animals, Westoby (1993) contrasted local diversity patterns of a variety of taxa there with that on other continents. He concluded that the divergent, local diversity of coral reef fishes, arthropod herbivores on bracken, and arid-zone lizards reflect differences in regional diversity. Furthermore, he interpreted these divergent diversity patterns to be direct manifestations of differences in evolutionary history. In contrast, the local diversity of vascular plants and possibly birds were not influenced by differences in evolutionary history (i.e, they were convergent). Convergence in vascular plant species was noted for temperate forests, semi-arid woodlands, desert shrublands and grasslands, sclerophyll woodlands, and tropical rain forests (see Rice 1985). Likewise, the number of bird species in Australian and North American forests and desert shrublands were convergent and thus independent of historical influences (see Recher 1969, Wiens 1991).

A different perspective regarding convergence in avifaunas and other taxa emerges from studies conducted on selected guilds rather than on entire faunas or from studies in particular regions rather than on continental scales. For example, Morton (1993) recognized some major divergent patterns in the avifaunas of Australia and North America particularly among granivorous species inhabiting arid regions. Although the total richness on these two continents are similar (140 and 130 species, respectively), the number of granivores are quite different (26 and 4 species, respectively). Morton (1993) speculated that history might explain the difference, but noted that “it will be difficult to untangle the relative roles of history and ecological opportunity in this case”. Cody (1993) specifically emphasized regional differences in Australia which he partially attributed to the influences of history and location. Some of the relatively high species richness characterizing woodland and forest birds in Queensland was attributed to “speciation... via faunal exchange with and repeated invasions from New Guinea”. Thus there is considerable variation in the standard one might use to evaluate convergence at large spatial scales (*i.e.*, there are large differences among taxa and geographic regions). Furthermore, there are both local and large-scale explanations for this variation. Cody (1983) invoked multiple processes operating within habitats, between habitats, and on regional scales to explain the richness patterns of the Australian avifauna.

Perhaps one of the best examples of the lack of convergence in local diversity (also referred to as a “diversity anomaly") was presented by Ricklefs and Latham (1993) in their evaluation of the number of plant species inhabiting mangroves. There are approximately four times as many genera and six times as many species in the Indo-West Pacific region as there are in the Atlantic, Caribbean, and eastern Pacific regions combined. Likewise, local species richness at hectare scales is at least two to three times higher in the Indo-West Pacific. This large anomaly is attributed to evolutionary "differences between the regions in the origin of mangrove clades and their subsequent diversification within the mangrove habitat".

In spite of the divergent patterns noted above, Schluter and Ricklefs (1993b) found "that at least some convergence is the rule rather than the exception". They evaluated the published literature on the number of species occurring within particular habitats in different regions. Their quantitative method permitted them to partition the variance in this richness data. If local richness converges within a given habitat type regardless of the differences among regions, then variation among habitats should be similar in different regions (*i.e.*, convergent). Thus the habitat component of variation should be large relative to the total variation in local richness. Conversely, if divergent regional factors are largely responsible for differences in local richness, then the habitat component of variation should be relatively small. Their objective then was to quantify the relative magnitude of the components of variation attributable to differences among habitats and regions.

Using thirty six examples in which terrestrial vertebrates predominated (i.e, there were 11, 8, 4, and 2 examples using birds, reptiles, mammals, and amphibians, respectively), the habitat component exceeded the regional component in twenty two cases (Schluter and Ricklefs 1993b). Thus, "it is clear that convergence in species diversity is not a rare or peculiar phenomenon in nature". The mean proportion of the total variance associated with convergent patterns of species richness among habitats was 42% (range = 0-98%). That attributed to regional factors was only 20% (range = 0-83%). These results were depicted by plotting each variance component against the total variance (Figure 8.2). While the regional component varied little as a function of total variance, the habitat component indicating convergence was strongly related to total variance. "Habitat or microhabitat differences were usually found to play a large role whenever sample units (sites, etc.) differed greatly in the numbers of species present. Conversely, chance and regional effects assumed the greatest relative importance when variation among sites in diversity was not large".

The most spectacular examples of convergence occurred in comparisons using the number of species in the same microhabitat (or foraging category within a habitat) on different continents. Virtually all of the variation (98%) in the number of lizard species occurring in tropical wetlands of Australia (fourteen species) and Zambia (six species) was attributed to differences between arboreal and terrestrial/aquatic microhabitats. None was attributed to a difference between regions and the total variance was relatively large (see Simbotwe and Friend 1985). In contrast, there was a large regional component of variation (83%) in the number of desert ant species occurring in multiple habitats at two sites in Argentina (thirty four species) and Arizona (twenty seven species). Only 11% of the total variation was attributed to habitat differences in this example. "The ant fauna at these two sites are remarkably similar at the generic level" (Orians and Solbrig 1977). Yet despite this taxonomic similarity, several ecological differences in the prevalence of particular guilds (*e.g.*, arboreal nesters, granivores, and fungus growers) appear to preclude convergent patterns of species richness.

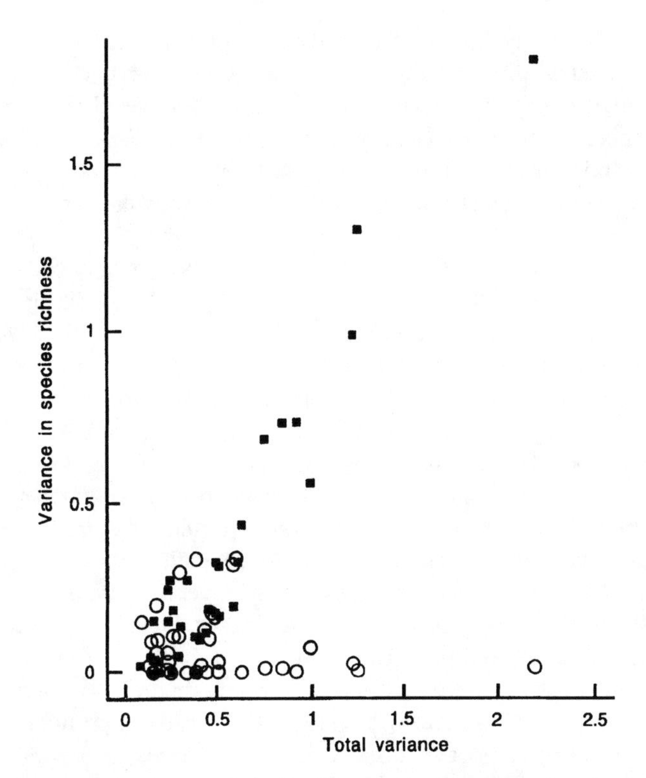

Figure 8.2 Variance components associated with differences in the number of species found in local habitats/microhabitats/climate zones (solid squares) and in different regions (open circles) plotted against the total variance. The relative importance of convergence (*i.e.*, predictable habitat differences) increases as the total variance in species richness among sites becomes large (redrawn using original data after figure 21.5 in Schluter and Ricklefs 1993b). © 1993 by The University of Chicago.

8.1.2 SATURATION OR REGIONAL ENRICHMENT

The notion that there is a limit to the number of species which can coexist together in a single habitat dates back at least to Elton (1933). Based on surveys of the number of animal species found in similar habitats in very different places, he inferred the existence of such limits. He referred to them as saturation points. Furthermore, he introduced the concept of limited membership. The number of species occurring together in a single habitat was noted to be far fewer than the number of species in the general region which could potentially occur together. The notions of saturation and limited membership invoke strong biological interactions within organized assemblages as the primary mechanism limiting local species richness. As a test of the notion of saturation, Cornell and Lawton (1992) advocated quantitative assessments of the relationship between local and regional species richness in order to determine if local richness reaches some upper limit in spite of increasing regional richness (see Figure 8.1).

Such a test would ideally involve the assessment of local richness within uniform habitats across a set of differentially rich regions. As a theoretical alternative to this saturation model, they hypothesized that local richness might merely be a linear function of the total number of species in the regional pool of species. This model is depicted as a proportional relationship between local and regional richness in Figure 8.1. Real assemblages of organisms were thought to fall between the extremes represented by these two models.

Quantitative empirical tests of these alternative models have only been conducted in a few assemblages. One of the first was conducted by Terborgh and Faaborg (1980) in their study of bird assemblages on twelve islands in the West Indies. They surveyed birds using mist-nets and direct observations in each of two habitat types representing coastal sclerophyll scrub and lower montaine rain forests. Regardless of habitat or method, local richness patterns appeared to match the saturation curves in Figure 8.1 when the total number of species per island was used as an estimate of regional richness. The upper limits to local richness (10-30 species) occurred across regional pools estimated at 40-80 species. Terborgh and Faaborg (1980) interpreted these patterns in terms of conventional competition theory. At saturation, these bird assemblages exhibited reduced habitat utilization patterns and more vertical stratification.

Wiens (1989) re-evaluated these same data to find equivocal support for saturation. Although statistical tests "confirm a saturation pattern for three of the four data sets", "the census data agree almost as well with a linear model" invoking proportional sampling (Figure 8.1). Furthermore, Wiens (1989) cautioned against inferring saturation based purely on an apparent equilibrial number of species in a local assemblage. Non-equilibrial processes can contribute to significant species turnover, yet still generate a stochastic equilibrial number of species. Thus he recommended consideration of alternative interpretations of species richness and the use of other measures of community structure. Although species richness can provide "an important starting point toward understanding how communities are put together and what factors influence that assembly", more attention should be paid to actual species composition, relative abundance patterns, the influence of abiotic environmental fluctuations, and a wider range of testable hypotheses than is generated by conventional equilibrial theory.

Cornell (1985a, 1985b) evaluated the relationship between local and regional richness in cynipid gall-forming wasps. The larvae of these wasps are specialized herbivores which feed exclusively on oaks (*Quercus*). In California, more than 150 species of wasps feed on 15 species of oaks. Given that host specificity occurs at the subgeneric level, "this oligophagous habit suggests a tight evolutionary linkage with the oaks" (Cornell 1985a). Local richness patterns might be strongly influenced by the unique history of the wasp fauna specialized on each host species. If such history is important, Cornell predicted that there should be a linear relationship between local and regional richness (called the "pool-enrichment" model). Alternatively, ecological and evolutionary processes at the local level might result in the convergence of local richness towards some saturated level.

Analysis of the data revealed an "eightfold increase in average local richness" (3-20 species) which was linearly related with a "ninefold increase in regional richness" (5-39 species) (Cornell 1985a). There was no evidence of a curvilinear relationship as one would expect if local cynipid assemblages were saturated. Thus local explanations invoking biological interactions and niche differentiation are insufficient explanations of local richness patterns in the wasp fauna. Cornell (1985b) recommended that large-scale biogeographic and evolutionary processes (*e.g.*, rates of speciation and extinction, habitat shifts, and changes in species distributions) be included in the conceptual framework for understanding the structure of these assemblages.

Ricklefs (1987) synthesized much of the early evidence regarding the relationship between local and regional richness to emphasize the potential importance of regional processes and history. He urged ecologists to expand the scale they use to evaluate richness patterns by incorporating data from systematics, biogeography, and paleontology [the same message echoed later in Ricklefs and Schluter (1993)]. As examples, he noted two cases in which there is a strong linear relationship between local and regional richness [*i.e.*, Caribbean birds (see Cox and Ricklefs 1977, Ricklefs and Cox 1977, 1978) and cynipid wasps (Cornell 1985a,b)]. In these unsaturated assemblages, local richness "is sensitive to such regional processes as geographical dispersal and the historical accumulation of herbivore species on host plants" (Ricklefs 1987). As a consequence, local richness is enriched by regional processes and historical events (as depicted for "Type I" assemblages in Figure 8.1).

Ricklefs (1987) referred to this theoretical relationship as the "regional enrichment" model. It is identical with the "pool-enrichment" model from Cornell (1985a) and the "proportional sampling" model from Cornell and Lawton (1992). The absence of saturation in local assemblages is attributed to differences in local richness, the degree of niche overlap or resource specialization, and biogeographical factors influencing regional richness. Ricklefs (1987) emphasized the link between this regional-historical perspective and conventional niche theory. Both perspectives are important. Cornell and Lawton (1992) further explored theoretical explanations for the lack of saturation. These included a non-interactive community model (see Caswell 1976 and section 7.1) as well as interactive models invoking spatiotemporal heterogeneity generated by predators, disturbances, aggregated resource use, and stochastic immigration and extinction. These processes represent putative mechanisms which may open local assemblages to immigrants from regional species pools and reduce the importance of competitive exclusion in local assemblages.

Cornell and Karlson (1997) examined the published literature on multiple taxa to determine the prevalence of linear and curvilinear relationships between local and regional richness. Significant linear relationships indicate proportional sampling (pool or regional enrichment). Curvilinear relationships are necessary (but not sufficient) evidence to establish saturation in accordance with the theoretical models (Figure 8.1). Data from eight of seventeen studies exhibited a significant linear relationship between local and regional richness with no evidence of curvilinearity (Table 8.1). Three studies exhibited significant linear and curvilinear relationships with no evidence supporting the existence of an upper limit to local richness (no saturation). Only six studies (presented in five publications) provided evidence consistent with the hypothesis that local assemblages are saturated. "However, this last result somewhat overstates the ubiquity of the Type II relationship since four of the regressions represent the same type of community (parasitic helminth worms in vertebrates)" (Cornell and Karlson 1997).

Furthermore, not all helminth assemblages show this pattern (Table 8.1). Our conclusion from this survey of the literature is that "unsaturated patterns are common and widespread in natural communities". A similar assessment of a variety of taxa throughout the world by Caley and Schluter (1997) yielded an even stronger conclusion in that they found virtually no quantitative evidence for saturation.

TABLE 8.1 Examples of studies documenting linear and curvilinear relationships between local and regional richness. Linear relationships are consistent with models invoking proportional sampling (Cornell and Lawton 1992), pool-enrichment (Cornell 1985a), and regional enrichment (Ricklefs 1987). Curvilinear relationships indicate saturation or an intermediate situation when local richness does not level off (after Cornell and Karlson 1997).

Relationship	Organism	Source
Linear	cynipid wasps	Cornell 1985a,b
	arthropods on bracken	Lawton 1990
	parasitoid wasps	Gaston and Gauld 1993
		Dawah *et al.* 1995
	freshwater fishes	Griffiths 1997
		Hugueny and Paugy 1995
	parasitic helminths	Kennedy and Guégan 1994
	mixed vertebrates	Caley and Schluter 1997
Curvilinear but no saturation	deep-sea gastropods	Stuart and Rex 1994
	Caribbean birds	Ricklefs 1987
	fig wasps	Hawkins and Compton 1992
Curvilinear and saturation	fig wasps	Hawkins and Compton 1992
	Banksia	Richardson *et al.* 1995
	parasitic helminths	Aho 1990
		Aho and Bush 1993
		Kennedy and Guégan 1994

8.2 Evidence from coral communities

The general evaluations of the published literature regarding convergence (Schluter and Ricklefs 1993b) and saturation (Cornell and Karlson 1997) focused primarily on well-studied organisms found in terrestrial and freshwater habitats. Marine organisms were represented in none of the thirty six examples used in the former analysis and were emphasized in only one of the seventeen examples used in the latter [*i.e.*, deep-sea gastropods (Stuart and Rex 1994)]. However, a broader search of the literature reveals that these subjects have been addressed using evidence from marine communities in general and coral reef communities in particular. For example, recent work by Cornell and Karlson (1996), Caley and Schluter (1997), and Caley (1997) indicate the existence of unsaturated assemblages of coral reef organisms. In what follows, I present evidence from coral communities using studies of reef fishes, decapod crustaceans, and corals. The reader should pay particular attention to the scale at which these studies have been conducted.

8.2.1 REEF FISHES

Sale (1980) considered both within- and between-habitat components of the species richness of coral reef fishes as well as the major global patterns of variation among regions. At the global scale, richness is highest in the central Indo-West Pacific especially in the Philippines [2177 species, see Goldman and Talbot (1976)]. Although this high richness might "have purely historical causes", there might also be important local ecological factors promoting the maintenance of this rich diversity. Sale (1980) contrasted local species richness in lagoonal patch reef habitats at One Tree Island with comparable data from multiple locations in the Caribbean Sea and other geographic regions. The data were plotted as the number of fish species in local samples against the total number of individuals in each census (Figure 8.3). Sale (1980) reasoned that "if the high diversity of the central Indo-West Pacific region is due to a greater alpha-diversity within habitats", then regressions for different regions should differ in terms of their slope and elevation. If, on the other hand, the high number of species found in the central Indo-West Pacific is due only to the between-habitat component (beta diversity), then these regressions should not differ. There were no significant differences in the slope or elevation of regressions of the data from One Tree Island and the Caribbean Sea. Thus, it would appear that between-habitat differences due to habitat specialization and species turnover across habitats may be primarily responsible for observed regional differences. However, Sale (1980) noted that collections from other regions fall well off the regression line (Figure 8.3). In addition, he acknowledged that it would be premature to conclude that within-habitat richness does not vary among regions.

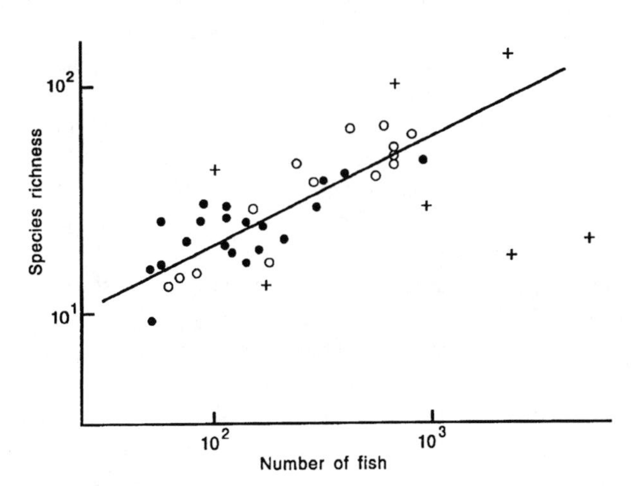

Figure 8.3 Species richness of fishes in censuses or collections from lagoonal patch reef and back reef habitats in several geographic regions. One Tree Island data (closed circles) are from twenty lagoonal patch reefs. Caribbean and western Atlantic data (open circles) are from thirteen sites in the Bahamas, Virgin Islands, Venezuela, and Florida. Other geographic regions (crosses) encompass sites in the Red Sea, Society Islands, Mafia Archipelago, and Enewetak Atoll. The regression line represents the relationship for the pooled data from One Tree Island and the Caribbean Sea (redrawn after Sale 1980). Reproduced with kind permission of UCL Press © 1980.

Bohnsack and Talbot (1980) explicitly tested the idea that within-habitat partitioning of space by reef fishes would differ in differentially rich regions. Standardized artificial reef habitats were constructed in the lagoon at One Tree Island and on the back reef at Big Pine Key, Florida. At each location, eight replicates comprised of 28 concrete blocks each were monitored at approximately monthly intervals for 32-39 months. "Colonization patterns were strikingly similar in both regions with mean saturation levels being reached in approximately one month". This reference to saturation refers to the equilibrial number of species predicted from colonization theory (MacArthur 1972), not the limit on local species richness predicted for saturated assemblages (Cornell and Lawton 1992). Bohnsack and Talbot (1980) reported almost exactly the same total number of fish species on the artificial reefs (89 and 88 species, respectively). This is in spite of there being approximately half as many species in the Florida Keys as there are in the southern region of the Great Barrier Reef. Furthermore, the mean species richness per artificial reef was significantly higher in Florida than in Australia (16.6 vs. 12.8 species). Thus there was a difference between regions in local richness, but not in the same direction as one would have predicted if these local assemblages were being regionally enriched. It would appear that there are significant regional differences in this fish fauna which require independent tests of saturation and regional enrichment across regional richness gradients in Australia and the Caribbean.

Although Sale (1980) recognized the general need for better data and more rigorous analyses to resolve the issue, the next major set of reviews of the ecology of coral reef fishes (see Sale 1991a) barely mentioned the notions of convergence and saturation. In fact, Thresher (1991) emphasized the persistent lack of adequate data for large-scale biogeographic comparisons. Furthermore, he noted that "it is premature to conclude that interoceanic differences in assemblage structure are slight". On the other hand, several contributors stressed the importance of habitat associations and specialization at the local scale. For example, Choat and Bellwood (1991) and Jones (1991) emphasized strong associations between fishes and particular habitats on coral reefs. Doherty (1991) mentioned the importance of habitat selection during settlement. Williams (1991) emphasized the high degree of heterogeneity exhibited by coral reef fish assemblages across several spatial scales (see section 2.3). It would appear that divergent patterns observed among reefs on the Great Barrier Reef are thought to be controlled primarily by larval transport processes and their influence on recruitment rates. At even larger biogeographic scales, broad differences in the structure of coral reef fish assemblages are typically considered in terms of evolutionary processes (*e.g.*, speciation and extinction), oceanographic transport processes, and the divergent histories of biogeographic regions. Consequently, habitat-based explanations for regional differences at this scale are generally "unsatisfying in two respects" (Thresher 1991). First, experimental tests are logistically difficult to conduct at such large scales. Second, such explanations are logically "impossible" to falsify; "there is always the possibility of some, as yet unidentified, habitat feature that dictates shifts in the composition of fish assemblages" (Thresher 1991).

Because biogeographic assemblages are not equivalent with smaller-scale ecological assemblages occurring within habitats, some clarification is needed regarding scale. Biogeography is the study of species distributions across large geographic scales.

Until recently, there have been two independent lines of inquiry within the subdisciplines of ecological and historical biogeography (Myers and Giller 1988). The former deals primarily with extant species and attempts to explain current distributional patterns in terms of local-scale processes, biological interactions, current environmental factors, and recent history. The latter explains broader-scale distributional patterns in terms of the origin, dispersal, and extinction of species as well as geological events. Historical biogeography is further split into dispersal and vicariance biogeography which explain distributional patterns based on the movement of organisms or events which fragment the biota of a region. Although ecological phenomena can be useful in explaining "the great parallels between community structure of different continental areas of similar climate and topography," they cannot "explain the great difference in the taxonomic composition which produces such structure nor intercontinental distribution patterns at higher taxonomic levels" (Myers and Giller 1988). Nevertheless, a full explanation of current distributional patterns of most organisms probably requires an understanding of both ecological and historical phenomena (Myers and Giller 1988).

In the present context, the relationship between local and regional species richness may indicate that the structure of reef fish assemblages is primarily determined by the local physical environment and biological interactions (if saturation is indicated), by larger-scale historical and geographical phenomena (if strong regional enrichment is indicated), or some combination of these influences. Thus one is required to blend one's knowledge of ecology, evolutionary biology, systematics, biogeography, and paleontology depending on the particular observed relationships. Two additional studies which have specifically attempted to evaluate local-regional relationships among coral reef fishes at local scales in different regions are presented below. Both indicate the importance of large-scale phenomena, but much work remains to verify the preliminary results, identify the causal factors for the observed relationships, and quantify the relative magnitude of these factors at the appropriate scales at which they operate.

Westoby (1985, 1993) contrasted the number of coral reef fishes found at One Tree Island with that at St. Croix in the Caribbean Sea to find significant regional differences in this fauna at a local scale. When species-area relationships were plotted, the difference in species richness at local (10^2 km^2) and regional (10^5-10^6 km^2) scales exhibited the same proportional relationship (Figure 8.4). In both cases, the number of species at the local scale was approximately 60% of the regional richness. Furthermore, there were 47% fewer species in St. Croix (420 species) than at One Tree Island (900 species) and 47% fewer in the Caribbean Sea [700 species, see Thresher (1980)] than on the Great Barrier Reef [1500 species, see Goldman and Talbot (1976)]. Thus Westoby concluded that local assemblages reflect the regional difference between the relatively depauperate Caribbean Sea and the more speciose Great Barrier Reef (*i.e.*, there was no convergence at the smaller scale).

The local scale used by Westoby (1985, 1993) encompasses many different habitats on these reefs. Thus it is not small enough to critically evaluate the saturation hypothesis for the number of species within habitats. In recognizing this limitation, Westoby stated his inferences regarding regional influences at this large local scale using the notion of between-habitat diversity (Whittaker 1977, see section 2.2).

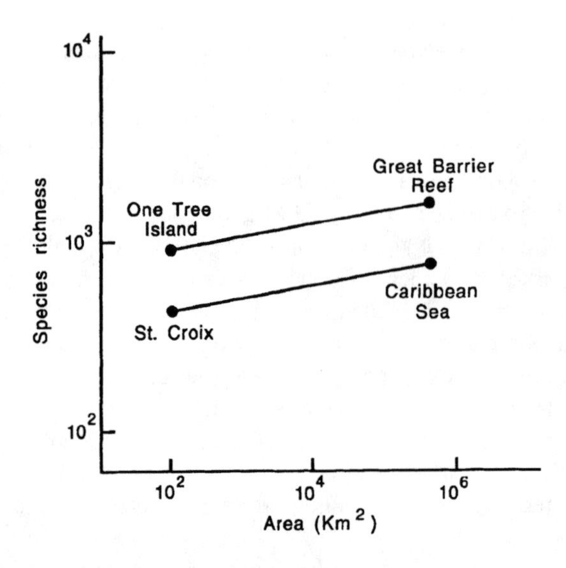

Figure 8.4 Local richness of fishes at One Tree Island and St. Croix and regional richness of the Great Barrier Reef and Caribbean Sea. The approximate area of the local scale is 10^2 km^2, while that of the two regions is 10^5-10^6 km^2 (redrawn after Westoby 1985).

As one examines species compositional changes over a range of habitats at One Tree Island and St. Croix, the degree of habitat specialization or species turnover appears to be directly related to differences in "evolutionary history determining the size of the regional biota" (Westoby 1993). Thus the speciose fish fauna on the Great Barrier Reef is strongly associated with high habitat heterogeneity, habitat specialization, and this between-habitat component of species diversity [see Sale (1980), Williams and Hatcher (1983), and section 2.3].

Caley (1997) examined the species richness of fishes occurring as residents in rubble patches at two differentially rich locations along the Great Barrier Reef. He reported significant regional enrichment at two, quite small local scales. In the north at Lizard Island, regional richness was estimated to be 40% higher than at One Tree Island in the southern Capricorn-Bunker Group (1200 vs. 859 species). Within 4-m^2 patches of rubble habitat, mean species richness was 10% higher at Lizard Island than at One Tree Island (13.2 vs. 12.0 species, a small but statistically significant difference). Mean species richness within the rubble habitat averaged over two locations at each site was 15% higher at Lizard Island. Thus the evidence is consistent with the notions of proportional sampling and regional enrichment. Caley (1997) concluded "that regional and historical factors exert considerable influence over" the local structure of these fish assemblages, yet also conceded that a more definitive test "will require estimation of species richness in a wider range of habitats and from more locations".

8.2.2 DECAPOD CRUSTACEANS

Decapod crustaceans are a particularly diverse group of marine organisms which flourish in a wide variety of habitats. Abele (1974) evaluated ten habitats along the coasts of Florida and Panama in terms of the species richness of this assemblage. He examined how this attribute varies with habitat complexity, temperature, salinity, tidal exposure, latitude, and longitude. In each habitat, samples were collected until cumulative species richness curves plotted against sampling effort "became asymptotic". Although the dispersal capabilities of these crustaceans were thought to make immigration into all habitats within each ocean possible (thus contributing to an approximately uniform regional pool of species), there were large differences among habitats. Abele (1974) reported a strong positive linear relationship between species richness and the number of types of substrate found in each habitat. None of the other environmental variables were related to richness patterns. The highest species richness estimates were generated from samples in rocky intertidal habitats in Panama (78 species along the Pacific coast and 67 species along the Caribbean coast associated with 9-10 types of substrate). The subtidal community dominated by pocilloporid corals off the Pacific coast of Panama was the third richest habitat with a total of 55 species associated with six types of substrate. Thus Abele stressed the importance of habitat complexity and substrate selection as important determinants of species richness.

Using Panamanian reef sites at Uva Island and in the Pearl Islands, Abele (1979) characterized the decapod assemblage specifically associated with live colonies of *Pocillopora damicornis.* This was one of the six types of substrate reported in Abele (1974). A total of 154 coral heads were found to harbor 74 decapod species (52 and 57 species at Uva and Pearl Islands, respectively, see section 2.3). Abele (1979) found that decapod species richness on individual coral heads was strongly related to coral-head size. There was no significant difference in this relationship between sites (Table 8.2). However, in spite of this similarity in species richness per sample, there were "significantly fewer species of decapod crustaceans" at Uva Island than in the Pearl Islands when the cumulative number of species was examined over multiple samples (*e.g.*, 55 vs. 32 species per 1000 sampled individuals). Although the two sites are located in the same biogeographic province less than 300 km apart, decapod species composition varied much more among coral heads sampled in the Pearl Islands than at Uva Island. Abele (1979) speculated that disturbances (upwelling of cool water) might enhance species coexistence more in the Pearl Islands. In addition, the lower species richness observed at Uva Island might also result from factors excluding fourteen species of habitat generalists on that reef.

Abele (1984) considered the decapod assemblage associated with *P. damicornis* at an even larger scale by contrasting regional species richness patterns across the Pacific Ocean. There are approximately five to eight times as many species in the central and western Pacific as there are in the eastern Pacific. If this regional difference is due primarily to higher species richness within this habitat, then comparisons of local richness should exhibit patterns consistent with the notion of regional enrichment. Alternatively, if regional differences mostly result from partitioning of this habitat at some intermediate scale, then local richness patterns might indicate saturation.

In addition to comparative studies, Abele (1984) also described the results of experiments in which decapods were removed from coral heads. Subsequent recolonization was monitored in one set of colonization experiments, while the persistence of species was followed after the introduction of "too many" decapods in "supersaturation" experiments.

TABLE 8.2 Linear regressions of the number of coral-associated decapod species per coral head (S) against coral-head size in terms of volume (V in cm^3). Y = $\log_{10}$ S, X = $\log_{10}$ V, *P* = the significance of the t-test for a non-zero slope, N = the number of coral heads sampled, T = the total number of decapod species sampled, and I = the mean number of individuals per species in each collection (after Abele 1979, 1984, Austin *et al.* 1980).
Information from Abele (1984); *Ecological Communities: Conceptual Issues and the Evidence.*

Location	Regression Equation	*P*	N	T	I
Uva Island, Panama	Y = 0.27 X + 0.01	<0.001	119	52	90.8
Pearl Islands, Panama	Y = 0.37 X - 0.32	<0.001	35	57	19.4
Palau, Micronesia	Y = 0.46 X - 0.65	<0.001	40	47	11.6
Lizard Island, Northern Great Barrier Reef	Y = 0.42 X - 0.64	<0.001	30	43	10.2
Heron and One Tree Islands, Southern Great Barrier Reef	Y = 0.36 X - 0.08	<0.05	40	65	12.3

As in earlier studies, the number of decapod species per coral head was strongly related to coral-head size (see linear regression relationships in Table 8.2). In spite of the regional differences in species richness, there were no significant differences among the five regressions determined for sites in Panama, Micronesia, and Australia. Thus, Abele (1984) concluded that there was no evidence indicating higher within-habitat species richness, increased habitat partitioning within individual coral heads, or more "species packing in the more species-rich regions" (*i.e.*, regional enrichment was not detected). On the other hand, variation in species composition among coral heads was higher in Micronesia and Australia than in Panama thus contributing to higher regional richness. "The faunal similarity among individual coral heads is higher in the Eastern Pacific" where any two coral heads "will share more species than any two coral heads in the Central or Western Pacific".

In order to understand the dynamics of these decapod assemblages in the more speciose regions, colonization and supersaturation experiments were conducted at Palau and Lizard Island. Abele (1984) reported that decapod colonization of defaunated coral heads was quite rapid. It occurred mostly at night and colonists were almost exclusively small adults and juveniles. Only a single larva (a crab megalopa) was observed to colonize a coral head at Lizard Island. In fact, colonization was so fast that the number of species on many of the experimental coral heads reached pre-defaunation levels in 4-6 days. The range (1-7 species per coral head) again varied with coral-head size. However, species composition was much less resilient in that several common species failed to colonize defaunated coral heads.

In supersaturation experiments, decapod species were rapidly lost from coral heads which had received up to 10 decapod species after defaunation. In Palau, 54 of 84 decapod species were lost from 11 coral heads. At Lizard Island, 36 of 60 decapod species were lost from 10 coral heads. Even though species were also lost from control coral heads which had undergone the defaunation and re-introduction treatments, the loss rate was considerably lower. Only 7 of 31 species were lost from 14 coral heads in Palau and 9 of 28 species were lost from 10 coral heads at Lizard Island. Based on these results, Abele (1984) concluded that there clearly was an upper limit to the number of decapod species coexisting on individual coral heads (*i.e.*, saturation). Apparently then, the high regional species richness observed at Palau and Lizard must be attributed to factors operating at larger spatial scales. As a partial explanation, Abele (1984) noted that in speciose regions "species are represented by fewer individuals (see Table 8.2) and are more patchily distributed". Thus, spatial heterogeneity within the pocilloporid habitat at scales encompassing multiple coral heads may substantially contribute to regional enrichment and the observed variation in decapod species composition among coral heads.

8.2.3 CORALS

In order to test for local saturation and regional enrichment of coral assemblages, Cornell and Karlson (1996) examined the published literature for quantitative estimates of both local and regional species richness. Prior to 1970, this literature contained very little quantitative data (see review in Stoddart 1969). Our initial search yielded 63 sources which had been published between 1918 and 1993. These provided 1329 local richness estimates from over 100 coral assemblages in the Indo-Pacific and western Atlantic Oceans (citations are given in Karlson and Cornell 1998). I present this work in two parts. In this chapter, I specifically examine the relationship between local and regional richness. In the final chapter, I address the relationships between local species richness and other independent variables. This separation is intended to simplify the presentation of a somewhat complicated set of analyses involving many statistical tests, a range of local scales of measurement, contrasts across different geographic scales, and consideration of multiple local and regional variables.

Local richness estimates for corals are reported in the literature in a variety of ways. The three most common field methods employed quadrats, line transects, and point intercepts. Quadrat sizes, transect lengths, and the number of points used in each sample are highly variable. Sometimes the data include a complete species list for each sample. Often, however, only summary data are presented in terms of the mean or cumulative total number of species for a set of samples. We included all data reported for individual samples. We used mean values only when data for individual samples were not reported and replicates were clearly taken in the same immediate vicinity. We did not use local richness data reported as means or cumulative totals for multiple depths, habitats, or reefs. Cumulative total data were included with means and individual samples only in a separate analysis directed at variation in the magnitude of regional enrichment across different local scales of measurement (Karlson and Cornell unpublished manuscript). Because of the highly variable nature of the data, we have focused three separate analyses on the following questions: 1) does the relationship between local and regional richness indicate saturation or regional enrichment? 2) does this relationship vary with the local scale of measurement? and 3) does this relationship vary with the geographic scale of the analysis?

• Although biological interactions in coral assemblages are known to be intense and we expected species membership to be limited, we concluded that these assemblages are, in fact, regionally enriched (Cornell and Karlson 1996). This is based on regression analyses of local richness against regional richness. We found no evidence for saturation when we analyzed quadrat, line-transect, and point-intercept data separately or when we pooled data collected using all three methods. In the former case, regional enrichment was strongest using line-transect data, weakest using the point-intercept data. Pooling of local richness data was done using the residual variation in log-log regressions of local species richness against sample size. The regression of these pooled local richness residuals against regional richness yielded a significant positive slope [0.21 ± 0.03 (1 S.E.), Table 8.3]. Thus local richness increases with the size of the regional pool of species. This suggests that there is no hard upper limit on the local richness of coral assemblages. The saturation model depicted in Figure 8.1 does not appear to be applicable to coral assemblages. However, since there is little currently available data from the most speciose regions of the Indo-Pacific, a more definitive assessment of saturation is still needed.

TABLE 8.3 The relationship between local and regional species richness of coral assemblages at multiple geographic scales. The regression coefficients for the slope are given for pooled log-transformed local richness residuals vs. regional richness as in Karlson and Cornell (1998). Untransformed richness values are given for the range in regional richness and for the mean local richness (± 1SE, n = sample size) found along 10-m line transects. Additional line-transect data for 15-m, 20-m, and 30-m samples collected on the Great Barrier Reef appear in Bull (1982), Done (1977), Morrissey (1980), and Pichon and Morrissey (1981).

Geographic scale	Range in regional richness	Mean local richness	Regression coefficient
Global	19-411	8.5 ± 0.5, n = 236	0.21***
Indo-Pacific	19-411	8.2 ± 0.6, n = 167	0.27***
Speciose Indo-Pacific	244-411	16.7 ± 1.0, n = 44	0.24**
Depauperate Indo-Pacific	19-88	3.0 ± 0.2, n = 102	0.60***
Great Barrier Reef	244-343	---	0.27*
Western Atlantic	19-50	9.5 ± 0.4, n = 69	0.06

***$p<0.00001$, **$p<0.00005$, *$p<0.0005$

• The local scale of measurement strongly influenced the relationship between local and regional richness (Karlson and Cornell unpublished manuscript). We ranked local spatial scale based on the linear extent of habitat traversed by four, commonly used sampling methods as follows: 1-m² quadrats < 5-m² quadrats < 10-m line transects < 30-m line transects. The published literature indicates that the mean number of species per sample increases with this ranking (3.2 < 6.1 < 8.5 < 10.2 species). In addition, spatial heterogeneity increases with spatial scale as indicated by increasing values for the variance and coefficient of variation associated with these estimates of local richness. There is also a corresponding increase in the degree of regional enrichment. Log-log regressions of local vs. regional richness yielded the highest slope estimates using data from 10-m line transects [0.81 ± 0.07 (1 S.E.)]. Much lower slope estimates were generated using data from small quadrats. The data from 10-m line transects also indicate a curvilinear relationship between local and regional richness (Figure 8.5).

• The relationship between local and regional richness also varied with the geographic scale used in our analyses (Karlson and Cornell 1998). We defined three geographic scales as follows: global (including all data from the Indo-Pacific and western Atlantic Oceans), provincial (Indo-Pacific and western Atlantic data considered separately), and sub-provincial (Indo-Pacific data considered for the speciose central Indo-Pacific with 244-411 species per region, the Great Barrier Reef with 244-343 species per region, and depauperate regions with 19-88 species per region). Our analysis indicates that Indo-Pacific coral assemblages are regionally enriched at all of these geographic scales (Karlson and Cornell 1998, Table 8.3). This is especially true in depauperate regions, but also applies to speciose regions where we found no evidence for saturation.

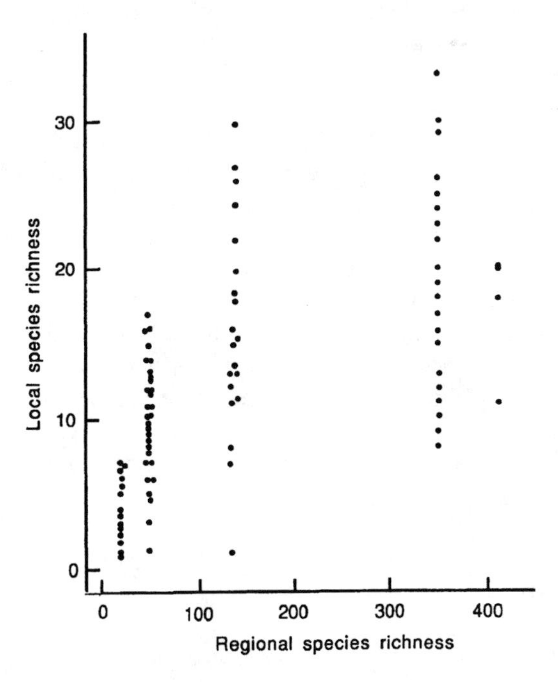

Figure 8.5 Local species richness along 10-m line transects plotted against regional richness in samples from the Indo-Pacific and western Atlantic Oceans (data from multiple sources cited in Karlson and Cornell 1998).

In contrast, coral assemblages in the western Atlantic (as represented by the published literature) do not appear to be regionally enriched or saturated. The inference of regional enrichment generated from the global scale of analysis does not apply to this provincial scale (*i.e.*, the inferences are scale-dependent). Much of the published data from the western Atlantic has been collected over relatively homogeneous regions which lack strong gradients in regional species richness (Liddell and Ohlhorst 1988, Karlson and Cornell 1998). However, more sampling may reveal regional enrichment along richness gradients extending out of the Gulf of Mexico or through the Lesser Antilles. If, on the other hand, western Atlantic assemblages are saturated, we would have expected higher local richness values than those observed. Current local richness in the western Atlantic is generally less than that observed in the speciose, central Indo-Pacific (*e.g.*, see Sy *et al.* 1982, Moll 1986). It is also probably less than that in the Caribbean four million years ago before the closure of the Isthmus of Panama. Species turnover since that time has been extensive and current regional richness is significantly lower (see Budd *et al.* 1994, 1996, Johnson *et al.* 1995). The unique history of this faunal province following its isolation from the Indo-Pacific warrants more detailed attention.

At this point, we believe much more empirical evidence is needed to confirm the results of our literature survey. Data from such surveys are often highly variable because it was collected for a wide range of reasons using multiple methods. We recommend the use of standardized methods for future work (*e.g.*, 10-m line transects), matched habitats across regions, and balanced sampling designs. Furthermore, we specifically note the need for more data and more rigorous tests for saturation and regional enrichment across speciose regions of the Indo-Pacific. Secondly, we note the potential importance of speciation across differentially rich regions of the western Atlantic. The recognition of sibling species in these coral assemblages suggests that a closer look at realized niches and habitat partitioning is needed (Knowlton and Jackson 1994, Rowan and Knowlton 1995). Lastly, the strong regional enrichment we detected in depauperate regions of the Indo-Pacific highlights the potential importance of disruptive regional factors (*i.e.*, marginal environmental conditions and major disturbances). Globally, these coral assemblages appear to be the most open to regional influences. Perhaps this inference is somehow related to the interruption of reef development and community assembly by these regional factors. Perhaps other regional phenomena also have strong influences on local assemblages in these depauperate regions. More comparative regional studies across the eastern Pacific like that by Glynn and Wellington (1983) appear to be needed. These might profitably be focused on dispersal across biogeographically important scales, the phylogenetic affinities of isolated taxa, and other large-scale phenomena relating oceanographic and evolutionary processes to local community structure.

8.2.4 PLEISTOCENE CORALS: TEMPORAL CONSTANCY AND LIMITED MEMBERSHIP

Pandolfi (1996) examined the species richness and composition of coral assemblages in Papua New Guinea in order to evaluate their temporal and spatial variability over a 95-ky late Pleistocene interval. Because of tectonic uplifting along the north coast of the Huon Peninsula, fossil corals from a series of nine reef-building episodes are distributed over a set of terraces dated at 30-125 ky. A total of 122 species were sampled in outcrops at three study sites along 35 km of this coast. While living, these corals occurred in intertidal, reef crest, reef slope, and fore-reef habitats. Because the intertidal and fore-reef habitats were less well represented in the fossil record at these sites, only reef crest and slope habitats were used for quantitative analyses. More species occurred on reef slopes (109) than on reef crests (77) and there was considerable overlap in the species composition in these two habitats. All of twelve common species on reef crests (defined as being present in at least 10 of 20 samples) were also common on reef slopes. Fifteen additional species were common only on reef slopes.

The 122 coral species were distributed among 48 genera. The most common were *Favia* and *Favites* with nine species each. These are typically massive forms which are well represented in the fossil record. Although *Acropora* is the most speciose genus of corals in the central Indo-Pacific [*e.g.*, seventy three species noted for eastern Australia (Veron 1986)], it was represented by only four species in these samples. Many *Acropora* spp. are not well represented in the fossil record because their fragile skeletons do not preserve well. They can also be generally rare or uncommon in mature, undisturbed assemblages (*e.g.*, see Tomascik *et al.* 1996 and section 4.2.1). Pandolfi (personal communication) has indicated that there were probably more *Acropora* spp. present in these fossil assemblages, but not many more. There was extensive coral rubble which could not be unequivocally identified to species, but only seventeen species in this genus are currently known to occur along this coast of the Huon Peninsula.

The analysis of variability in the species composition of the fossil assemblages revealed a high degree of temporal constancy over the 95-ky interval, but significant variation over two spatial scales (Pandolfi 1996). This assessment was based on comparisons among the nine reef-building episodes, two well-sampled habitats, and three study sites. The Bray-Curtis index of dissimilarity was determined for each pair of samples using presence/absence data. This index varies between 1 (the most dissimilar samples with no species in common) and 0 (with all species in common). Using two different non-parametric tests on these indices, Pandolfi (1996) found no significant differences among reef-building episodes. Between-episode dissimilarity was not significantly greater than that within episodes.

In contrast, the species composition of reef crest and slope habitats was significantly different in spite of the large degree of overlap noted above. This was attributed to the "diverse massive Favidae and *Fungia* spp." which were quite common on the reef slopes, but not on the reef crests. The mean species richness per assemblage for reef crests and slopes was 18.3 and 31.4 species, respectively. Species contributing to the highest between-habitat differences in composition included *Gardineroseris planulata, Montastrea curta,* and *M. annuligera.* These species occurred in 60-70% of the samples from reef slopes, but in only 20-30% of those from reef crests. Significant between-site differences in species composition were detected only in samples from reef crests. Several species, which were quite common in this habitat at one or both of two sites at Sialum and Hubegong, were absent or uncommon at Kanzarua. The mean richness of common species in samples from reef crests at these three sites was 17.5, 18.3, and 10.7 species per sample, respectively. Also contributing to the significant difference among sites were a few species which were more common at Kanzarua (most notably *Acropora cuneata* and *A. hyacinthus*). These differences among sites were largely attributed to riverine influences which are quite strong at Kanzarua and much weaker at the other sites.

Based on the absence of significant temporal variation in species composition and richness over the nine reef-building episodes, Pandolfi (1996) suggested that species membership in these coral assemblages might be limited (after Elton 1933). He noted that "during any one reef-building episode, only a portion of the available species pool actually occupied a particular reef" habitat. The same few species appeared to dominate assemblages spatially segregated among sites and habitats, while almost half the species were rare or uncommon in this set of samples (53 of 122 species). However, in order to conclude that Elton's notion of limited membership actually applies to these coral assemblages, Pandolfi (1996) recognized two further questions which "must be addressed".

The first is to resolve whether species composition is significantly different from that predicted by the null hypothesis that assembly is a random sampling of the available species pool. Pandolfi (1996) used a statistical null model from Connor and Simberloff (1978, 1979) to predict the expected number of shared species in paired samples from different times at the same sites. Because there were significantly more species shared in the paired samples than predicted by the null model, he rejected the hypothesis that these assemblages represent random samples from the pool of available species. The high temporal similarity of these assemblages would therefore appear to result from a more organized nonrandom process.

The second is to determine if the realized niches of these corals are narrower than their fundamental niches. In other words, were these corals restricted to a narrower range of reef environments than they could potentially occupy? Certainly not all common species occurred in every assemblage sampled at a site. On the other hand, the majority of species occurred in both of the well-sampled habitats and there was considerable overlap in the species composition among sites. However, this particular data set does not provide a good basis for evaluating niche dimensions in any detail. Consequently, Pandolfi (1996) made a general observation regarding corals on reefs in Jamaica and on the Great Barrier Reef. Although many of these corals occur over broad depth ranges, the Jamaican reefs are characterized by multiple zones which are dominated by only one or a few species (see section 2.2.6). "The same is also probably true for Indo-Pacific reefs where zonation has been studied (Done 1983, 1992)." Thus Pandolfi (1996) concluded that realized niches among corals are smaller than fundamental niches.

The combined results led Pandolfi to conclude that the membership in coral assemblages is limited. Temporal constancy, nonrandom colonization of habitats, and narrow realized niches all support this conclusion. However, because regional enrichment has been detected even in speciose regions of the central Indo-Pacific (section 8.2.3), the degree of this limitation remains in question. These assemblages appear to be open to some proportion of the regional fauna, but certainly not to all species. The results are consistent with some degree of partial limited membership, greater restrictions on membership in speciose assemblages, and partial openness of local assemblages to the regional pool of species.

8.3 Middle ground

Roughgarden (1989) considered the notions of limited vs. unlimited membership in ecological communities to conclude that these dated theoretical perspectives [from Elton (1933) and Gleason (1926), respectively] are only partially true representations of real systems. "After half a century of research, the need for a middle ground between the classically stated polarities was never clearer" (Roughgarden 1989). The same has been said for the contrasting notions of saturation vs. regional enrichment. Real assemblages fall between these hypothetical extremes (Cornell and Lawton 1992). Explanations for species richness patterns in local assemblages are likely to include elements of each extreme (Karlson and Cornell 1997, 1998). However, it is not quite clear how we should proceed to integrate these polarized notions into a single conceptual framework. Some degree of confusion along the way is probably inevitable as we use similar terms and concepts in different ways. For example, identification of important processes controlling the structure of real assemblages is likely to depend on whether we use local or biogeographical scales to designate assemblages. In addition, designations referring to saturation (Bohnsack and Talbot 1980), oversaturation (Cornell and Lawton 1992), supersaturation (Abele 1984), and undersaturation (Roughgarden 1989) can be confusing. They are sometimes used to refer to quite different phenomena (*e.g.*, colonization at a single site vs. local species richness across a range of differentially rich regions).

Consider the notion of undersaturation. "When communities are undersaturated, they become structured as much or more by the dynamics of the transport processes governing the introduction of propagules as by interactions among the residents" (Roughgarden 1989). Undersaturation then is relative to the predicted richness when all available niche space is occupied. Since undersaturated assemblages have fewer species than predicted for saturated assemblages, they seemingly would fall outside the continuum hypothesized by Cornell and Lawton (1992) for most real assemblages. This, however, is not the case. These notions of saturation use different standards as a basis for comparison. The standard used by Roughgarden (1989) is based on the amount of available niche space rather than the number of species in the available pool. Thus undersaturation implies that some species are actually not part of the available pool. The standard used by Cornell and Lawton (1992) sets a lower limit on local richness again based on niche space, but also on the notion of competitive exclusion. There is little or no species coexistence in a saturated assemblage and several putative mechanisms are posited to generate oversaturation.

Undersaturation represents one potential consequence of recruitment limitation. As transport processes constrain recruitment into benthic habitats (sections 1.4.1 and 5.2.3), some proportion of the species in the region may be lost from the pool of species colonizing a particular habitat. Recruitment limitation also includes cases in which species may be present in the regional pool and capable of successful colonization, but limited in terms of abundance. Low abundances of important species in ecological assemblages are likely to reduce their capacity to limit species membership. Hence competitive exclusion might be diminished in some cases and more species may co-occur than predicted by a saturation model (*i.e.*, oversaturation). Thus transport processes may restrict or enhance local species richness depending on the degree of recruitment limitation. In order to evaluate the different notions of saturation and recruitment limitation in coral communities, we will need to integrate knowledge regarding processes operating in the water column as well as in benthic habitats. Transport processes directly influence the size of the available pool of species, colonization of the benthos, and subsequent recruitment into benthic assemblages. These may represent sufficient explanations of observed community structure in some cases. However, much of the coral community literature strongly emphasizes the importance of post-recruitment competition and various sources of spatiotemporal heterogeneity as primary determinants of benthic community structure. These should not be ignored.

Hubbell (1997) recently considered local, niche-based and larger, dispersal-based perspectives on the assembly of species assemblages. His objective was to unify the theories of island biogeography (MacArthur and Wilson 1963, 1967) and of relative species abundances (*e.g.*, MacArthur 1957, 1960, 1972). He noted that from the biogeographical perspective, the theory of island biogeography "is incomplete because it embodies no mechanism of speciation". According to this theory, species cannot originate on islands or in source areas of potential colonists. Furthermore, the number of species per island is determined purely on the basis of island size, relative isolation from sources, and the balance between immigration and local extinction rates. From a local ecological perspective, conventional theory predicts species richness patterns, but fails to "predict the abundances of species".

Hubbell's new theory "is a direct generalization of the equilibrium theory of island biogeography. It rests on a key first principle, namely that the interspecific dynamics of ecological communities are a stochastic zero-sum game". That is to say that colonization can occur only until resources become limited. At equilibrium, there can be no births or successful immigration unless these new individuals replace vacancies created by death. This death process is assumed to be stochastic and the new theory posits that there are no niche differences among species. Yet new species can "arise in the theory like rare point mutations, and they may spread and become more abundant, or more likely, die out quickly". Furthermore, they may originate on isolated islands, in fragmented landscapes (metacommunities), or in large source areas.

Using this theory, Hubbell (1997) generated several predicted patterns describing community structure. These included the number of species in hypothetical assemblages, species-area relationships, species incidence curves, and dominance-diversity relationships. Predictions also included a set of Bray-Curtis similarity indices for paired assemblages separated spatially, yet linked by dispersal. "This theory then shows how species richness and relative species abundances will evolve and equilibrate over a metacommunity landscape. Remarkably this happens in the perfectly homogeneous environment inhabited by perfectly identical species in terms of per capita probabilities of birth, death, and migration". Even more remarkable is the fact that some of the predictions of this theory closely match the observed patterns exhibited by "a latitudinally broad range of closed canopy tree" assemblages.

The relevance of this to coral assemblages is twofold. First of all, Hubbell (1997) recognized the similarities among space-limited assemblages found in closed-canopy forests, the rocky intertidal zone, and coral reefs. The assumption that new individuals enter these assemblages as in a zero-sum game may be justified. Secondly, the capacity of many corals to disperse over large distances links local coral assemblages into metacommunities. Thus the structure and dynamics of coral assemblages at large biogeographical scales may be controlled more by dispersal than by species-specific niche differences.

Hubbell (1997) also argued that it is premature to conclude that local coral assemblages have limited membership and niche-based assembly rules control their structure. He pointed out that according to his unified theory, “even moderate rates of dispersal will ensure that these species are nearly everywhere nearly all the time”. Dispersal increases the proportion of species from the metacommunity present in individual local assemblages. Therefore, species membership in local assemblages can be easily explained by dispersal phenomena.

As a concluding comment on the divergent perspectives provided by dispersal and niche-based explanations of assemblage structure, Hubbell (1997) suggested that this “long standing” debate “is probably here to stay”. There are at least two good reasons to support this view. The first is that each notion is partially correct, yet they differ in relative importance based on the scale at which one defines the assemblage and the degree of dispersal linking local assemblages. At biogeographical scales, the unified theory suggests that “most of the detail about niche structure is lost or becomes ineffective at controlling community structure”. Yet “real species are not identical, and they have niches”. In that both dispersal and niche differentiation contribute to the structure of real communities, we require a theoretical framework incorporating both. Hubbell’s unified theory is a step in this direction because it incorporates both local and metacommunity scales. Yet it does not include several important aspects of coral communities (*e.g.*, niche differences, local spatial heterogeneity, competitive exclusion, or limited membership by predation). A truly unified theory would bring the divergent perspectives together.

The second reason for the continuing debate is that the dispersal-based perspective “is likely to be a lot more difficult to falsify than might have been thought” (Hubbell 1997). This certainly appears to be the case for closed-canopy forests. However, rigorous tests conducted across large spatial scales (*e.g.*, the fragmented landscape in Indonesia or across a set of differentially rich regions in the central Indo-Pacific) may be both feasible and successful in testing for the influence of dispersal on local coral assemblages. In that the published literature already provides support for both limited membership (section 8.2.4) and regional enrichment in coral communities (section 8.2.3), it is likely that both niche-based and dispersal phenomena are important.

8.4 Overview

Comparisons of ecological communities across large scales have been used to explore the notions of convergence, saturation, limited membership, and regional enrichment. Conventional niche theory predicts that competition restricts the number of co-occurring species. Thus species richness is predicted to be convergent among similar organisms exploiting similar habitats and resources in different places. Furthermore, an upper limit is predicted for both species membership and richness as local assemblages become saturated with species. Studies examining the number of species in particular habitats on different continents confirm that convergence is common, but highly variable. This variation occurs among taxa and guilds and with the regional scale of the analysis. Divergent patterns are also quite common. These are generally interpreted as evidence for historical influences on species richness patterns. When examined across a range of differentially rich regions, most studies of local assemblages support the conclusion that they are not saturated. Generally, local assemblages in speciose regions are richer than those in depauperate regions and there are no hard upper limits to local richness.

The evidence from coral communities is not extensive, but does support the view that we need to integrate the local niche-based perspective with larger-scale phenomena. Among fishes, habitat heterogeneity strongly influences local species richness. In addition, there is growing evidence that regional factors also influence local richness. Among decapod crustaceans associated with corals, regional differences in richness do not appear to influence local richness on individual coral heads. However, regional influences may be detected at local scales encompassing multiple coral heads. Indo-Pacific coral assemblages appear to be regionally enriched rather than saturated. Local species richness increases significantly over regional richness gradients. This pattern is strongest in depauperate regions, yet is also evident across regional gradients in the most speciose, central Indo-Pacific. The evidence from species-rich coral assemblages supports the notions of regional enrichment and limited membership. At the extreme, these properties are mutually exclusive. Thus each property is likely to represent only a partial explanation for the observed patterns. Speciose coral assemblages may be open to some, but not all species in the regional pool. Likewise, the notion of limited membership may apply to many species, but not to all.

9 INTEGRATION ACROSS SCALES

9.1 The local environment

In this final chapter, I bring together the local niche-based perspective on the dynamics of coral communities with larger scale perspectives invoking the effects of geography and history. This integration requires the crossing of disciplinary boundaries into fields where historical phenomena are central (*e.g.*, biogeography, paleontology, evolutionary biology). This crossing requires the community ecologist to face the challenge of blending history with a largely non-historical perspective regarding community dynamics. Harbingers of this trend emerged over a decade ago with the growing appreciation of the importance of nonequilibrial phenomena (section 3.4) and scale (sections 1.3 and 8.1). This presentation provides a preliminary view of a set of current ideas, observations, and a field in the midst of change. Several years will be required for such efforts to emerge as mature scientific endeavors. This is because we are only just beginning to appreciate the nature of the association of large-scale phenomena with coral community patterns. Rigorous evaluation of the hypothetical processes responsible for these patterns largely remain for future investigations.

To begin, I briefly review some of the main points made previously regarding the influence of local factors on the structure of coral communities. In chapter 2, zonation and diversity patterns were described for corals and fishes across depth and habitat gradients at multiple locations. Among corals, these patterns are thought to result largely from the interactive influences of interspecific competition and disturbances (Connell 1978, Huston 1985a). Furthermore, competitive success or reductions in population size due to disturbances are affected by a wide variety of abiotic and biotic factors. These factors include wave action, tidal exposure, sedimentation, consumer-resource interactions, and specialized species and habitat associations. Several instances were noted in which individual coral species exhibited some degree of habitat specialization over depth or habitat gradients. Most species with wide depth distributions reach high abundance levels in only a limited portion of their depth range. Hence, critical evaluation of the factors restricting coral distributions and abundance patterns is warranted. Among fishes and invertebrates other than corals, local diversity patterns also appear to be strongly influenced by depth and habitat heterogeneity. In addition, topographic complexity and other local environmental attributes related to cross-shelf location and current regimes are important (*e.g.*, see Williams and Hatcher 1983).

In chapter 3, keystone species were highlighted as determinants of coral community structure. By definition, these species control the stability of local communities. This keystone status is spatiotemporally variable, dependent on factors operating at multiple scales, and is typically given to consumer species involved in very strong biological interactions. For example, the radical shift of Jamaican coral reefs from coral to algal domination has been attributed to overfishing, hurricanes, and the regional mass mortality of the herbivore *Diadema antillarum* (Hughes 1994a). Thus man's activities, other consumer-resource interactions, and the devastation generated by Hurricane Allen were implicated as important local factors contributing to keystone species status of the sea urchin. The major phase shift of this entire community followed the mass mortality of this keystone species. Other keystone species include territorial damselfishes and the corallivorous *Acanthaster planci.* They influence species composition, richness, and zonation patterns in coral communities as they interact with a variety of resource species, competitors, and other consumers. Examples of local factors contributing to variation in these effects include topographic relief (Wellington 1982) and biological interactions. The abundance and distribution of keystone species can be controlled by interactions with predators (*e.g.*, Glynn 1977, 1982) and symbiotic crustaceans living with corals (*e.g.*, Glynn 1976).

In chapters 3 and 4, a number of local factors contributing to successional processes in coral communities were noted. Biological interactions are a central component of most models of ecological succession and the rules governing the assembly of natural communities. Colgan (1987) documented coral species replacements following the devastation of coral communities by *Acanthaster planci* in Guam. The actual successional mechanisms included herbivory, inhibition of coral settlement, spatial competition, and differential settlement and growth rates among corals. At Heron Island, spatial competition, physical disturbances, and recruitment processes all contribute to temporal changes in the abundance of corals (Tanner *et al.* 1994, Connell *et al.* 1997). In Hawaii, local disturbance regimes and recruitment processes are thought to influence the rate and direction of coral succession (Grigg and Maragos 1974, Grigg 1983). At smaller spatial scales, damselfishes control algal successional patterns within their territories (Hixon and Brostoff 1996). Evidence is currently accumulating from a variety of organisms found in coral communities that larval settlement and metamorphosis are controlled by highly species-specific biological interactions with benthic plants and animals. Lastly, there is preliminary evidence that planktivory on the larvae of benthic organisms may also influence the successional process.

In chapters 5 and 6, interspecific competition and consumer-resource interactions were explicitly emphasized as important fundamental processes in coral communities. The relative "community importance" of individual species (Power *et al.* 1996) is often expressed through a variety of direct and indirect effects of these interactions. The magnitude of the effects of particular types of interactions varies strongly among species and typically varies with the densities of each interacting species. These effects are further modified by a variety of abiotic and biotic factors characterizing the local environment. Interspecific competition involving corals and other colonial invertebrates takes many forms and the importance of any particular competitive mechanism varies in space and time (Lang and Chornesky 1990). For example, Alino *et al.* (1992) suggested that the outcome of coral interactions with soft corals on the Great Barrier Reef depends on water clarity (light), nutrients, and corallivory by fishes. Likewise, overtopping by fast growing corals depends on physical conditions favoring relatively fragile branching corals rather than more resistant and slower growing massive forms.

Spatial heterogeneity in consumer-resource interactions in coral communities varies strongly with depth, habitat, and availability of sheltered sites.

The influence of disturbances on the dynamics of coral communities are also highly variable in space and time (see chapter 7). Jackson (1991) identified predators, damselfishes, agents of bioerosion, and storms as important disturbances operating at small spatial and/or short temporal scales (Table 7.1). Although catastrophic disturbances can result in major extinctions on a global scale (section 2.1), most natural and human-caused disturbances known to have significant impacts on coral communities occur at local or regional scales (Table 8.2). In the 30-year study at Heron Island, Connell *et al.* (1997) implicated important disturbances operating at scales of 10-100s of meters. The regional enrichment of coral communities reported by Cornell and Karlson (1996) appears to be influenced by factors generating spatial heterogeneity at scales of less than 10 meters. Probable local factors contributing to this heterogeneity include predators, small-scale disturbances, local "neighborhood" competition, and intraspecific aggregation (Cornell and Lawton 1992, Karlson and Cornell unpublished manuscript).

9.1.1 LOCAL RICHNESS OF CORALS ACROSS DEPTH AND HABITAT GRADIENTS

As part of our analysis of local saturation and regional enrichment of coral assemblages (section 8.2.3), we also evaluated the sensitivity of local richness to environmental variation across depth and habitat gradients (Cornell and Karlson 1996). Our intent was to quantify the relative importance of local and much larger scale phenomena on local richness. As emphasized above, much of the variation in the local environment occurs along depth and habitat gradients. For example, Huston (1985a) subsumed most variation in coral species diversity on coral reefs to depth and disturbance gradients representing shifts in the influence of abiotic factors and grazing by fish and sea urchins (as depicted in Figure 7.1). Disturbance gradients can clearly occur across intertidal reef flats or subtidally along exposure gradients at uniform depths (section 2.2). We used depth relative to mean low water and identified specific habitats which we ranked by relative distance from shore (*i.e.*, inner flat, mid- and outer flat, reef crest and upper slope, mid-slope, and lower slope) (Cornell and Karlson 1996). Most of the data (89%) extracted from the published literature could be categorized in this way.

Variation in the local richness of coral assemblages was strongly associated with depth and habitat (Table 9.1). Using simple and stepwise regression procedures, we determined that local richness was most sensitive to depth. The vast majority of published estimates of local richness have been collected in shallow water from intertidal reef flats to depths of approximately 20 m (Cornell and Karlson 1996). Over this depth range, local species richness increases towards some maximum value which Huston (1985a) estimated to occur at 15-30 m. Considerable variation in local richness also occurs along habitat gradients; it increases across reef flats towards the crest and reaches a maximum on the mid-to-upper slope (Cornell and Karlson 1996). When independent variables were considered alone in simple regressions, the linear effect of habitat variation on local richness was almost as strong as that due to depth, but stronger than that associated with a nonlinear, quadratic term for habitat (Table 9.1). However, as a consequence of covariation between depth and habitat, only depth and the quadratic term for habitat entered our stepwise regression models. These two local variables explain approximately twice as much variation in local species richness as does regional species richness (Cornell and Karlson 1996). Thus regional enrichment (see chapter 8) and local environmental factors jointly contribute to observed local richness patterns.

TABLE 9.1 The relationship between the local richness of coral assemblages (pooled residuals as described in section 8.2.3) and several independent variables used in simple and stepwise regressions (after Cornell and Karlson 1996[1], Karlson and Cornell 1998[2]). Species richness and depth were log-transformed prior to analysis and all variables were standardized to reduce differences in measurement scale. Both linear and quadratic terms for depth and habitat were evaluated as local environmental variables. Regional variables included regional species richness (linear and quadratic terms), number of genera, distance to nearest high-diversity region, distance to equator, a simple contrast between the Indo-Pacific and western Atlantic provinces, and the average age of genera.

	Regression Coefficient		
	Simple regressions	Stepwise Model 1[1]	Stepwise Model 2[2]
Local variables			
Depth	0.39*****	0.46***	0.44***
Depth (quadratic)	0.06*	--	--
Habitat	0.34*****	--	--
Habitat (quadratic)	-0.25*****	-0.44***	-0.42***
Regional variables			
Regional species	0.21*****	0.32***	0.16***
Regional species (quadratic)	-0.16*****	-0.14**	-0.04
Regional genera	0.18*****	--	0.23**
Distance to high-diversity region	-0.30*****	--	-0.08***
Distance to equator	0.11****	--	0.24**
Provincial contrast	0.06*	--	--
Age of genera	0.02	--	--

*****$p<0.00001$, ****$p<0.00005$, ***$p<0.0001$, **$p<0.0005$, *$p<0.05$

The sensitivity of local richness to depth and habitat depended on the geographic scale of the analysis (as discussed previously for the relationship between local and regional richness in section 8.2.3). At global and provincial scales, local richness consistently increases with depth. The regression coefficient for this relationship is the same using data from the Indo-Pacific or western Atlantic Oceans (Table 9.2). However, this relationship differs substantially when one considers depauperate regions of the Indo-Pacific. Depth did not enter the stepwise model in this case, whereas it was the most important independent variable at all other scales (Karlson and Cornell 1998). In similar fashion, the term for habitat entered the stepwise model at all geographic scales, but was notably weak in the regression for the depauperate Indo-Pacific

(Table 9.2). The variation in local richness explained by both depth and habitat combined was 24.3 - 33.9% at global and provincial scales, 25.3% for the speciose Indo-Pacific, but only 2.1% for the depauperate Indo-Pacific (Table 9.2). These results are consistent with a general assessment that local species richness among corals does not vary to any great extent across depth and habitat gradients in relatively impoverished regions (*e.g.*, see section 2.2.5).

TABLE 9.2 The relationship between the local richness of coral assemblages and local environmental variables representing depth and the quadratic term for habitat at multiple geographic scales. Regression coefficients were generated using the stepwise regression model identified in Table 9.1 as Model 2. R^2 values (in parentheses) are given as percentages. The terms for depth and habitat were both significant at all but one geographic scale. In depauperate regions of the Indo-Pacific, the former was not significant while the latter was weak (after Karlson and Cornell 1998).

Geographic scale	Regression coefficients			
	Depth	(R^2)	Habitat	(R^2)
Global	0.44***	(13.8%)	-0.42***	(10.5%)
Indo-Pacific	0.45***	(12.3%)	-0.51***	(14.4%)
Speciose Indo-Pacific	0.64***	(19.9%)	-0.40***	(5.4%)
Depauperate Indo-Pacific	--	--	-0.11*	(2.1%)
Western Atlantic	0.45***	(30.5%)	-0.21**	(3.4%)

***$p<0.0001$, **$p<0.0005$, *$p<0.005$

9.2 The regional setting and cross-scale linkage

In chapter 1, a variety of large-scale geographical and historical phenomena were introduced as significant determinants of the structure of ecological communities. At the largest scales, global and regional climate, atmospheric-oceanographic coupling, geologically significant extinction events, and unique radiations of particular taxa were noted to influence local aquatic communities. At regional spatial scales, the contribution of oceanographic processes to restricted larval transport and recruitment limitation was highlighted for rocky intertidal communities (section 1.4.1). Recruitment limitation may be severe primarily constraining local species richness. Alternatively, it may primarily constrain abundance levels and the intensity of competitive processes without restricting local species richness. Under this less severe form of recruitment limitation, transport processes can contribute to regional enrichment (section 8.1.2). Disturbances were also noted to operate at large scales. Temporal fluctuations in the frequency of storms and El Niño events in the eastern Pacific were linked to periodic atmospheric phenomena operating at multi-year scales. Other long term temporal phenomena include the influence of past recruitment or disturbance events on the structure of benthic communities, especially those dominated by long-lived clonal invertebrates (section

1.4.3). Lastly, the evolution of a variety of biological attributes of benthic organisms was used to illustrate their dynamic response to selection by disturbances, consumers, and competitors.

It should not be surprising that most of the large-scale phenomena known to influence ecological communities also have important effects on patterns studied by biologists in other disciplines (*e.g.*, biogeographers, systematists, and paleontologists). Living systems are obviously influenced by climatic fluctuations, geological catastrophes, and other large-scale phenomena. As a consequence of different disciplinary perspectives, there can be some degree of semantic confusion as we use similar terms to refer to different biological attributes. For example, Rosen (1988a) noted the ambiguity in the meaning of "dispersal" by biogeographers studying reef corals. Dispersal represents one of the major factors controlling distributional limits and membership in biogeographical assemblages. At the community level, Dayton (1994) suggested that dispersal represents a major unifying theme in the marine environment because virtually all of the biological order we see regardless of scale "is a result of coupling of dispersal and internal dynamics". Just as dispersal links local communities into metacommunities (*e.g.*, Hubbell 1997), it also links populations into metapopulations (Levins 1969, 1970). Thus dispersal (and the limits thereof) contribute to spatial distributions and the dynamics of populations, communities, and entire biotas.

Other sources of ambiguity arise within ecological subdisciplines like ecosystem and community ecology. For example, Chapin *et al.* (1997) noted "the apparent conflict between the perspectives that each species is important and that there is ecological redundancy among species". The notion of redundant species emerges as one considers regional and ecosystem processes. "Regional processes include trace gas fluxes to the atmosphere and nutrient fluxes from terrestrial to aquatic systems". "Ecosystem processes include productivity and nutrient cycling". The notion that individual species are important (*e.g.*, keystone species) comes from the study of biological interactions and the stability of ecological communities. While a species might be extremely important at the community level, it might not be recognized as important at the regional or ecosystem level. Chapin *et al.* (1997) concluded that we need to consider "functional types of organisms and their environmental responses". In particular, they highlighted the importance of identifying and protecting both keystone species and functional groups because the stability of both communities and ecosystems are linked. We are only just beginning to understand the extent of this linkage, yet it is emerging "as a problem of fundamental concern" (also see Schulze and Mooney 1994).

Among the consequences of considering ecological patterns at large-scale has been the emergence of macroecology as a new discipline (Brown and Maurer 1989, Brown 1995, Blackburn and Gaston 1998). It embraces the geographical and historical perspective and relies on inter-disciplinary integration of the more traditional fields of biogeography, paleontology, and evolutionary biology. The macroecologist seeks to promote the understanding of the distribution, abundance, and diversity of organisms comprising regional or continental biotas in terms of resource partitioning and the evolutionary history of particular taxa. Thus it represents another attempt to integrate conventional notions from community ecology with large-scale phenomena. In that most work in this relatively new area has focused on terrestrial mammals, birds, and insects, it is not yet clear how it will contribute to our understanding of the dynamics of coral communities. At the same time this is happening, other developments in more traditional disciplines are integrating notions from community ecology. Thus the integration across disciplines involves efforts in multiple directions and may eventually result in the coalescence of fields (or at least the blurring of disciplinary boundaries).

For example, Myers (1994) integrated the marine biogeography of shallow-water organisms with elements of community ecology by examining patterns of endemism and species richness across a range of spatiotemporal scales. Among historical biogeographers, perhaps the "biggest frontier" left to be explored are the events which have led to endemic distribution patterns. "Endemicity is relative because, although endemicity implies geographical restrictedness, it is actually a continuum such that all species are endemic at some scale". The important focus for this research is on the factors generating endemism. These include colonization over ecological time scales as well as vicariance processes occurring in the historical past. The former occurs following "jump-dispersal" across biogeographic barriers. The latter involves the splitting of distributional ranges into multiple "discontinuous ranges or populations". Successful colonization depends on a number of factors influencing dispersal, invasion of uncolonized sites, and the establishment of viable populations. Speciation over evolutionary time and colonization success over ecological time are important among recently evolved taxa (neoendemics) which may be expanding their ranges. Older palaeoendemics largely owe their endemicity to extinction events due to large-scale processes (*e.g.*, geological events and changes in climate, sea level, and current systems). Of course, in the face of periodic climate changes, taxa regardless of age may also undergo range expansions and contractions rather than extinctions (see section 3.4).

In recognizing the dual importance of historical and ecological processes on patterns of endemicity, Myers (1994) considered factors contributing to biogeographic dispersal (*i.e.*, current systems and speciation) and post-dispersal colonization success separately. This mirrors the way community ecologists often consider the separate effects of larval settlement and post-settlement mortality on recruitment success (section 1.4.1). While historical phenomena directly influence the size and composition of "the immigrating pool of taxa", local "ecological factors determine colonization success" (Myers 1994). Despite the large number of larvae produced by many marine organisms and their ability to travel long distances in the sea, "there does not appear to be a close correlation, in general, between the immigration rates by larvae of shallow-water marine organisms and subsequent colonization". This may result from the fact that many marine organisms are capable of extensive dispersal as adults. Even sessile corals are known to raft across the Pacific Ocean attached to floating pumice (Jokiel 1990a,b). While successful immigration is a prerequisite for successful colonization, important ecological factors operating at local scales influence the invasion of new species and the establishment of viable populations.

Myers (1994) recognized that species richness patterns in marine systems are also influenced by a range of ecological and historical processes operating at multiple spatiotemporal scales (Table 9.3). "The extent to which historical processes influence species richness in ecological time is currently a matter of debate (Cornell and Lawton 1992) and depends upon whether communities are saturated or near saturation" (see chapter 8). At the smallest spatiotemporal scales (within habitats) and under conditions of saturation, resource partitioning is thought to dictate species richness patterns. Several factors contributing to local spatiotemporal heterogeneity can enhance the number of coexisting species within habitats (*e.g.*, disturbances, patchy resources, aggregated resource utilization patterns). Consequently, local assemblages are not saturated, but are open to the influences of larger-scale historical phenomena (as described in chapter 8). At an intermediate spatial scale, Myers (1994) highlighted the role of habitat heterogeneity on patterns of species richness. This between-habitat component of diversity is "probably strongly influenced by regional richness (Cornell and Lawton 1992)" and several historical processes (Table 9.3). These include immigration, colonization, population extinctions, and variation in the local climate. I would add that all of these processes also can influence richness patterns at the smallest spatial scale within habitats because most communities are unsaturated (section 8.1.2). At the largest scale (pan-continental), Myers (1994) emphasized the effects of speciation, global extinction events, and a wide range of other phenomena influencing species richness patterns. These included the movement of the earth's crust (tectonics) and changes in sea level (eustasy), global climate, and major ocean current systems. These historical processes are thought to strongly control richness patterns at this scale. In addition, they contribute to the size and composition of regional species pools and to the regional enrichment of local species assemblages within habitats. Likewise, the effects of ecological processes on richness patterns are probably strongest at smaller spatial scales, yet they cannot be ignored at larger scales. Local spatial heterogeneity, immigration, colonization, and speciation all contribute to the cumulative number of species observed at large spatial scales.

TABLE 9.3 Processes influencing species richness patterns in marine systems across a range of spatiotemporal scales. Ecologically important factors including resource partitioning and disturbances operate at the smallest scale, while speciation, global extinction, and "TECO" processes operate at the largest scales (T = Tectonic, E = Eustatic, C = Climatic, O = Oceanographic) (modified after Myers 1994).

Temporal Scale	Spatial Scale		
	Small (within habitats)	Intermediate (between habitats)	Large (pancontinental scales)
Short	Resource partitioning Disturbances	Habitat heterogeneity Immigration Local climate changes	
Intermediate		Colonization Population extinctions Local climate trends	
Long			Speciation Global extinction TECO events / processes

9.2.1 CROSS-SCALE LINKAGE IN BIOGEOGRAPHIC ASSEMBLAGES: A CORAL EXAMPLE

Here I describe a recent study which relates variation in the richness of biogeographic assemblages of corals to a variety of variables representing the effects of ecological, geographical, and historical processes (Fraser and Currie 1996). The primary objective of this study was to evaluate a set of seven competing hypotheses for global patterns of coral diversity (Table 9.4). These patterns include well known latitudinal gradients as well as centers of high diversity [see chapter 2, Stehli and Wells (1971), and Veron (1995)]. The set of hypothetical explanations for these biogeographical patterns invoked processes operating at multiple spatiotemporal scales. Fraser and Currie (1996) used correlation and regression analyses on the number of coral genera occurring at 130 sites spread throughout the Indo-Pacific and Atlantic Oceans. Some sites were as small as those at Clipperton Atoll (where they estimated shelf area at 42 km^2) and others as large as all of Indonesia (1,075,044 km^2) (personal communication). This range of spatial scales permits the evaluation of global-scale variation along a variety of large-scale gradients. However, it is too large to permit the sort of analysis recommended by Westoby (1985, 1993) (see section 8.2.1 regarding the evaluation of richness patterns across a range of spatial scales). While recognizing that causal relationships cannot be inferred using this approach, Fraser and Currie (1996) hoped to eliminate inappropriate hypotheses when no statistical relationships could be detected between generic richness and the putative causal factors.

The first hypothesis evaluated by Fraser and Currie (1996) invoked the well known and pervasive relationship between taxonomic richness and area. Typically richness increases with the size of the area sampled along with increases in habitat diversity and environmental heterogeneity. Not surprisingly, the generic richness of corals was highly correlated with the size of the continental shelf (area to a depth of 100 m) at each site. Three of the largest sites (namely Indonesia, Malaysia, and the Philippines) are situated in the richest diversity region in the world where there are 64-78 genera (Veron 1993, 1995). In order to evaluate the other hypotheses and to correct for the fact that the shelf areas "varied by nearly six orders of magnitude", Fraser and Currie (1996) conducted correlation analyses before and after correcting for the area effect. If the variation in coral richness were determined solely by the effect of area, they could eliminate all of the remaining hypotheses. Prior to this correction, 9 of 15 other independent variables (representing 5 of 6 remaining hypotheses) were significantly correlated with generic richness (Table 9.4). After correction, 10 of 15 variables (again representing 5 of 6 remaining hypotheses) were significantly correlated with generic richness. Thus shelf area alone provided an inadequate predictive framework for understanding richness patterns and none of the competing hypotheses could be excluded.

Fraser and Currie (1996) concluded that a combination of ecological and historical factors act as determinants of coral richness patterns. However, their presentation emphasized some of these factors more than others. For example, "the best environmental predictors of diversity" were "mean annual ocean temperature and an estimate of regional coral biomass" (namely the length of coastline with coral reefs at a site). The former was considered as a possible correlate of energy available to corals, while the latter was a "relative measure of total coral biomass or energy accumulation". Both variables were highly correlated with generic richness as predicted by the hypothesis that the availability of energy limits taxonomic richness (Table 9.4). The highest number of coral genera were located at multiple sites in the tropical Indo-Pacific with mean annual ocean temperatures in excess of 28°C. The lowest richness was reported at sites with temperatures below 18°C (*e.g.*, Honshu, Japan and Esperance Bay, Australia).

TABLE 9.4 Seven general hypotheses for the global richness patterns of reef-building corals. Factors were evaluated using correlation analysis before and after correction for the effect of area on generic richness (after Fraser and Currie 1996).© 1996 by The University of Chicago.

Hypothesis	Argument	Factors tested
Area	Richness reflects sampling effort and environmental heterogeneity	Continental shelf area[1]
Available energy	Partitioning of energy among species limits richness	Mean ocean temperature[1,3] Solar irradiation[1,3] Length of reef[1,3]
Environmental stress	Physiological tolerances of harsh conditions limit richness	Mean annual salinity Ocean temperature extremes[1,3] Secchi depth (water clarity)[4]
Environmental stability	Physiological tolerances of variable environments limit richness	Annual variation in temperature[2,4], salinity, and cloud cover
Disturbance	Disturbance prevents competitive exclusion	Frequency of tropical cyclones (number per 20 years) originating in the area[1]
Biological interactions	Competition and predation affect niche partitioning	Phytoplankton productivity[4]
Historical factors	Richness is influenced by speciation, extinction, and "TECO" events/ processes like ocean currents and glaciation	Density of up-current islands[1,3] Contrast between Indo-Pacific and Atlantic sites Latitude[2,4]

[1]correlation significant at $p<0.0005$ or [2]$p<0.005$ before correction for shelf area; [3]significant after correction at $p<0.0005$ or [4]$p<0.005$

Furthermore, three geographical variables were noted for their significant relationships with generic richness and their association with historical explanations for diversity patterns. First, there were significantly more genera at Indo-Pacific than at Atlantic sites. Second, more genera were represented at sites near the equator rather than at higher latitudes. Third, more genera occurred at sites with high island densities located up-current (within 5000 km of a site) than at sites associated with low island densities. This last result "is consistent with vicariance models of speciation and theories of coral dispersal". The weakest ecological predictors of coral generic richness were those supporting the hypotheses that disturbances, biological interactions, and environmental variability limit taxonomic richness.

One should note that these weak correlations were detected at quite large biogeographic scales. At smaller spatial scales, local disturbances and a variety of biological interactions are well known to strongly influence richness patterns. Furthermore, these ecological factors typically involve species-specific responses to physical and biological aspects of the environment. For example, disturbance is generally invoked as a mechanism promoting species coexistence (chapter 7). How disturbance should influence diversity at higher taxonomic levels is unclear. Fraser and Currie (1996) recognized that disturbances typically are predicted to promote richness at local spatial scales within ecological communities (Connell 1978), yet their examination of this relationship at large spatial scales is also valid. For example, Huston (1994) extended his dynamic equilibrium model to explain global diversity patterns (section 7.1). Furthermore, some types of disturbance actually operate at very large scales (Table 7.1). Using data from Simpson and Riehl (1981), Fraser and Currie (1996) found that there were 0-30 cyclones originating in the vicinity of their 130 sites during 1958-1978. Cyclone frequency was significantly correlated with generic richness before the richness data were corrected for shelf area, but not afterwards (Table 9.4). This result probably is indicative of a strong correlation between cyclone frequency and shelf area, though cyclones are also well known to vary with other geographical variables (*e.g.*, latitude).

Fraser and Currie (1996) conducted stepwise regression analyses in order to reduce the number of variables one might use to predict the generic richness of corals. They noted that several of the variables they used were not only correlated with generic richness, but also they were highly correlated with one another (most notably reef length, mean ocean temperature, and shelf area). In their best regression model, two geographical variables and two supporting the energy limitation hypothesis explained 71% of the variation in coral generic richness. The contrast between Indo-Pacific and Atlantic sites was highly significant thus implicating the importance of historical differences in these two faunas. The density of up-current islands was also significant thus implicating potential effects of dispersal and/or speciation on richness. The mean ocean temperature and coral biomass (reef length) again were significant predictors of generic richness.

None of the ecological variables representing disturbances, biological interactions, or environmental stability entered two alternative stepwise regression models. Interestingly, shelf area also did not enter these two models. The effects of sampling effort and environmental heterogeneity on generic richness were masked by correlations with reef length and ocean temperatures. This illustrates the care required in interpreting the results of multivariate analyses of competing hypotheses. Clearly shelf area was an important variable to consider in developing their predictive model of generic richness. Fraser and Currie (1996) retained it in their best stepwise regression model even though it did not contribute to any additional significant variation in generic richness.

9.2.2 CROSS-SCALE LINKAGE IN ECOLOGICAL ASSEMBLAGES: A CORAL EXAMPLE

In contrast to the biogeographical assemblages evaluated by Fraser and Currie (1996), Cornell and Karlson (1996) analyzed the coral richness patterns of ecological assemblages using unequivocally local samples (*i.e.*, within quadrats or along transects) (sections 8.2.3 and 9.1.1). Thus the data provided quantitative estimates of the number of coral species occurring within habitats. We purposefully avoided data collected at larger spatial scales which might include both the within and between-habitat components of species richness (see section 2.2 and chapter 8). In order to determine the relative magnitude of local and regional influences on coral assemblages, we used stepwise regression analysis to jointly consider depth, habitat, and regional richness as predictors of local richness. Our first analysis determined that local richness was just as sensitive to regional richness as it was to each of the two local variables (Cornell and Karlson 1996, 1997). Thus we detected significant regional enrichment of coral assemblages as well as significant variation in richness over local environmental scales.

In a second analysis, we expanded the number of large-scale variables to consider the potential influences of other regional factors (*e.g.*, evolutionary history, dispersal, and geographic location). We were interested in finding significant variation in local richness which could be attributed to variables operating independently of regional richness (Karlson and Cornell 1998). We reasoned that the regional enrichment of coral assemblages might be primarily controlled at the local level by factors contributing to spatiotemporal heterogeneity (Cornell and Lawton 1992. If true, little additional variation in local richness would be attributable to large-scale variables other than regional richness. Alternatively, large-scale phenomena might be the primary determinants of local richness patterns regardless of local constraints on the number of coexisting species. In this case, the observed regional enrichment of coral assemblages would largely result from regional rather than local processes and we would expect to find considerable additional variation in local richness due to other regional variables.

We considered five additional regional variables in this second analysis (Karlson and Cornell 1998). They were as follows:

- We used the distance of sampled sites from the equator as a potentially important geographic variable. It corresponds with several environmental variables which vary across latitudinal gradients (*e.g.*, solar irradiation, ocean temperatures, etc.). In addition, it permits an independent evaluation of the geographic displacement of high-diversity centers north of the equator (Stehli and Wells 1971).
- As a second geographic variable, we used the distance of sampled sites from the nearest high-diversity centers in the Indian, Pacific, and Atlantic Oceans (as depicted by Stehli and Wells 1971: figure 2). This term corresponds to the degree of regional isolation of each site and has the potential to implicate processes of speciation or oceanic transport as determinants of local species richness.
- We used the number of genera reported by Stehli and Wells (1971: figure 2 and Appendix A) for a variety of regional locations as a measure of richness above the species level. This variable is likely to be highly correlated with regional species richness and may provide a second estimate of the degree of regional enrichment.

In addition, generic richness might implicate evolutionary processes influencing the differential success or failure of particular clades across regional scales. In the latter case, we were again looking for a statistical relationship independent of that due to regional enrichment.
• As another potentially important evolutionary attribute, we used the average age of genera reported by Stehli and Wells (1971: figure 8 and Appendix A).
• Finally, we used the same simple contrast between the Indo-Pacific and Atlantic faunas as did Fraser and Currie (1996). Thus these new variables offered an opportunity to evaluate a mix of geographical and evolutionary attributes as predictors of local species richness.

Examination of the set of regression coefficients describing the simple relationships between local richness and each of nine independent variables (two local and seven large-scale terms) reveals that local richness was sensitive to all but one variable (the average age of genera in a region) (Table 9.1). As noted in section 9.1.1, local richness was most sensitive to depth and habitat in both simple and stepwise regressions. Among the large-scale variables used in simple regressions, local richness was most sensitive to the distance between each site and the nearest high-diversity region (Table 9.1). This is indicative of the importance of regional isolation and the comparatively low local richness observed for coral assemblages at highly isolated sites, especially in the eastern Pacific (Karlson and Cornell 1998). Local richness was also sensitive to the two variables representing taxonomic richness at regional scales, the distance to the equator, and the contrast between the Indo-Pacific and western Atlantic. The terms for taxonomic richness are indicative of significant regional enrichment, yet may also reflect independent influences of evolutionary history (see below).

The geographic variables corresponding to latitudinal gradients in environmental conditions and differences between the faunal provinces indicated two surprising results. First, local richness increased with distance from the equator. While Fraser and Currie (1996) found that generic diversity decreased with increasing latitude (section 9.2.1), local species richness actually increased to reach maximal values within the diversity centers displaced north of the equator (*e.g.*, in the Philippines and Red Sea). Second, local richness was slightly higher on average in the western Atlantic than in the Indo-Pacific (Karlson and Cornell 1998). This is just the opposite of what Fraser and Currie (1996) reported for generic richness. Below I interpret this result as a spurious consequence of averaging data from the relatively depauperate Indo-Pacific and much richer assemblages of the central Indo-Pacific. The high observed variation across the entire Indo-Pacific is contrasted with the relatively homogeneous western Atlantic.

Just as Fraser and Currie (1996) had attempted to reduce the number of variables used to predict generic richness, we also used stepwise regressions to screen all of the large-scale variables (along with depth and habitat) as predictors of local richness (Karlson and Cornell 1998). At the global scale of analysis, depth, habitat, and regional richness remained as strong predictors of local richness (Tables 9.1). Additional variation in local richness beyond that due to regional species richness was explained by the number of genera in a region, distance to the nearest high-diversity center, and distance to the equator (Tables 9.1 and 9.5). Thus geographical and historical phenomena other than regional enrichment are implicated as significant determinants of local richness. The insignificant term representing the average age of genera in a region and the weak term for the provincial contrast did not enter the stepwise regression model. These two terms were very highly correlated with one another and their weak-to-insignificant relationships with local richness were superseded by the more significant variation explained by the other large-scale variables (Table 9.1, Karlson and Cornell 1998).

Using the expanded regression model as a basis for evaluating the relative importance of local and large-scale variables, we explored the influence of the geographic scale of the analysis on our inferences regarding local richness. There were two primary reasons for altering geographic scale. First, the mixing of biotas with different evolutionary histories can introduce errors in the assessment of regional enrichment and saturation (Hugueny *et al.* 1997). This is particularly true when historical phenomena influence the relationship between local and regional richness. Hugueny *et al.* (1997) illustrated this point by blending data from two isolated fish faunas sampled in African and South American rivers. The blended data appeared to exhibit a pattern consistent with the saturation model (Figure 8.1), but separate analyses revealed strong regional enrichment [as indicated in Table 8.1, Hugueny and Paugy (1995)]. Therefore these local fish assemblages were unsaturated and local richness patterns on the two continents were significantly influenced by their different evolutionary histories. The severity of this particular problem is uncertain at the present time as only a few inter-provincial studies have been conducted and saturated patterns appear to be generally uncommon (Cornell and Karlson 1997). The second reason for considering the influence of geographic scale was our realization that regional enrichment can only be detected across a set of heterogeneous regions (chapter 8). Thus as one reduces the geographic scale, regional homogeneity becomes more likely and one is eventually precluded from detecting regional influences on local assemblages.

As described in section 8.2.3, we used global, provincial, and sub-provincial scales to examine the local species richness of coral assemblages (Karlson and Cornell 1998). A strong contrast between Indo-Pacific and western Atlantic corals emerged at the provincial scale of analysis. While depth and habitat remained as significant predictors of local richness in both provinces (Table 9.2), none of the large-scale variables entered the stepwise model for the western Atlantic (Table 9.5). This result is consistent with the assessment that species distributions within this small, isolated province are relatively homogeneous (section 8.2.3). This province lacks the large gradients so typical of the vast expanses of the Indian and Pacific Oceans. We found that local richness in the Indo-Pacific was sensitive to the distance to the nearest high-diversity region, distance to the equator, and the number of genera in a region (Table 9.5). The first of these variables largely reflects the difference in local richness between speciose regions in the western Pacific and more depauperate regions in the eastern Pacific. As noted above, the second is indicative of the geographic displacement of rich local assemblages north of the equator. The third indicates that the number of genera in a region was a better predictor of local richness than was the number of species.

Since regional species richness is correlated with all three of the above variables (Karlson and Cornell 1998), the significant relationship between local and regional richness alone is masked in the stepwise model (Tables 8.3 and 9.5). Furthermore, there is considerable additional variation in local richness attributable to these other variables which is independent of the influence of regional richness. Regional richness alone explained only 7.4% of the variation in local richness, while these other large-scale variables explained at least 17.0% (Karlson and Cornell 1998). Provincial differences among coral faunas in terms of niche partitioning (Ricklefs 1987), relative competitive and dispersal abilities (Cornell and Lawton 1992), and evolutionary history (Hugueny *et al.* 1997) are among the potential factors influencing local richness patterns. In so doing, these factors also contribute to variation in the relationship between local and regional richness (*e.g.*, Hugueny *et al.* 1997).

TABLE 9.5 The relationship between the local richness of coral assemblages and regional variables representing number of species and genera, distance to the nearest high-diversity region, and distance to the equator. Regression coefficients for significant terms were generated using the stepwise regression model identified in Table 9.1 as Model 2. R^2 values (percentages in parentheses) indicate the contribution of each variable to the cumulative total explained variation (after Karlson and Cornell 1998).

Geographic scale	Independent variable	Regression coefficient	(R^2)
Global	Regional species	0.16***	(10.9%)
	Distance to nearest high-diversity region	-0.08***	(2.7%)
	Distance to equator	0.24**	(1.0%)
	Regional genera	0.23**	(0.8%)
Indo-Pacific	Distance to nearest high-diversity region	-0.24***	(14.1%)
	Distance to equator	0.28***	(2.5%)
	Regional genera	0.19*	(0.4%)
Speciose Indo-Pacific	Distance to nearest high-diversity region	-0.50***	(10.7%)
	Distance to equator	0.59***	(4.5%)
Depauperate Indo-Pacific	Regional genera	0.67***	(46.9%)
Western Atlantic	--	--	--

***$p<0.0001$, **$p<0.0005$, *$p<0.05$

Given the large difference noted above between the Indo-Pacific and western Atlantic provinces, we considered the possibility that the lack of regional enrichment in the latter might merely be a statistical consequence of using excessively narrow ranges for the regional variables. For example, regional richness in the western Atlantic spanned only 19-50 species, whereas there were 19-411 species in regions located in the Indo-Pacific. As a test of this possibility, we considered speciose regions of the Indo-Pacific (with 244-343 species) separately from depauperate regions (with 19-88 species) as described in section 8.2.3. The former includes coral assemblages located in the central Indo-Pacific (Figure 2.2), while the latter includes most notably the eastern Pacific along with a few other depauperate regions in the central Pacific and Indian Oceans.

We detected significant regional enrichment in both speciose and depauperate regions of the Indo-Pacific (Table 8.3). Therefore, the lack of regional enrichment in the western Atlantic cannot be attributed to the narrow range of values used for regional richness.

Furthermore, we determined that three other large-scale variables were significant predictors of local richness at these reduced geographic scales in the Indo-Pacific. In the speciose Indo-Pacific, the best large-scale predictors of local richness were the distances to the nearest high-diversity region and to the equator (Table 9.3). These variables implicate strong geographic influences on local richness which appear to be greater than that due purely to regional enrichment, but less than that due to local environmental variation (Karlson and Cornell 1998). In depauperate regions, the best predictors of local richness were the regional number of species (Table 8.3) and genera (Table 9.3). As the regression coefficients associated with these two large-scale variables are quite comparable and those associated with all other variables are either insignificant or relatively weak, we infer that there is strong regional enrichment in these regions.

In summary, our analysis of the published literature suggests that the local richness patterns of coral assemblages are influenced by a variety of ecological, geographical, and historical phenomena. The following specific inferences are offered here not as conclusions, but as predictions to be tested empirically:

• Local and regional factors influence local richness simultaneously. They do not represent mutually exclusive effects.

• The relative magnitude of these effects varies with the regional setting and the geographic scale of analysis.

• Local richness is strongly influenced by local environmental factors everywhere except in the most depauperate regions.

• Coral assemblages throughout the Indo-Pacific are regionally enriched. Forthcoming evidence from the most speciose regions of the central Indo-Pacific may very well challenge this prediction.

• The magnitude of regional enrichment is inversely related to species richness (*i.e.*, assemblages in depauperate regions appear to be quite open to the regional pool of species). This may reflect the influence of regional disturbances on depauperate faunas (*e.g.*, as in the eastern Pacific) or other factors promoting immigration and colonization processes. In addition, the resistance of resident species to invasions by non-residents (or exclusion due to other biological interactions like predation) may contribute to the limited membership of assemblages in speciose regions.

• The influence of evolutionary history and other geographical and historical phenomena on local richness patterns are strongest in speciose regions of the Indo-Pacific where local environmental factors also strongly influence local richness. This implicates a high degree of niche partitioning within local assemblages in accordance with conventional theory, yet also suggests that large-scale regional factors play an important role in determining local richness patterns.

Perhaps, the most general inference we can make regarding coral assemblages is that the relative influence of local and regional factors varies from place to place and with the scale of the analysis. Particularly relevant to community ecologists is the notion that observational and experimental studies conducted at the local scale may yield quite different results depending on the regional setting (see section 1.4.1 for rocky intertidal examples). While it has been generally appreciated that there may be strong environmental influences on local assemblages studied within regions [*e.g.*, across the continental shelf on the Great Barrier Reef (section 2.3, Williams and Hatcher 1983)], variation among regions or provincial faunas is only just beginning to be appreciated [*e.g.*, among freshwater fishes (Hugueny *et al.* 1997)]. As we explore ecological communities across these enormous scales, it will be necessary to progress beyond the use of our simplest community attribute, species richness. We will need to explore more detailed patterns of community structure and the processes which generate them. In particular, I reiterate the message from chapter 8 that we will need to consider the actual species composition of coral assemblages and relative abundance patterns of these species (Wiens 1989). This is exactly the kind of information Pandolfi (1996) used to infer limited membership in Pleistocene coral assemblages (section 8.2.4). In addition to more thoroughly describing the structure of coral assemblages, we should also begin to evaluate the putative explanations for limited membership and regional enrichment (see chapter 8). These invoke evolutionary phenomena influencing habitat and resource utilization patterns, strong biological interactions, various sources of spatiotemporal heterogeneity at the local scale, and a range of ecologically important attributes which may vary among clades with different evolutionary histories (*i.e.*, attributes influencing dispersal, colonization, and competition).

9.3 A biogeographical context for coral community ecology

As a consequence of recognizing that ecological assemblages differ across large biogeographic scales and evolutionary history can influence both local and regional richness patterns, it is appropriate to explore the faunal differences among the major provinces. For the community ecologist, this means returning to the biogeographical literature to develop a context for studying community dynamics. Here I address the following simple questions: What are the major faunal differences among provinces and how does evolutionary history differ among them? For the sake of brevity, I refer mostly to corals below. Many different taxa exhibit similar patterns of distribution, richness, and endemicity among provinces, yet one cannot deny that every taxon has a unique evolutionary history. In that this is certainly not an exhaustive treatment of the subject and it is well outside my own areas of expertise, I encourage the interested reader to explore the biogeographical literature directly. Since I began the development of the conceptual framework for this book by considering the influence of scale on the analysis of community dynamics (section 1.3), it is fitting that these final sections place coral communities in this large-scale context.

Virtually all analyses of the biogeographical affinities of corals clearly recognize the distinction between the Atlantic and Indo-Pacific faunas (Stehli and Wells 1971, Rosen 1988b, Veron 1995, Paulay 1997). This emphasizes the importance of the Isthmus of Panama as a major geographic barrier and the complete isolation of these faunas 3.5 my ago (Jones and Hasson 1985, Coates *et al.* 1992). Two other recognized barriers further subdivide the shallow-water fauna of the tropics. "The East Pacific Barrier is the formidable stretch of deep water that lies between Polynesia and America" isolating the East Pacific Region from the Indo-Pacific Region (Briggs 1974). Likewise, "the Mid-Atlantic Barrier is the broad, deep-water region that separates the Western Atlantic tropics from those of the West African coast" (Briggs 1974). These are designated as the West and East Atlantic Regions, respectively (Briggs 1974, Paulay 1997). Although each of these four regions are "characterized by having largely endemic biotas" (Paulay 1997), finer-scale provincial distinctions have also been recognized on the basis of endemicity among all the major taxa (Briggs 1974) and among corals alone (Rosen 1988b, Veron 1995). Briggs (1974) subdivided the Indo-West Pacific Region into eight faunal provinces (Western Indian Ocean, Red Sea, Indo-Polynesian, Northwestern Australia, Lord Howe-Norfolk, Hawaiian, Easter Island, Marquesas) and further recognized three provinces in the East Pacific Region (Mexican, Panamanian, Galápagos). In the Atlantic Ocean, he recognized the Caribbean, Brazilian, West Indian Provinces in the west and the West African and St. Helena-Ascension Provinces in the east.

More recent analyses of the geographic distributions of corals match some of the provincial distinctions recognized by Briggs (1974), while others do not. Using parsimony analysis of endemicity among coral genera, Rosen (1988b) separated the Atlantic fauna into Caribbean and eastern Atlantic faunas and the Indo-Pacific into four subregions (Indo-West Pacific, West Indian Ocean, Mid Pacific, and East Pacific). Using the geographic ranges of genera, Veron (1995) recognized four distinct provinces in the Atlantic (Gulf of Mexico, Caribbean, Brazil, Eastern Atlantic) and six in the Indo-Pacific (Indo-West Pacific, Central Pacific, Southeast Pacific, Far Southeast Pacific, Far East Pacific, Hawaiian Islands).

If one ignores all the problems associated with taxonomic uncertainties and poor sampling, patterns of endemicity can also be examined at the species level. Most coral species in the highly diverse central Indo-Pacific are widely distributed both latitudinally and longitudinally over much of the Indo-Pacific (Veron 1995). Nevertheless, regional endemism characterizes up to 21% of the central Indo-Pacific fauna (> 100 species). However, "single regions within the central Indo-Pacific have less than 10 per cent endemic coral species" and most regions have less than 5% (Veron 1995). Patterns of endemism are much more variable in other Indo-Pacific provinces. A high level of endemism was noted among 46 species in the Hawaiian Islands, low levels in the Far East Pacific (with only one endemic species), and no endemics at all among 29 species at Johnston Atoll in the Central Pacific. Johnston Atoll is one of the world's most isolated atolls (Maragos and Jokiel 1986), yet the corals there have strong affinities with those in the western Pacific (Veron 1995). In the West Atlantic, endemism is generally absent among Caribbean corals because species distributions are relatively uniform throughout the region (see section 8.2.3). In contrast, the few corals occurring off Brazil (19 species) are highly endemic. These variable patterns of endemism among corals appear to emerge as a direct consequence of different dispersal regimes and degrees of isolation among biogeographic regions and provinces.

The West Atlantic

Among the four major tropical regions of the world recognized by Briggs (1974), the West Atlantic is unusual in terms of its history of regional extinctions. Near the Oligocene-Miocene boundary (24 mya), the Indo-Pacific and Atlantic faunas began to separate with the rise of the isthmus which would eventually connect North and South America (Jones and Hasson 1985). During the Miocene 16-24 mya, approximately half of the genera in the Caribbean went regionally extinct presumably as a consequence of changes in oceanic currents, temperature regimes, and other environmental factors (Edinger and Risk 1994, 1995). Yet despite these extinctions, "Pliocene reef communities were dominated by *Stylophora*, *Goniopora*, and a suite of agariciid and poritid species that more closely resemble modern Indo-Pacific species than modern Caribbean species" (Budd *et al.* 1996). Following the complete closure of the isthmus during the Pliocene, rates of both extinction and origination of coral species accelerated. Over 70% of Pliocene coral species (49-53 species) went extinct during the Plio-Pleistocene (1-4 mya), while over half of the species now living in the Caribbean originated during this interval (Budd *et al.* 1994, 1996, Johnson *et al.* 1995). Thus the dominance of a few species of *Acropora* and *Montastrea* in Caribbean coral communities today (sections 2.2.6 and 3.4) dates back only into the mid-Pleistocene (Jackson 1992).

At the present time, it is uncertain why both origination and extinction rates accelerated in the Caribbean during the Plio-Pleistocene. "Although the overall pattern of faunal change was generally punctuated, it cannot be ascertained whether extinction and origination events were steady or occurred in pulses" (Budd *et al.* 1996). Furthermore, it is not clear whether these two processes are linked. Budd *et al.* (1996) suggested that better evidence from successional studies are needed to evaluate the role of biotic factors on this high rate of species turnover. However, it appears that "a number of interrelated regional environmental agents" were directly responsible (namely fluctuations in temperature, salinity, siltation, and nutrients). Faunal turnover rates were reported to be geographically patchy and variable among habitats. "Extinction was heaviest in seagrass communities" relative to that noted for reefal communities. A full understanding of these events would appear to require knowledge of the regional environmental setting, ecologically important life history attributes of corals (especially those influencing dispersal capabilities at regional scales), their tolerances to extreme environmental conditions, and local variation among habitats in terms of both species composition and the physical environment.

A similar comment can be made regarding coral communities in the Caribbean today. Understanding potentially catastrophic consequences of natural and anthropogenic disturbances on these communities (see section 3.3.1 and chapter 7) will require knowledge of dynamic processes at the regional scale. Sustainable population dynamics for many species (corals and non-corals alike) as well as the resilience of coral communities partially depends on dispersal capabilities and recruitment from regional sources. "The great majority of species have a dispersive pelagic larval stage, and many also disperse as eggs" (Roberts 1997). Furthermore, some major sources of mortality like the disease which killed *Diadema antillarum* also disperse over regional scales. Roberts (1997) used the major current patterns in the Caribbean to estimate the scale of the interconnections among 18 coral reef sites in the Caribbean. Maps of one and two-month "transport envelopes" were used to approximate the total reef area available to passively dispersed propagules upstream and downstream of each location. At some locations characterized by strong "current fields", 1000-km dispersal distances were attainable in a single month. All but a few locations within the Caribbean would appear to have large upstream sources of immigrants (Roberts 1997). Sites in the Bahamas and Florida Keys were estimated to have the largest available pool of immigrants based on the area of upstream coral reefs, while Banco de Serranilla and Barbados had the smallest. Likewise, sites representing the largest potential sources of immigrants with a large area of downstream coral reefs included Banco de Serranilla and Mexico. The smallest downstream reef area was reported for the Flower Gardens in the Gulf of Mexico and Bonaire.

Given the large scale of these interconnections, it is clear that any evaluation of the dynamics of Caribbean coral communities should include consideration of dispersal at the regional scale. Coral communities at all but a few "self-seeded" locations [*e.g.*, Barbados (Hunte and Younglao 1988, Roberts 1997)] may be unstable in the absence of a supply of larvae from upstream sources. In accordance with this "supply-side" notion, the resilience of these communities is predicted to vary with the availability of potential immigrants from the regional pool (see sections 1.4.1 and 3.1). Although some have questioned the details of the passive dispersal model used by Roberts (1997) to characterize the immigrant pool of particular sites (Bellwood *et al.* 1998, Sale and Cowen 1998), Roberts (1998) defended its use as a management tool based on its broad general features. The regional stability of coral reef communities may very well depend on large-scale dispersal among individual reefs.

The East Pacific

Although the East Pacific Region "is biologically very diverse, little of this diversity is associated with reef communities" (Paulay 1997). The shallow-water biota other than reef-associated organisms is generally "endemic and distinctive". In contrast, many species comprising the reef biota of the East Pacific Region share affinities with the Indo-West Pacific and endemism is quite low. However, the genera representing the impoverished coral fauna of this region are distinctive being dominated by *Pavona* and *Pocillopora* (Table 9.6). The low coral diversity has been attributed to 1) the extreme isolation of the region, 2) oceanographic conditions generally limiting coral reef development while favoring upwelling of cool, nutrient-rich water, and 3) relatively frequent major environmental disturbances associated with El Niño events (section 7.2.4). Furthermore, Paulay (1997) noted the importance of historical events on current regional patterns of coral species composition, richness, and endemism. In particular, he noted affinities between the extant eastern Pacific fauna and "the older amphi-American fauna" throughout "much of the Cenozoic". With the closure of the Isthmus of Panama, there was "even greater turnover in the reef biota of the East Pacific than in the West Atlantic". Nevertheless, some extinct Caribbean genera "are important components of modern eastern Pacific reefs" (Budd *et al.* 1994).

The regional setting in the East Pacific appears to severely restrict the development of speciose coral communities so much so that regional endemism is extremely low. Most extant species are thought to be very widely distributed as a consequence of exceptional dispersal capabilities. Consequently, the influence of dispersal on community dynamics may very well provide a suitable focus for community studies in this region. I reiterate three points made previously as support for this view:

• Oceanographic transport mechanisms are thought to be primarily responsible for a latitudinal gradient in recruitment rates of intertidal organisms extending from the Gulf of California to Panama (section 1.4.1). Recruitment limitation is thought to be most severe along the Panamanian coast. Given this large spatial scale, recruitment limitation may also have significant effects on coral community structure in this same region. As noted in chapter 8, the degree of this limitation can influence local communities by restricting the number of species reaching a site and by diminishing the importance of competitive exclusion.

• Local upwelling was thought to contribute to finer scale differences in the decapod crustacean fauna associated with pocilloporid corals in Panama (section 8.2.2). Although predictable patterns of upwelling might very well contribute to predictable recruitment patterns, Abele (1979) actually speculated that upwelling generates unpredictable disturbances to this fauna and represents a significant source of spatiotemporal heterogeneity. Thus this fauna may be undersaturated in accordance with the view that recruitment processes limit the richness of local communities (Roughgarden 1989) or oversaturated relative to available niche space due to disturbance (Cornell and Lawton 1992).

• The analysis of local coral richness patterns in depauperate regions like the East Pacific indicates that these communities are regionally enriched and thus also implicates the potential importance of oversaturation (sections 8.2.3 and 9.2.2). This may seem counter-intuitive in that these are not speciose regions, yet the relative magnitude of regional enrichment in coral assemblages is 2-3 times stronger in depauperate regions than in the speciose, central Indo-Pacific (based on the regression coefficients presented in Table 8.3).

The Indo-West Pacific

This enormously large region is by far the most diverse. Over 80% of all living coral species and genera occur in the Indo-West Pacific (Table 9.6). Among living corals, the most speciose family in the region is the Acroporidae with at least 248 species represented by four genera (Veron 1995). In terms of the number of living species per genus, the coral fauna is dominated by six genera (*Acropora*, *Montipora*, *Porites*, *Goniopora*, *Fungia*, and *Favia*). These include over half of all living coral species in the Indo-West Pacific (Table 9.6). *Goniopora* and *Favia* are old genera having originated in the middle Cretaceous, while *Acropora*, *Montipora*, and *Porites* date back to the Eocene (Jokiel and Martinelli 1992). However, the average age of genera in different regions of the Indo-West Pacific is only 23-44 my in contrast to an average age of 47-127 my for Atlantic corals (Stehli and Wells 1971). This is indicative of the young age of many less speciose genera in the Indo-West Pacific. Twenty three genera are known only from the Recent; they are unknown in the fossil record. Five genera date back into the Plio-Pleistocene and 13 genera including the speciose genus *Fungia* date back into the Miocene (5-25 mya) (Jokiel and Martinelli 1992). Although "no fossil sequences comparable with the Caribbean are known from the Indo-Pacific", Veron (1995) estimated that the species turnover rates in the latter are approximately half of those reported for the Neogene in the Caribbean by Budd *et al.* (1994). Species turnover rates among all regions are estimated at 4-10% per million years (Veron 1995), yet "no extant coral genus is known to have undergone regional extinction" in the Indo-West Pacific (Paulay 1997).

Within the Indo-West Pacific, the highly speciose region encompassing the Philippines, Malaysia, and Indonesia (section 2.2.1) can be contrasted with marginal areas with much lower diversity. The latter have been long recognized as regions where environmental factors become limiting (Wells 1954) and there appear to be only "minor differences in the order in which genera drop out" along these gradients (Stehli and Wells 1971). The Persian Gulf is one example in which corals must contend with high salinity and algal competition (*e.g.*, Coles 1988). High-latitude coral communities represent another in which low temperatures and competition with algae and other sessile invertebrates become severe problems (section 2.2.7). In that the regional species richness of corals is likely to be strongly correlated with these large-scale environmental gradients, perhaps these regions may provide opportunities to test the notions of recruitment limitation and regional enrichment as suggested above for the East Pacific. Within the highly speciose Indo-West Pacific, the focus should shift to jointly consider local niche-based explanations of community patterns with the regional perspective emphasizing geographical and historical phenomena.

TABLE 9.6 Zooxanthellate Scleractinian corals of the Indo-West Pacific (IWP), East Pacific (EP), West Atlantic (WA), and East Atlantic (EA) Regions. Twenty two genera known from amphi-American fossils predate the isolation of the EP from the WA and are indicated below under WA. Likewise, fifty genera known from fossils in Europe, Africa, and islands of the EA predate the isolation of the EA from the proto-IWP and are indicated below as part of the west Tethyan (WT) fauna (after Paulay 1997).

	IWP	EP	WA	EA	WT
Number of living genera occurring in modern assemblages[1]	91	13	25	8	-
Approximate number of living species	700[2]	30[1]	65[1]	14[1]	
Number of living genera known only as fossils in a region[1]	0	-	22	-	50
Family and genus	Approximate minimum number of living species in selected speciose genera				Global Total
Acroporidae					
Acropora	147	0	3[3,4]	0	150[2]
Montipora	80	0	0	0	80[2]
Poritidae					
Porites	73-75	3[5]	5[3,4]	3[6]	80[2]
Goniopora	30	0	0	0	30[2]
Fungiidae					
Fungia	33	0	0	0	33[2]
Faviidae					
Favia	27	0	3[3,4,6]	2[6]	30[2]
Agariciidae					
Pavona	22	7[2]	0	0	22[2]
Pocilloporidae					
Pocillopora	9-10	6-7[2,5]	0	0	10[2]

[1] Paulay (1997), [2] Veron (1995), [3] Goreau and Wells (1967), [4] Wells and Lang (1973), [5] Glynn and Wellington (1983), [6] Laborel (1974)

"Since the Middle Miocene, the earth's climate has been in glacial mode" (Veron 1995) and the world's oceans have been exposed to major fluctuations in temperature, salinity, sediment-load, sea level, and current systems. At least seventeen glacial cycles dominated the global environment during the Pleistocene (Grigg and Epp 1989). The last major glaciation was only 18,000 years ago and was associated with a sea level minimum (low stand) estimated at 120-165 m below the present level. Between 18,000 and 9,000 years ago, sea level rose rapidly at a rate estimated at 10-12 mm yr^1. This rate "outpaced the ability of all reefs in the world submerged below critical depth (30 to 40 m or less) to build upward" (Grigg and Epp 1989). This rapid rise emphasizes the importance of larval transport processes on the recolonization of shallow water environments. The present high stand has been stable for only about 6000 years! Veron (1995) speculated that the high diversity center of the Indo-West Pacific "was much more affected by glacio-eustatic changes than other parts of the world". This is because of the proximity of continental land masses and aerial exposure of the large continental shelves. Although "the evolutionary consequences of sea level changes have generally been overstated" (Veron 1995), the associated shifts in "surface circulation" have undoubtedly been important in maintaining the "genetic continuity" of coral populations and the low level of endemism currently observed in the region.

The consequences of Pleistocene glacial cycles on the structure and dynamics of coral communities in the Indo-West Pacific appears to be open to question. The fossil assemblages within the local habitats examined by Pandolfi (1996) in Papua New Guinea were remarkably stable over 95 ky of the late Pleistocene (section 8.2.4). Sea level fluctuated extensively over this time period (Potts 1984). Nevertheless, Pandolfi (1996) concluded that "the fossil reefs are characterized by preservation of the same coral zonation patterns and dominant taxa over the same range of reef environments observed on the adjacent living reef". Thus he stressed the ecological stability of local assemblages in the face of radical changes in the global environment. This provides a fascinating counter point to the notion that ecological communities are nonequilibrial systems (section 3.4). Is there evidence to refute the idea that Indo-West Pacific coral assemblages are so stable? Might not coral communities be just as open and flexible as most other communities exposed to Pleistocene climate changes (see Valentine and Jablonski 1993)? Is there any evidence supporting the notion that species assemblages were "reshuffled" in response to glacial cycles?

9.4 Closing comments

I offer three relevant final points as food for thought in this ongoing development of a biogeographical context for coral community ecology.

• Analysis of the published literature indicates that coral assemblages throughout the Indo-Pacific and even in the most speciose regions of the Indo-West Pacific are regionally enriched. Thus they appear to be open to regional sources of immigration (Karlson and Cornell 1998). This result needs confirmation. It should be tested in the field using standardized methods and balanced sampling designs.

• The degree of enrichment appears to be much reduced in speciose regions.

Perhaps the notion of limited membership applies to all species in these rich assemblages (Pandolfi 1996). Alternatively, the membership in these assemblages may be only partially limited. Further exploration of this "middle ground" should include the simultaneous evaluation of local and regional factors influencing the structure of these ecological communities.

• For the community ecologist, understanding the within and between-habitat components of variation in community structure is of fundamental importance. While such an understanding may not be essential when considering purely biogeographical phenomena (Veron 1995), the integration of local and regional perspectives in community ecology requires closer attention to local factors (*e.g.*, physical factors, the recruitment regime, and biological interactions) as well as to larger-scale processes.

Veron (1995) made the following generalizations regarding variation in the abundance of individual coral species between habitats:

"Species occurring in a particular habitat in one region may be absent from similar habitats in another. Species that are almost always associated with a particular type of habitat in one region may be found in a completely different habitat in another. Species that usually occur in a wide range of habitats may, in some regions, be restricted to one habitat type".

In that these habitat associations are geographically variable, the causal factors restricting them are likely to include both large-scale and local phenomena. Perhaps some of this variation is due to "reshuffling" in response to Pleistocene glacial cycles. Constraints due to oceanographic transport processes may also be important. At the local scale, particular biological interactions may strongly control species composition. Yet spatial variation in these interactions may contribute significantly to the observed variation in habitat associations. It is unlikely that simple explanations will suffice, but the challenge is to blend the patently ecological phenomena with those operating at larger scales.

REFERENCES

Abele, L.G. (1974) Species diversity of decapod crustaceans in marine habitats. *Ecology* **55**, 156-161.

Abele, L.G. (1979) The community structure of coral-associated decapod crustaceans in variable environments, in *Ecological Processes in Coastal and Marine Systems*, (ed R.J. Livingston), Plenum Press, New York, pp. 265-287.

Abele, L.G. (1984) Biogeography, colonization, and experimental community structure of coral-associated crustaceans, in *Ecological Communities: Conceptual Issues and the Evidence*, (eds D.R. Strong, Jr., D. Simberloff, L.G. Abele and A.B. Thistle), Princeton University Press, Princeton, pp.123-137.

Abrams, P. (1987) Indirect interactions between species that share a predator: varieties of indirect effects, in *Predation: Direct and Indirect Impacts on Aquatic Communities*, (eds W.C. Kerfoot and A. Sih), University Press of New England, Hanover, pp. 38-54.

Abrams, R.W., Abrams, M.D. and Schein, M.W. (1983) Diurnal observations on the behavioral ecology of *Gymnothorax moringa* (Cuvier) and *Muraena miliaris* (Kaup) on a Caribbean coral reef. *Coral Reefs* **1**, 185-192.

Aerts, L.A.M. and van Soest, R.W.M. (1997) Quantification of sponge/coral interactions in a physically stressed reef-community, NE Columbia. *Marine Ecology Progress Series* **148**, 125-134.

Aho, J.M. (1990) Helminth communities of amphibians and reptiles: comparative approaches to understanding patterns and processes, in *Parasite Communities: Patterns and Processes*, (eds G.W. Esch, A.O. Bush and J.M. Aho), Chapman and Hall, London, pp. 157-195.

Aho, J.M. and Bush, A.O. (1993) Community richness in parasites of some freshwater fishes from North America, in *Species Diversity in Ecological Communities: Historical and Geographical Perspectives*, (eds R.E. Ricklefs and D. Schluter), The University of Chicago Press, Chicago, pp. 185-193.

Alino, P.M., Sammarco, P.W. and Coll, J.C. (1992) Competitive strategies in soft corals (Coelenterata, Octocorallia). IV. Environmentally induced reversals in competitive superiority. *Marine Ecology Progress Series* **81**, 129-145.

Andrewartha, H.G. and Birch, L.C. (1954) *The Distribution and Abundance of Animals*. The University of Chicago Press, Chicago.

Armstrong, R.A. (1976) Fugitive species: experiments with fungi and some theoretical considerations. *Ecology* **57**, 953-963.

Aronson, R.B. (1994) Scale-independent biological processes in the marine environment. *Oceanography and Marine Biology: an Annual Review* **32**, 435-460.

Attaway, D.H. and Ciereszko, L.S. (1974) Isolation and partial characterization of a Caribbean palytoxin. *Proceedings of the Second International Coral Reef Symposium* **1**, 497-504.

Austin, A.D., Austin, S.A. and Sale, P.F. (1980) Community structure of the fauna associated with the coral *Pocillopora damicornis* (L.) on the Great Barrier Reef. *Australian Journal of Marine and Freshwater Research* **31**, 163-174.

Bak, R.P.M. (1977) Coral reefs and their zonation in Netherlands Antilles. *The American Association of Petroleum Geologists Studies in Geology* **4**, 3-16.

Bak, R.P.M. (1985) Recruitment patterns and mass mortalities in the sea urchin *Diadema antillarum* *Proceedings of the Fifth International Coral Reef Congress, Tahiti* **5**, 267-272.

Bak, R.P.M., Carpay, M.J.E. and de Ruyter van Steveninck, E.D. (1984) Densities of the sea urchin *Diadema antillarum* before and after mass mortalities on the coral reefs of Curaçao. *Marine Ecology Progress Series* **17**, 105-108.

Bak, R.P.M. and Engle, M.S. (1979) Distribution, abundance and survival of juvenile hermatypic corals

(Scleractinia) and the importance of life history strategies in the parent coral community. *Marine Biology* **54**, 341-352.

Bak, R.P.M., Lambrechts, D.Y.M., Joenje, M., Nieuwland, G. and Van Veghel, M.L.J. (1996) Long-term changes on coral reefs in booming populations of a competitive colonial ascidian. *Marine Ecology Progress Series* **133**, 303-306.

Bak, R.P.M., Termaat, R.M. and Dekker, R. (1982) Complexity of coral interactions: influence of time, location of interaction and epifauna. *Marine Biology* **69**, 215-222.

Bak, R.P.M. and van Eys, G. (1975) Predation of the sea urchin *Diadema antillarum* Philippi on living coral. *Oecologia* **20**, 111-115.

Bakus, G.J. (1981) Chemical defense mechanisms on the Great Barrier Reef, Australia. *Science* **211**, 497-499.

Barnes, J., Bellamy, D.J., Jones, D.J. *et al.* (1971) Morphology and ecology of the reef front of Aldabra. *Symposia of the Zoological Society of London* **28**, 87-114.

Begon, M. and Mortimer, M. (1986) *Population Ecology: A Unified Study of Animals and Plants*, 2nd edn, Blackwell Scientific Publications, Oxford.

Bellwood, D.R., Leis, J.M. and Stobutzki, I.C. (1998) Fishery and reef management. *Science* **279**, 2021-2022.

Berquist, P.R. (1978) *Sponges*, University of California Press, Berkeley.

Birkeland, C. (1997) Implications for resource management, in *Life and Death of Coral Reefs*, (ed. C. Birkeland), Chapman and Hall, New York, pp. 411-435.

Birkeland, C. and Lucas, J.S. (1990) *Acanthaster planci: major management problem of coral reefs*, CRC Press, Boca Raton.

Birkeland, C., Randall, R.H., Wass, R.C. *et al.* (1987) Biological resource assessment of the Fagatele Bay National Marine Sanctuary. NOAA Technical Memorandum NOS MEMD **3**, U.S. Department of Commerce, Washington, D.C.

Blackburn, T.M. and Gaston, K.J. (1998) Some methodological issues in macroecology. *The American Naturalist* **151**, 68-83.

Bohnsack, J.A. and Talbot, F.H. (1980) Species-packing by reef fishes on Australian and Caribbean reefs: an experimental approach. *Bulletin of Marine Science* **30**, 710-723.

Bolser, R.C. and Hay, M.E. (1996) Are tropical plants better defended? Palatability and defenses of temperate vs. tropical seaweeds. *Ecology* **77**, 2269-2286.

Bouchon, C. (1978) Étude quantitative des peuplements a base de scléractiniaires d'un récif frangeant de l'ile de la Réunion (Océan Indien). Thése Doctorat de Spécialité Océanographie Université Aix-Marseille 2.

Bouchon, C. (1981) Quantitative study of the scleractinian coral communities of a fringing reef of Reunion Island (Indian Ocean). *Marine Ecology Progress Series* **4**, 273-288.

Bouchon, C. (1985) Quantitative study of scleractinian coral communities of Tiahura Reef (Moorea Island, French Polynesia). *Proceedings of the Fifth International Coral Reef Congress, Tahiti* **6**, 279-284.

Bouchon-Navaro, Y. (1980) Quantitative distribution of the Chaetodontidae on a fringing reef of the Jordanian coast (Gulf of Aqaba, Red Sea). *Tethys* **9**, 247-251.

Bouchon-Navaro, Y. (1981) Quantitative distribution of the Chaetodontidae on a reef of Moorea Island (French Polynesia). *Journal of Experimental Marine Biology and Ecology* **55**, 145-157.

Branch, G.M., Harris, J.M., Parkins, C. *et al.* (1992) Algal 'gardening' by marine grazers: a comparison of the ecological effects of territorial fish and limpets, in *Plant-Animal Interactions in the Marine Benthos*, (eds D.M. John, S.J. Hawkins and J.H. Price), The Systematics Association Special Volume No. 46, Clarendon Press, Oxford, pp. 405-423.

Branham, J.M., Reed, S.A., Bailey, J.H. and Caperon, J. (1971) Coral-eating sea stars *Acanthaster planci* in Hawaii. Science **172**, 1155-1157.

Brawley, S.H. (1992) Mesoherbivores, in *Plant-Animal Interactions in the Marine Benthos*, (eds D.M. John, S.J. Hawkins and J.H. Price), The Systematics Association Special Volume No. 46, Clarendon Press, Oxford, pp. 235-263.

Brawley, S.H. and Adey, W.H. (1977) Territorial behavior of threespot damselfish (*Eupomacentrus planifrons*) increases reef algal biomass and productivity. *Environmental Biology of Fishes* **2**, 45-51.

Briand, F. and Cohen, J.E. (1987) Environmental correlates of food chain length. *Science* **238**, 956-960.

Briggs, J.C. (1974) *Marine Zoogeography*, McGraw-Hill, New York.

Bright, T.J., Kraemer, G.P., Minnery, G.A. and Viada, S.T. (1984) Hermatypes of the Flower Garden banks, northwestern Gulf of Mexico: a comparison to other western Atlantic reefs. *Bulletin of Marine Science* **34**, 461-476.

Brown, B.E. (1997) Disturbances to reefs in recent times, in *Life and Death of Coral Reefs*, (ed C. Birkeland), Chapman and Hall, New York, pp. 354-379.

Brown, B.E., Le Tissier, M.D., Dunne, R.D. and Scoffin, T.P. (1993) Natural and anthropogenic disturbances on intertidal reefs of S.E. Phuket, Thailand 1979-1992, in *Proceedings of the Colloquium on Global Aspects of Coral Reefs: Health, Hazards and History*, (ed R.N. Ginsburg), Rosenstiel School of Marine and Atmospheric *Science*, University of Miami, Miami, pp. 278-285.

Brown, B.E., Le Tissier, M.D.A., Scoffin, T.P. and Tudhope, A.W. (1990) Evaluation of the environmental impact of dredging on intertidal coral reefs at Ko Phuket, Thailand, using ecological and physiological parameters. *Marine Ecology Progress Series* **65**, 273-281.

Brown, B.E. and Suharsono (1990) Damage and recovery of coral reefs affected by El Niño related seawater warming in the Thousand Islands, Indonesia. *Coral Reefs* **8**, 163-170.

Brown, J.H. (1995) *Macroecology*. The University of Chicago Press, Chicago.

Brown, J.H. and Maurer, B.A. (1989) Macroecology: the division of food and space among species on continents. *Science* **243**, 1145-1150.

Bryan, P.G. (1973) Growth rate, toxicity and distribution of the encrusting sponge *Terpios* sp. (Hadromerida: Suberitidae) in Guam, Marianas Islands. *Micronesica* **9**, 237-242.

Budd, A.F., Johnson, K.G. and Stemann, T.A. (1996) Plio-Pleistocene turnover and extinctions in the Caribbean reef coral fauna, in *Evolution and Environment in Tropical America*, (eds J.B.C. Jackson, A.F. Budd and A.G. Coates), University of Chicago Press, Chicago, pp. 168-204.

Budd, A.F., Stemann, T.A. and Johnson, K.G. (1994) Stratigraphic distributions of genera and species of Neogene to Recent Caribbean reef corals. *Journal of Paleontology* **68**, 951-977.

Bull, G.D. (1982) Scleractinian coral communities of two inshore high island fringing reefs at Magnetic Island, North Queensland. *Marine Ecology Progress Series* **7**, 267-272.

Buss, L.W. (1976) Better living through chemistry: the relationship between allelochemical interactions and competitive networks, in *Aspects of Sponge Biology* (eds F.W. Harrison and R.R. Cowden), Academic Press, New York, pp. 315-327.

Buss, L.W. (1979) Habitat selection, directional growth, and spatial refuges: why colonial animals have more hiding places, in *Biology and Systematics of Colonial Organisms*, (eds G. Larwood and B.R. Rosen), Academic Press, New York, pp. 459-497.

Buss, L.W. (1980) Competitive intransitivity and size-frequency distributions of interacting populations. *Proceedings of the National Academy of Sciences USA* **77**, 5355-5359.

Buss, L.W. (1986) Competition and community organization on hard surfaces in the sea, in *Community Ecology*, (eds J. Diamond and T.J. Case), Harper & Row, Publishers, Inc., New York, pp. 517-536.

Buss, L.W. and Jackson, J.B.C. (1979) Competitive networks: nontransitive competitive relationships in cryptic coral reef environments. *The American Naturalist* **113**, 223-234.

Buss, L.W. and Jackson, J.B.C. (1981) Planktonic food availability and suspension-feeder abundance: evidence of depletion. *Journal of Experimental Marine Biology and Ecology* **49**, 151-161.

Butler, A.J. (1991) Effect of patch size on communities of sessile invertebrates in Gulf St Vincent, South Australia. *Journal of Experimental Marine Biology and Ecology* **153**, 255-280.

Caley, M.J. (1997) Are local patterns of reef fish diversity related to patterns of diversity at a larger scale? *Proceedings of the Eighth International Coral Reef Symposium* **1**, 993-998.

Caley, M.J. and Schluter, D. (1997) The relationship between local and regional diversity. *Ecology* **78**, 70-80.

Carpenter, R.C. (1981) Grazing by *Diadema antillarum* (Philippi) and its effects on the benthic algal community. *Journal of Marine Research* **39**, 749-765.

Carpenter, R.C. (1986) Partitioning herbivory and its effects on coral reef algal communities. *Ecological Monographs* **56**, 345-363.

Carpenter, R.C. (1988) Mass mortality of a Caribbean sea urchin: Immediate effects on community metabolism and other herbivores. *Proceedings of the National Academy of Sciences USA* **85**, 511-514.

Carpenter, R.C. (1990) Mass mortality of *Diadema antillarum* I. Long-term effects on sea urchin population dynamics and coral reef algal communities. *Marine Biology* **104**, 67-77.

Carpenter, R.C. (1997) Invertebrate predators and grazers, in *Life and Death of Coral Reefs*, (ed C. Birkeland), Chapman and Hall, London, pp. 198-229.

Caswell, H. (1976) Community structure: a neutral model analysis. *Ecological Monographs* **46**, 327-354.

Caswell, H. and Cohen, J.E. (1991) Disturbance and diversity in metapopulations. *Biological Journal of the Linnean Society* **42**, 193-218.

Caswell, H. and Cohen, J.E. (1993) Local and regional regulation of species-area relations: a patch-occupancy model, in *Species Diversity in Ecological Communities: Historical and Geographical Perspectives*, (eds R.E. Ricklefs and D. Schluter), The University of Chicago Press, Chicago, pp. 99-107.

Chapin III, F.S., Walker, B.H., Hobbs, R.J., Hooper, D.U. *et al.* (1997) Biotic control over the functioning of ecosystems. *Science* **277**, 500-504.

Chaplin, C.C.G. and Scott, P. (1972) *Fishwatchers Guide to West Atlantic Coral Reefs*. Livingston Publishing Company, Wynnewood.

Chesser, R.H. (1969) Destruction of Pacific corals by the sea star *Acanthaster planci*. *Science* **165**, 280-283.

Chesson, P.L. and Case, T.J. (1986) Overview: Nonequilibrium community theories: chance, variability, history, and coexistence, in *Community Ecology*, (eds J. Diamond and T.J. Case), Harper and Row, New York, pp. 229-239.

Chia, F.-S. and Rice, M.E. (eds) (1978) *Settlement and Metamorphosis of Marine Invertebrate Larvae*, Elsevier, New York.

Chiappone, M. and Sullivan, K.M. (1996) Distribution, abundance and species composition of juvenile scleractinian corals in the Florida Reef Tract. *Bulletin of Marine Science* **58**, 555-569.

Choat, J.H. and Bellwood, D.R. (1991) Reef fishes: their history and evolution, in *The Ecology of Fishes on Coral Reefs*, (ed P.F. Sale), Academic Press, New York, pp. 39-66.

Chornesky, E.A. (1983) Induced development of sweeper tentacles on the reef coral *Agaricia agaricites*: a response to direct competition. *Biological Bulletin* **165**, 569-581.

Chornesky, E.A. (1989) Repeated reversals during spatial competition between corals. *Ecology* **70**, 843-855.

Clark, R.D. (1977) Habitat distribution and species diversity of chaetodontid and pomacentrid fishes near Bimini, Bahamas. *Marine Biology* **40**, 277-289.

Clements, F.E. (1936) Nature and structure of the climax. *Journal of Ecology* 24, 252-284.

Coates, A.G., Jackson, J.B.C., Collins, L.S., Cronin, T.M. *et al.* (1992) Closure of the Isthmus of Panama: the nearshore marine record of Costa Rica and western Panama. *Geological Society of America Bulletin*

104, 814-828.

Cody, M.L. (1993) Bird diversity components within and between habitats in Australia, in *Species Diversity in Ecological Communities: Historical and Geographical Perspectives*, (eds R.E. Ricklefs and D. Schluter), The University of Chicago Press, Chicago, pp. 147-158.

Cohen, J.E. (1978) *Food Webs and Niche Space*, Princeton University Press, Princeton.

Cohen, J.E. (1989a) Food webs and community structure, in *Perspectives in Ecological Theory*, (eds J. Roughgarden, R.M. May and S.A. Levin), Princeton University Press, Princeton, pp. 181-202.

Cohen, J.E. (1989b) Ecologist's Co-Operative Web Bank (ECOWeB). Version 1.0. Machine-readable data base of food webs. Rockefeller University, New York.

Cohen, J.E. (1994) Marine and continental food webs: three paradoxes? *Philosophical Transactions of the Royal Society of London, Series B, Biological Sciences* **343**, 57-69.

Cohen, J.E., Briand, F. and Newman, C.M. (1990) Community food webs: data and theory. *Biomathematics*, Vol. 20, Springer, Berlin.

Coles, S.L. (1988) Limitations on coral reef development in the Arabian Gulf: temperature or algal competition? *Proceedings of the Sixth International Coral Reef Symposium* **3**, 211-216.

Coley, P.D., Bryant, J.P. and Chapin III, F.S. (1985) Resource availability and plant antiherbivore defense. *Science* **230**, 895-899.

Colgan, M.W. (1982) Succession and recovery of a coral reef after predation by *Acanthaster planci* (L.). *Proceedings of the Fourth International Coral Reef Symposium* **2**, 333-338.

Colgan, M.W. (1987) Coral reef recovery on Guam (Micronesia) after catastrophic predation by *Acanthaster planci*. *Ecology* **68**, 1592-1605.

Colgan, M.W. (1990) El Niño and the history of eastern Pacific reef building, in *Global Ecological Consequences of the 1982-83 El Niño-Southern Oscillation*, (ed P.W. Glynn), Elsevier, Amsterdam, pp. 183-232.

Colin, P.I. (1978) *Caribbean Reef Invertebrates and Plants. A Field Guide to the Invertebrates and Plants Occurring on Coral Reefs of the Caribbean, the Bahamas and Florida*. T.F.H. Publications, Inc. Ltd., Hong Kong.

Coll, J.C., La Barre, S., Sammarco, P.W., Williams, W.T. *et al.* (1982) Chemical defences in soft corals (Coelenterata: Octocorallia) of the Great Barrier Reef: a study of comparative toxicities. *Marine Ecology Progress Series* **8**, 271-278.

Connell, J.H. (1961a) The influence of interspecific competition and other factors on the distribution of the barnacle *Chthamalus stellatus*. *Ecology* **42**, 710-723.

Connell, J.H. (1961b) Effects of competition, predation by *Thais lapillus*, and other factors on natural populations of the barnacle *Balanus balanoides*. *Ecological Monographs* **31**, 61-104.

Connell, J.H. (1972) Community interactions on marine rocky intertidal shores. *Annual Review of Ecology and Systematics* **3**, 169-192.

Connell, J.H. (1973) Population ecology of reef-building corals, in *Biology and Geology of Coral Reefs*, Volume II: Biology 1, (eds O.A. Jones and R. Endean), Academic Press, New York, pp. 205-245.

Connell, J.H. (1976) Competitive interactions and the species diversity of corals, in *Coelenterate Ecology and Behavior* (ed G.O. Mackie), Plenum Press, New York, pp. 51-58.

Connell, J.H. (1978) Diversity in tropical rain forests and coral reefs. *Science* **199**, 1302-1310.

Connell, J.H. (1980) Diversity and coevolution of competitors, or the ghost of competition past. *Oikos* **35**, 131-138.

Connell, J.H. (1983) On the prevalence and relative importance of interspecific competition: evidence from field experiments. *The American Naturalist* **122**, 661-696.

Connell, J.H. (1985) The consequences of variation in initial settlement vs. post-settlement mortality in rocky intertidal communities. *Journal of Experimental Marine Biology and Ecology* **93**, 11-45.

Connell, J.H. (1987) Change and persistence in some marine communities, in *Colonization, Succession and Stability*, (eds A.J. Cray, M.J. Crawley and P.J. Edwards), Blackwell Scientific Publications, Oxford., pp. 339-352.

Connell, J.H. (1997) Disturbance and recovery of coral assemblages. *Coral Reefs* **16**, S101-S113.

Connell, J.H., Hughes, T.P. and Wallace, C.C. (1997) A 30-year study of coral abundance, recruitment, and disturbance in space and time. *Ecological Monographs* **67**, 461-488.

Connell, J.H. and Keough, M.J. (1985) Disturbance and patch dynamics of subtidal marine animals on hard substrata, in *The Ecology of Natural Disturbance and Patch Dynamics*, (eds S.T.A. Pickett and P.S. White), Academic Press, Orlando, pp. 125-151.

Connell, J.H. and Orias, E. (1964) The ecological regulation of species diversity. *The American Naturalist* **98**, 399-414.

Connell, J.H. and Slatyer, R.O. (1977) Mechanisms of succession in natural communities and their role in community stability and organization. *The American Naturalist* **111**, 1119-1144.

Connell, J.H. and Sousa, W.P. (1983) On the evidence needed to judge ecological stability and persistence. *The American Naturalist* **121**, 789-824.

Connor, E.F. and Simberloff, D.S. (1978) Species number and compositional similarity of the Galàpagos flora and avifauna. *Ecological Monographs* **48**, 219-248.

Connor, E.F. and Simberloff, D.S. (1979) The assembly of species communities: chance or competition? *Ecology* **60**, 1132-1140.

Cornell, H.V. (1985a) Local and regional richness of cynipine gall wasps on California oaks. *Ecology* **66**, 1247-1260.

Cornell, H.V. (1985b) Species assemblages of cynipid gall wasps are not saturated. *The American Naturalist* **126**, 565-569.

Cornell, H.V. and Lawton, J.H. (1992) Species interactions, local and regional processes, and limits to the richness of ecological communities: a theoretical perspective. *Journal of Animal Ecology* **61**, 1-12.

Cornell, H.V. and Karlson, R.H. (1996) Species richness of reef-building corals determined by local and regional processes. *Journal of Animal Ecology* **65**, 233-241.

Cornell, H.V. and Karlson, R.H. (1997) Local and regional processes as controls of species richness, in *Spatial Ecology: The Role of Space in Population Dynamics and Interspecific Interactions*, (eds D. Tilman and P. Kareiva), Princeton University Press, Princeton, pp. 250-268.

Cox, E.F. (1986) The effects of a selective corallivore on growth rates and competition for space between two species of Hawaiian corals. *Journal of Experimental Marine Biology and Ecology* **101**, 161-174.

Cox, G.W. and Ricklefs, R.E. (1977) Species diversity and ecological release in Caribbean land bird faunas. *Oikos* **28**, 113-122.

Dana, T.F. (1979) Species-number relationships in an assemblage of reef-building corals: McKean Island, Phoenix Islands. *Atoll Research Bulletin* **228**, 1-27.

Davies, P.S. (1991) Effect of daylight variations on the energy budgets of shallow-water corals. *Marine Biology* **108**, 137-144.

Davies, P.S. (1992) Endosymbiosis in marine cnidarians, in *Plant-Animal Interactions in the Marine Benthos*, (eds D.M. John, S.J. Hawkins and J.H. Price), The Systematics Association Special Volume No. 46, Clarendon Press, Oxford, pp. 511-540.

Davies, P.S., Stoddart, D.R. and Sigee, D.C. (1971) Reef forms of Addu Atoll, Maldive Islands. *Symposia of the Zoological Society of London* **28**, 217-259.

Davis, M.B. (1986) Climatic instability, time lags, and community disequilibrium, in *Community Ecology*, (eds J. Diamond and T.J. Case), Harper and Row, New York, pp. 269-284.

Dawah, H.A., Hawkins, B.A. and Claridge, M.F. (1995) Structure of the parasitoid communities of grass-feeding chalcid wasps. *Journal of Animal Ecology* **64**, 708-720.

Dayton, P.K. (1971) Competition, disturbance and community organization: the provision and subsequent utilization of space in a rocky intertidal community. *Ecological Monographs* **41**, 351-389.

Dayton, P.K. (1972) Toward an understanding of community resilience and the potential effects of enrichments to the benthos at McMurdo Sound, Antarctica, in *Proceedings of the Colloquium on Conservation Problems in Antarctica,* (ed B.C. Parker), Allen Press, Lawrence, pp. 81-95.

Dayton, P.K. (1985a) The structure and regulation of some South American kelp communities. *Ecological Monographs* **55**, 447-468.

Dayton, P.K. (1985b) Ecology of kelp communities. *Annual Review of Ecology and Systematics* **16**, 215-245.

Dayton, P.K. (1989) Interdecadal variation in an Antarctic sponge and its predators from oceanographic climate shifts. *Science* **245**, 1484-1486.

Dayton, P.K. (1994) Community landscape: scale and stability in hard bottom marine communities, in *Aquatic Ecology: Scale, Pattern and Process,* (eds P.S. Giller, A.G. Hildrew and D.G. Raffaelli), Blackwell Scientific Publications, Oxford, pp. 289-332.

Dayton, P.K., Currie, V., Gerrodette, T. *et al.* (1984) Patch dynamics and stability of some California kelp communities. *Ecological Monographs* **54**, 253-289.

Dayton, P.K. and Hessler, R.R. (1972) Role of biological disturbance in maintaining diversity in the deep sea. *Deep-Sea Research* **19**, 199-208.

Dayton, P.K., Robilliard, G.A. and DeVries, A.L. (1969) Anchor ice formation in McMurdo Sound, Antarctica, and its biological effects. *Science* **163**, 273-274.

Dayton, P.K., Robilliard, G.A., Paine, R.T. and Dayton, L.B. (1974) Biological accommodation in the benthic community at McMurdo Sound, Antarctica. *Ecological Monographs* **44**, 105-128.

Dayton, P.K. and Tegner, M.J. (1984a) Catastrophic storms, El Niño, and patch stability in a southern California kelp community. *Science* **224**, 283-285.

Dayton, P.K. and Tegner, M.J. (1984b) The importance of scale in community ecology: a kelp forest example with terrestrial analogs, in *A New Ecology: Novel Approaches to Interactive Systems,* (eds P.W. Price, C.N. Slobodchikoff and W.S. Gaud), John Wiley & Sons, Inc., New York, pp. 457-481.

Dayton, P.K., Tegner, M.J., Parnell, P.E. *et al.* (1992) Temporal and spatial patterns of disturbance and recovery in a kelp forest community. *Ecological Monographs* **62**, 421-445.

Dayton, P.K., Tegner, M.J., Edwards, P.B. and Riser, K.L. (1998) Sliding baselines, ghosts, and reduced expectations in kelp forest communities. *Ecological Applications,* **8**, 323-325.

De Angelis, D.L. (1975) Stability and connectance in food web models. *Ecology* **56**, 238-243.

den Boer, P.J. (1968) The spreading of risk and stabilization of animal numbers. *Acta Biotheoretica* **18**, 165-194.

den Boer, P.J. (1991) Seeing the trees for the wood: random walks or bounded fluctuations of population size? *Oecologia* **86**, 484-491.

den Hartog, J.C. (1977) The marginal tentacles of *Rhodactis sanctithomae* (Corallimorpharia) and the sweeper tentacles of *Montastrea cavernosa* (Scleractinia); their cnidom and possible function. *Proceedings of the Third International Coral Reef Symposium* **1**, 464-469.

Diaz, H.F. and Markgraf, V. (eds) (1992) *El Niño. Historical and Paleoclimatic Aspects of the Southern Oscillation,* Cambridge University Press, Cambridge.

Diamond, J.M. (1978) Niche shifts and the rediscovery of interspecific competition. *American Scientist* **66**, 322-331.

Dobzhansky, T. (1950) Evolution in the tropics. *American Scientist* **38**, 208-221.

Dodge, R.E., Logan, A. and Antonius, A. (1982) Quantitative reef assessment studies in Bermuda: a comparison of methods and preliminary results. *Bulletin of Marine Science* **32**, 745-760.

Doherty, P.J. (1991) Spatial and temporal patterns in recruitment, in *The Ecology of Fishes on Coral Reefs*, (ed P.F. Sale), Academic Press, New York, pp. 261-293.

Doherty, P.J. and Williams, D.McB. (1988) The replenishment of coral reef fish populations. *Oceanography and Marine Biology: an Annual Review* **32**, 487-551.

Dollar, S.J. and Tribble, G.W. (1993) Recurrent storm disturbance and recovery: a long-term study of coral communities in Hawaii. *Coral Reefs* **12**, 223-233.

Done, T.J. (1977) A comparison of units of cover in ecological classifications of coral communities. *Proceedings of the Third International Coral Reef Symposium* **1**, 10-14.

Done, T.J. (1982) Patterns of the distribution of coral communities across the Central Great Barrier Reef. *Coral Reefs* **1**, 95-107.

Done, T.J. (1983) Coral zonation: its nature and significance, in *Perspectives on Coral Reefs*, (ed D.J. Barnes), The Australian Institute of Marine Science by Brian Clouster, Manuka, pp. 107-147.

Done, T.J. (1985) Effects of two *Acanthaster* outbreaks on coral community structure. The meaning of devastation. *Proceedings of the Fifth International Coral Reef Congress, Tahiti* **5**, 315-320.

Done, T.J. (1992) Constancy and change in some Great Barrier Reef coral communities: 1980-1990. *American Zoologist* **32**, 655-662.

Drake, J.A. (1990) The mechanics of community assembly and succession. *Journal of Theoretical Biology* **147**, 213-233.

Edinger, E.N. and Risk, M.J. (1994) Oligocene-Miocene extinction and geographic restriction of Caribbean corals: roles of turbidity, temperature, and nutrients. *Palaios* **9**, 576-598.

Edinger, E.N. and Risk, M.J. (1995) Preferential survivorship of brooding corals in a regional extinction. *Paleobiology* **21**, 200-219.

Edmunds, P.J. and Davies, P.S. (1986) An energy budget for *Porites porites* (Scleractinia). *Marine Biology* **92**, 339-347.

Elton, C. (1927) *Animal Ecology* (New impression with additional notes, 1935.), Sidgwick & Jackson, Ltd., London.

Elton, C. (1933) *The Ecology of Animals*, Methuen & Co. Ltd, London.

Elton, C. (1948) Competition and the structure of ecological communities. *Journal of Animal Ecology* **15**, 54-68.

Elton, C. and Miller, R.S. (1954) The ecological survey of animal communities: with a practical system of classifying habitats by structural characters. *Journal of Ecology* **42**, 460-496.

Endean, R. (1973) Population explosions of *Acanthaster planci* and associated destruction of hermatypic corals in the Indo-West Pacific region, in *Biology and Geology of Coral Reefs*, Volume II: Biology 1, (eds O.A. Jones and R. Endean), Academic Press, New York, pp. 389-438.

Endean, R. (1974) *Acanthaster planci* on the Great Barrier Reef. *Proceedings of the Second International Coral Reef Symposium* **1**, 563-576.

Estes, J.A. and Steinberg, P.D. (1988) Predation, herbivory, and kelp evolution. *Paleobiology* **14**, 19-36.

Fabricius, K.E., Benayahu, Y. and Genin, A. (1995) Herbivory in asymbiotic soft corals. *Science* **268**, 90-92.

Faulkner, D.J. (1993) Marine natural products. *Natural Products Report* **10**, 497-539.

Fauth, J.E., Bernardo J., Camara M., Resetarits, Jr., W. J. *et al.* (1996) Simplifying the jargon of community ecology: a conceptual approach. *The American Naturalist* **147**, 282-286.

Feeney, P.P. (1975) Biochemical evolution between plants and their herbivores, in *Coevolution of Animals and Plants*, (eds L.E. Gilbert and P.H. Raven), University of Texas Press, Austin, pp. 3-19.

Feeney, P.P. (1976) Plant apparency and chemical defense. *Recent Advances in Phytochemistry* **10**, 1-40.

Fischer, A.G. (1960) Latitudinal variations in organic diversity. *Evolution* **14**, 64-81.

Fisher, S.G. (1994) Pattern, process and scale in freshwater systems: some unifying thoughts, in *Aquatic Ecology: Scale, Pattern and Process*, (eds P.S. Giller, A.G. Hildrew and D.G. Raffaelli), Blackwell Scientific Publications, Oxford, pp. 575-591.

Fisk, D.A. and Done, T.J. (1985) Taxonomic and bathymetric patterns of bleaching in corals, Myrimidon Reef (Queensland). *Proceedings of the Fifth International Coral Reef Congress, Tahiti* **6**, 149-154.

Fowler, A.J. (1990) Spatial and temporal patterns of distribution and abundance of chaetodontid fishes at One Tree Reef, southern GBR. *Marine Ecology Progress Series* 64, 39-53.

Fox, J.F. (1979) Intermediate-disturbance hypothesis. *Science* **204**, 1344-1345.

Fraser, R.H. and Currie, D.J. (1996) The species richness-energy hypothesis in a system where historical factors are thought to prevail: coral reefs. *The American Naturalist* **148**, 138-159.

Gaines, S. and Roughgarden, J. (1985) Larval settlement rate: A leading determinant of structure in an ecological community of the marine intertidal zone. *Proceedings of the National Academy of Sciences USA* **82**, 3707-3711.

Gaines, S.D. and Roughgarden, J. (1987) Fish in offshore kelp forests affect recruitment to intertidal barnacle populations. *Science* **235**, 479-481.

Galzin, R. (1987) Potential fisheries of a Moorea fringing reef (French Polynesia) by the analysis of three dominant fishes. *Atoll Research Bulletin* **305**, 1-21.

Gaston, K.J. and Gauld, I.D. (1993) How many species of pimplines (Hymenoptera: Ichneumonidae) are there in Costa Rica? *Journal of Tropical Biology* **9**, 491-499.

Gause, G.F. (1935) *Vérifications Expérimentales de la Théorie Mathématique de la Lutte Pour la Vie.* Hermann et Cie, Paris.

Gerhart, D.J. (1984) Prostaglandin A_2: an agent of chemical defense in the Caribbean gorgonian *Plexaura homomalla. Marine Ecology Progress Series* **19**, 181-187.

Gerhart, D.J. (1986) Gregariousness in the gorgonian-eating gastropod *Cyphoma gibbosum*: tests of several possible causes. *Marine Ecology Progress Series* **31**, 255-263.

Giller, P.S., Hildrew, A.G. and Raffaelli, D.G. (eds) (1994) *Aquatic Ecology: Scale, Pattern and Process*, (The 34th Symposium of the British Ecological Society), Blackwell Scientific Publications, Oxford.

Ginsburg, R.N. (ed)(1993) *Proceedings of the Colloquium on Global Aspects of Coral Reefs: Health, Hazards and History.* Rosenstiel School of Marine and Atmospheric *Science*, University of Miami, Miami.

Gleason, H.A. (1917) The structure and development of the plant association. *Bulletin of the Torrey Botanical Club* **44**, 463-481.

Gleason, H.A. (1926) The individualistic concept of the plant association. *Bulletin of the Torrey Botanical Club* **53**, 7-26.

Gleibs, S., Mebs, D. and Werding, B. (1995) Studies on the origin and distribution of palytoxin in a Caribbean coral-reef. *Toxicon* **33**, 1531-1537.

Glynn, P.W. (1973) Aspects of the ecology of coral reefs in the western Atlantic region, in *Biology and Geology of Coral Reefs*, Volume II: Biology 1, (eds O.A. Jones and R. Endean), Academic Press, New York, pp. 271-324.

Glynn, P.W. (1974) Rolling stones among the Scleractinia: mobile coralliths in the Gulf of Panamá. *Proceedings of the Second International Coral Reef Symposium* **2**, 183-198.

Glynn, P.W. (1976) Some physical and biological determinants of coral community structure in the eastern

Pacific. *Ecological Monographs* **46**, 431-456.

Glynn, P.W. (1977) Interactions between *Acanthaster* and *Hymenocera* in the field and laboratory. *Proceedings of the Third International Coral Reef Symposium* **1**, 209-215.

Glynn, P.W. (1980) Defense by symbiotic crustacea of host corals elicited by chemical cues from predator. *Oecologia* **47**, 287-290.

Glynn, P.W. (1982) *Acanthaster* population regulation by a shrimp and a worm. *Proceedings of the Fourth International Coral Reef Symposium* **2**, 607-612.

Glynn, P.W. (1983) Crustacean symbionts and the defense of corals: coevolution on the reef?, in *Coevolution* (ed M.H. Nitecki), The University of Chicago Press, Chicago, pp. 111-178.

Glynn, P.W. (1985) El Niño-associated disturbance to Coral Reefs and post disturbance mortality by *Acanthaster planci. Marine Ecology Progress Series* **26**, 295-300.

Glynn, P.W. (1988a) Predation on coral reefs: some key processes, concepts and research directions. *Proceedings of the Sixth International Coral Reef Symposium* **1**, 51-62.

Glynn, P.W. (1988b) El Niño-Southern Oscillation 1982-1983: nearshore population, community, and ecosystem responses. *Annual Review of Ecology and Systematics* **19**, 309-345.

Glynn, P.W. (1990a) Feeding ecology of selected coral-reef macroconsumers: patterns and effects on coral community structure, in *Ecosystems of the World 25 Coral Reefs*, (ed Z. Dubinsky), Elsevier, Amsterdam, pp. 365-400.

Glynn, P.W. (ed)(1990b) *Global Ecological Consequences of the 1982-83 El Niño-Southern Oscillation*, Elsevier, Amsterdam.

Glynn, P.W. (1990c) Coral mortality and disturbances to coral reefs in the tropical eastern Pacific, in *Global Ecological Consequences of the 1982-83 El Niño-Southern Oscillation*, (ed P.W. Glynn), Elsevier, Amsterdam, pp. 55-126.

Glynn, P.W. and Colgan, M.W. (1992) Sporadic disturbances in fluctuating coral reef environments: El Niño and coral reef development in the eastern Pacific. *American Zoologist* **32**, 707-718.

Glynn, P.W. and De Weerdt, W.H. (1991) Elimination of two reef-building hydrocorals following the 1982-83 El Niño warming event. *Science* **253**, 69-71.

Glynn, P.W., Wellington, G.M. and Birkeland, C. (1979) Coral reef growth in the Galápagos: limitation by sea urchins. *Science* **203**, 47-49.

Glynn, P.W. and Wellington, G.M. (1983) *Corals and Coral Reefs of the Galápagos Islands*, University of California Press, Berkeley.

Glynn, P.W., Veron, J.E.N and Wellington, G.M. (1996) Clipperton Atoll (eastern Pacific): oceanography, geomorphology, reef-building coral ecology and biogeography. *Coral Reefs* **15**, 71-99.

Goldberg, W.M. (1973) The ecology of the coral-octocoral communities off the southeast Florida coast: geomorphology, species composition, and zonation. *Bulletin of Marine Science* **23**, 465-488.

Goldman, B. and Talbot, F.H. (1976) Aspects of the ecology of coral reef fishes, in *Biology and Geology of Coral Reefs,* Volume III: Biology 2, (eds O.A. Jones and R. Endean), Academic Press, New York, pp. 125-154.

Goldwasser, L. and Roughgarden, J. (1993) Construction of a large Caribbean food web. *Ecology* **74**, 1216-1233.

Goldwasser, L. and Roughgarden, J. (1997) Sampling effects and the estimation of food-web properties. *Ecology* **78**, 41-54.

Goreau, T.F. (1959) The ecology of Jamaican coral reefs. I. Species composition and zonation. *Ecology* **40**, 67-90.

Goreau, T.F. and Goreau, N.I. (1973) The ecology of Jamaican coral reefs. II. Geomorphology, zonation, and sedimentary phases. *Bulletin of Marine Science* **23**, 399-464.

Goreau, T.F. and Wells, J.W. (1967) The shallow-water scleractinia of Jamaica: revised list of species and their vertical distribution range. *Bulletin of Marine Science* **17**, 442-453.

Graham, R.W. (1986) Responses of mammalian communities to environmental changes during the late Quaternary, in *Community Ecology*, (eds J. Diamond and T.J. Case), Harper and Row, New York, pp. 300-313.

Grassle, J.F. (1973) Variety in coral reef communities, in *Biology and Geology of Coral Reefs*, Volume II: Biology 1, (eds O.A. Jones and R. Endean), Academic Press, New York, pp. 247-270.

Grassle, J.F. and Grassle, J.P. (1994) Notes from the abyss: the effects of a patchy supply of organic material and larvae on soft-sediment benthic communities, in *Aquatic Ecology: Scale, Pattern and Process*, (eds P.S. Giller, A.G. Hildrew and D.G. Raffaelli), Blackwell Scientific Publications, Oxford, pp. 499-515.

Green, A.L. (1996) Spatial, temporal and ontogenetic patterns of habitat use by coral reef fishes (Family Labridae). *Marine Ecology Progress Series* **133**, 1-11.

Griffiths, D. (1997) Local and regional diversity in North American lacustrine fish species. *Journal of Animal Ecology* **66**, 49-56.

Grigg, R.W. (1983) Community structure, succession and development of coral reefs in Hawaii. *Marine Ecology Progress Series* **11**, 1-14.

Grigg, R.W. and Dollar, S.J. (1990) Natural and anthropogenic disturbance on coral reefs, in *Ecosystems of the World 25 Coral Reefs*, (ed Z. Dubinsky), Elsevier, Amsterdam, pp. 439-452.

Grigg, R.W. and Epp, D. (1989) Critical depth for the survival of coral islands: effects on the Hawaiian Archipelago. *Science* **243**, 638-641.

Grigg, R.W. and Maragos, J.E. (1974) Recolonization of hermatypic corals on submerged lava flows in Hawaii. *Ecology* **55**, 387-395.

Grime, J.P. (1973) Competitive exclusion in herbaceous vegetation. *Nature* **242**, 344-347.

Grosberg, R.K. (1981) Competitive ability influences habitat choice in marine invertebrates. *Nature* **290**, 700-702.

Grubb, P.J. (1977) The maintenance of species-richness in plant communities: the importance of the regeneration niche. *Biological Reviews* **52**, 107-145.

Gu, D. and Philander, S.G.H. (1997) Interdecadal climate fluctuations that depend on exchanges between the tropics and extratropics. *Science* **275**, 805-807.

Guzmán, H.M. and Cortés, J. (1989) Coral reef community structure at Caño Island, Pacific Costa Rica. *Marine Ecology - Pubblicazioni della Stazione Zoologica di Napoli I* **10**, 23-41.

Hadfield, M.G. (1978) Metamorphosis in marine molluscan larvae: an analysis of stimulus and response, in *Settlement and Metamorphosis of Marine Invertebrate Larvae*, (eds F.-S. Chia and M.E. Rice), Elsevier, New York, pp. 165-175.

Hadfield, M.G. and Scheuer, D. (1985) Evidence for a soluble metamorphic inducer in *Phestilla*: ecological, chemical and biological data. *Bulletin of Marine Science* **37**, 556-566.

Hairston, Jr., N.G. and Hairston, Sr., N.G. (1997) Does food web complexity eliminate trophic-level dynamics. *The American Naturalist* **149**, 1001-1007.

Hairston, Sr., N.G. (1989) *Ecological Experiments: Purpose, Design, and Execution*, Cambridge University Press, Cambridge.

Hall, S.J. and Raffaelli, D. (1991) Food-web patterns: lessons from a species-rich web. *Journal of Animal Ecology* **60**, 823-842.

Hallam, A. (ed) (1973) *Atlas of Palaeobiogeography*, Elsevier Scientific Publishing Company, Amsterdam.

Hanski, I. and Gilpin, M. (1991) Metapopulation dynamics: brief history and conceptual domain. *Biological Journal of the Linnean Society* **42**, 3-16.

Hardin, G. (1960) The competitive exclusion principle. *Science* **131**, 1292-1297.

Harrison, P.L., Babcock, R.C., Bull, G.D., Oliver, J.K. *et al.* (1984) Mass spawning in tropical reef corals. *Science* **223**, 1186-1189.

Harrison, S. (1991) Local extinction in a metapopulation context. *Biological Journal of the Linnean Society* **42**, 73-88.

Harvell, C.D. (1990) The ecology and evolution of inducible defenses. *The Quarterly Review of Biology* **65**, 323-340.

Harvell, C.D., Fenical W., Roussis, V., Ruesink, J.L. *et al.* (1993) Local and geographical variation in the defensive chemistry of a West Indian gorgonian coral (*Briareum asbestinum*). *Marine Ecology Progress Series* **93**, 165-173.

Hashimoto, Y., Fusetani, N. and Kimura, S. (1969) Aluterin: a toxin of filefish, *Alutera scripta*, probably originating from a zoantharian, *Palythoa tuberculosa*. *Bulletin of the Japanese Society of Scientific Fisheries* **35**, 1086-1093.

Hassell, M.P., Latto, J. and May, R.M. (1989) Seeing the wood for the trees: detecting density dependence from existing life-table studies. *Journal of Animal Ecology* **58**, 883-892.

Hastings, A. (1980) Disturbance, coexistence, history, and competition for space. *Theoretical Population Biology* **18**, 363-373.

Hastings, A. and Harrison, S. (1994) Metapopulation dynamics and genetics. *Annual Review of Ecology and Systematics* **25**, 167-188.

Hawkins, B.A. and Compton, S.G. (1992) African fig wasp communities: vacant niches and latitudinal gradients in species richness. *Journal of Animal Ecology* **61**, 361-372.

Hay, M.E. (1981) Herbivory, algal distribution, and the maintenance of between-habitat diversity on a tropical fringing reef. *The American Naturalist* **118**, 520-540.

Hay, M.E. (1984) Patterns of fish and urchin grazing on Caribbean coral reefs: are previous results typical. *Ecology* **65**, 446-454.

Hay, M.E. (1986) Associational plant defenses and the maintenance of species diversity: turning competitors into accomplices. *The American Naturalist* **128**, 617-641.

Hay, M.E. (1991) Fish-seaweed interactions on coral reefs: effects of herbivorous fishes and adaptations of their prey, in *The Ecology of Fishes on Coral Reefs*, (ed P.F. Sale), Academic Press, New York, pp. 96-119.

Hay, M.E. and Fenical, W. (1988) Marine plant-herbivore interactions: the ecology of chemical defense. *Annual Review of Ecology and Systematics* **19**, 111-145.

Hay, M.E. and Fenical, W. (1992) Chemical mediation of seaweed-herbivore interactions, in *Plant-Animal Interactions in the Marine Benthos*, (eds D.M. John, S.J. Hawkins and J.H. Price), The Systematics Association Special Volume No. 46, Clarendon Press, Oxford, pp. 319-337.

Hay, M.E., Paul, V.J., Lewis, S.M., Gustafson, K. *et al.* (1988) Can tropical seaweeds reduce herbivory by growing at night? Diel patterns of growth, nitrogen content, herbivory, and chemical versus morphological defenses. *Oecologia* **75**, 233-245.

Hiatt, R.W. and Strasburg, D.W. (1960) Ecological relationships of the fish fauna on coral reefs of the Marshall Islands. *Ecological Monographs* **30**, 65-127.

Hixon, M.A. (1997) Effects of reef fishes on corals and algae, in *Life and Death of Coral Reefs*, (ed C. Birkeland), Chapman and Hall, London, pp. 230-248.

Hixon, M.A. and Brostoff, W.N. (1983) Damselfish as keystone species in reverse: intermediate disturbance and diversity of reef algae. *Science* **220**, 511-513.

Hixon, M.A. and Brostoff, W.N. (1996) Succession and herbivory: effects of differential fish grazing on

Hawaiian coral-reef algae. *Ecological Monographs* **66**, 67-90.

Hixon, M.A. and Menge, B.A. (1991) Species diversity: prey refuges modify the interactive effects of predation and competition. *Theoretical Population Biology* **39**, 178-200.

Hobson, E.S. (1974) Feeding relationships of teleostean fishes on coral reefs in Kona, Hawaii. *Fisheries Bulletin* **72**, 915-1031.

Hodgson, G. (1994) Coral reef catastrophe. *Science* **266**, 1930-1931.

Holling, C.S. (1973) Resilience and stability of ecological systems. *Annual Review of Ecology and Systematics* **4**, 1-23.

Holt, R.D. and Polis, G.A. (1997) A theoretical framework for intraguild predation. *The American Naturalist* **149**, 745-764.

Horn, H.S. (1975) Markovian properties of forest succession, in *Ecology and Evolution of Communities*, (eds M.L. Cody and J.M. Diamond), The Belknap Press of Harvard University Press, Cambridge, pp. 196-211.

Horn, H.S. (1981) Succession, in *Theoretical Ecology: Principles and Applications*, 2nd edn, (ed R.M. May), Sinauer Associates, Inc., Sunderland, pp. 253-271.

Horn, M.H. (1992) Herbivorous fishes: feeding and digestive mechanisms, in *Plant-Animal Interactions in the Marine Benthos*, (eds D.M. John, S.J. Hawkins and J.H. Price), The Systematics Association Special Volume No. 46, Clarendon Press, Oxford, pp. 339-362.

Howe, H.F. and Westley, L.C. (1988) *Ecological Relationships of Plants and Animals*, Oxford University Press, New York.

Hubbell, S.P. (1979) Tree dispersion, abundance, and diversity in a tropical dry forest. *Science* **203**, 1299-1309.

Hubbell, S.P. (1997) A unified theory of biogeography and relative species abundance and its application to tropical rain forests and coral reefs. *Coral Reefs* **16**, S9-S21.

Hubbell, S.P. and Foster, R.B. (1986) Biology, chance, and history and the structure of tropical rain forest tree communities, in *Community Ecology*, (eds J. Diamond and T.J. Case), Harper and Row, New York, pp. 314-329.

Hughes, T.P. (1985) Life histories and population dynamics of early successional corals. *Proceedings of the Fifth International Coral Reef Congress, Tahiti* **4**, 101-106.

Hughes, T.P. (ed)(1993) Disturbance: effects on coral reef dynamics. *Coral Reefs* **12**, 115-233.

Hughes, T.P. (1994a) Catastrophes, phase shifts, and large-scale degradation of a Caribbean coral reef. *Science* **265**, 1547-1551.

Hughes, T.P. (1994b) Coral reef catastrophe. *Science* **266**, 1932-1933.

Hughes, T.P. (1996) Demographic approaches to community dynamics: a coral reef example. *Ecology* **77**, 2256-2260.

Hughes, T.P., Keller, B.D., Jackson, J.B.C. and Boyle, M.J. (1985) Mass mortality of the echinoid *Diadema antillarum* Philippi in Jamaica. *Bulletin of Marine Science* **36**, 377-384.

Hughes, T.P., Reed, D.C. and Boyle, M.J. (1987) Herbivory on Coral Reefs: community structure following mass mortalities of sea urchins. *Journal of Experimental Marine Biology and Ecology* **113**, 39-59.

Hugueny, B. and Paugy, D. (1995) Unsaturated fish communities in African rivers. *The American Naturalist* **146**, 162-169.

Hugueny, B., Tito de Morais, L., de Mérona, B. and Ponton, D. (1997) The relationship between local and regional species richness: comparing biotas with different evolutionary histories. *Oikos* **80**, 583-587.

Hunte, W., Côté, I. and Tomascik, T. (1986) On the dynamics of the mass mortality of *Diadema antillarum* in Barbados. *Coral Reefs* **4**, 135-139.

Hunte, W. and Younglao, D. (1988) Recruitment and population recovery of *Diadema antillarum* (Echinodermata; Echinoidea) in Barbados. *Marine Ecology Progress Series* **45**, 109-119.

Hurd, L.E., Mellinger, M.V., Wolf, L.L. and McNaughton, S.J. (1971) Stability and diversity at three trophic levels in terrestrial successional ecosystems. *Science* **173**, 1134-1136.

Hurd, L.E. and Wolf, L.L. (1974) Stability in relation to nutrient enrichment in arthropod consumers of old-field successional ecosystems. *Ecological Monographs* **44**, 465-482.

Huston, M.A. (1979) A general hypothesis of species diversity. *The American Naturalist* **113**, 81-101.

Huston, M.A. (1985a) Patterns of species diversity on coral reefs. *Annual Review of Ecology and Systematics* **16**, 149-177.

Huston, M.A. (1985b) Patterns of species diversity in relation to depth at Discovery Bay, Jamaica. *Bulletin of Marine Science* **37**, 928-935.

Huston, M.A. (1994) *Biological Diversity. The Coexistence of Species on Changing Landscapes*, Cambridge University Press, Cambridge.

Hutchinson, G.E. (1961) The paradox of the plankton. *The American Naturalist* **95**, 137-145.

Imbrie, J., Hays, J.D., Martinson, D.G. *et al.* (1984) The orbital theory of Pleistocene climate: support from a revised chronology of the marine δ^{18} O record, in *Milankovitch and Climate, Part I*, (eds A.L. Berger, J. Imbrie, J. Hays, G. Kukla and B. Saltzman), D. Reidel Publishing Co., Dordrecht, pp. 269-305.

Irvine, G.V. (1983) Fish as farmers: an experimental study of herbivory by a coral reef damselfish. Ph.D. dissertation, University of California, Santa Barbara.

Jackson, J.B.C. (1977a) Competition on marine hard substrata: the adaptive significance of solitary and colonial strategies. *The American Naturalist* **111**, 743-767.

Jackson, J.B.C. (1977b) Habitat area, colonization, and development of epibenthic community structure, in *Biology of Benthic Organisms*, (eds B.F. Keegan, P.O. Ceidigh and P.J.S. Boaden), Pergamon Press, Oxford, pp.349-358.

Jackson, J.B.C. (1979a) Morphological strategies of sessile animals, in *Biology and Systematics of Colonial Organisms*, (eds G. Larwood and B.R. Rosen), Academic Press, New York, pp. 499-555.

Jackson, J.B.C. (1979b) Overgrowth competition between encrusting cheilostome ectoprocts in a Jamaican cryptic reef environment. *Journal of Animal Ecology* **48**, 805-823.

Jackson, J.B.C. (1981) Interspecific competition and species' distributions: the ghosts of theories and data past. *American Zoologist* **21**, 889-901.

Jackson, J.B.C. (1984) Ecology of cryptic coral reef communities. III. Abundance of encrusting organisms with particular reference to cheilostome bryozoans. *Journal of Experimental Marine Biology and Ecology* **75**, 37-57.

Jackson, J.B.C. (1991) Adaptation and diversity of reef corals. *Bioscience* **41**, 475-482.

Jackson, J.B.C. (1992) Pleistocene perspectives on coral reef community structure. *American Zoologist* **32**, 719-731.

Jackson, J.B.C. and Buss, L.W. (1975) Allelopathy and spatial competition among coral reef invertebrates. *Proceedings of the National Academy of Sciences USA* **72**,5160-5163.

Jackson, J.B.C., Goreau, T.F. and Hartman, W.D. (1971) Recent brachiopod-coralline sponge communities and their paleoecological significance. *Science* **173**, 623-625.

Jackson, J.B.C. and Hughes, T.P. (1985) Adaptive strategies of coral-reef invertebrates. *American Scientist* **73**, 265-274.

Jackson, J.B.C. and Kaufmann, K.W. (1987) *Diadema antillarum* was not a keystone predator in cryptic reef environments. *Science* **235**, 687-689.

Jackson, J.B.C. and Winston, J.E. (1982) Ecology of cryptic coral reef communities. I. Distribution and abundance of major groups of encrusting organisms. *Journal of Experimental Marine Biology and Ecology* **57**, 135-147.

Jennings, S. and Polunin, N.V.C. (1996) Impacts of fishing on tropical reef ecosystems. *Ambio* **25**, 44-49.

Johannes, R.E., Coles, S.L. and Kuenzel, N.T. (1970) The role of zooplankton in the nutrition of some scleractinian corals. *Limnology and Oceanography* **15**, 579-586.

Johannes, R.E., Wiebe, W.J., Crossland, C.J. *et al.* (1983) Latitudinal limits of coral reef growth. *Marine Ecology Progress Series* **11**, 105-111.

Johnson, C.R., Sutton, D.C., Olson, R.R. and Giddens, R. (1991) Settlement of crown-of-thorn starfish: role of bacteria on surfaces of coralline algae and a hypothesis for deepwater recruitment. *Marine Ecology Progress Series* **71**, 143-162.

Johnson, K.G., Budd, A.F. and Stemann, T.A. (1995) Extinction selectivity and ecology of Neogene Caribbean reef corals. *Paleobiology* **21**, 52-73.

Johnson, K.H., Vogt, K.A., Clark, H.J. *et al.* (1996) Biodiversity and the productivity and stability of ecosystems. *Trends in Ecology and Evolution* **11**, 372-377.

Jokiel, P.L. (1990a) Long-distance dispersal by rafting: reemergence of an old hypothesis. *Endeavour* **14**, 66-73.

Jokiel, P.L. (1990b) Transport of reef corals into the Great Barrier Reef. *Nature* **347**, 665-667.

Jokiel, P.L. and Martinelli, F.J. (1992) The vortex model of coral reef biogeography. *Journal of Biogeography* **19**, 449-458.

Jones, D.S. and Hasson, P.F. (1985) History and development of the marine invertebrate faunas separated by the central American Isthmus, in *The Great American Biotic Exchange*, (eds F.G. Stehli and S.D. Webb), Plenum Press, New York, pp. 325-355.

Jones, G.P. (1991) Postrecruitment processes in the ecology of coral reef fish populations: a multifactorial perspective, in *The Ecology of Fishes on Coral Reefs*, (ed P.F. Sale), Academic Press, New York, pp. 294-328.

Jones, G.P., Ferrell, D.J. and Sale, P.F. (1991) Fish predation and its impact on the invertebrates of coral reefs and adjacent sediments, in *The Ecology of Fishes on Coral Reefs*, (ed P.F. Sale), Academic Press, New York, pp. 156-179.

Kaplan, E.H. (1982) *A Field Guide to Coral Reefs of the Caribbean and Florida*. The Peterson Field Guide Series **27**, Houghton Mifflin Co., Boston.

Karieva, P.M., Kingsolver, J.G. and Huey, R.B. (1993) *Biotic Interactions and Global Change*, Sinaeur Associates Inc., Sunderland.

Karlson, R.H. (1978) Predation and space utilization patterns in a marine epifaunal community. *Journal of Experimental Marine Biology and Ecology* **31**, 225-239.

Karlson, R.H. (1980) Alternative competitive strategies in a periodically disturbed habitat. *Bulletin of Marine Science* **30**, 894-900.

Karlson, R.H. (1983) Disturbance and monopolization of a spatial resource by *Zoanthus sociatus* (Coelenterata, Anthozoa). *Bulletin of Marine Science* **33**, 118-131.

Karlson, R.H. (1985) Competitive overgrowth interactions among sessile colonial invertebrates: a comparison of stochastic and phenotypic variation. *Ecological Modelling* **27**, 299-312.

Karlson, R.H. (1988) Size-dependent growth in two zoanthid species: a contrast in clonal strategies. *Ecology* **69**, 1219-1232.

Karlson, R.H. and Cornell, H.V. (1997) Stability and invasibility of coral assemblages. *Trends in Ecology and Evolution* **12**, 195.

Karlson, R.H. and Cornell, H.V. (1998) Scale-dependent variation in local vs. regional effects on coral species richness. *Ecological Monographs* **68**, 259-274.

Karlson, R.H., Hughes, T.P. and Karlson, S.R. (1996) Density-dependent dynamics of soft coral aggregations: the significance of clonal growth and form. *Ecology* **77**, 1592-1599.

Karlson, R.H. and Hurd, L.E. (1993) Disturbance, coral reef communities, and changing ecological paradigms. *Coral Reefs* **12**, 117-125.

Karlson, R.H. and Jackson, J.B.C. (1981) Competitive networks and community structure: a simulation study. *Ecology* **62**, 670-678.

Karlson, R.H. and Levitan, D.R. (1988) Recruitment-limitation in open populations of *Diadema antillarum*: an evaluation. *Oecologia* **82**, 40-44.

Karplus, I. (1978) A feeding association between the grouper *Epinephelus fasciatus* and the moray eel *Gymnothorax griseus*. *Copeia* 1978, 164.

Kauffman, E.G. and Fagerstrom, J.A. (1993) The Phanerozoic evolution of reef diversity, in *Species Diversity in Ecological Communities: Historical and Geographical Perspectives*, (eds R.E. Ricklefs and D. Schluter), The University of Chicago Press, Chicago, pp. 315-329.

Kaufman, L.S. (1983) Effects of Hurricane Allen on reef fish assemblages near Discovery Bay, Jamaica. *Coral Reefs* **2**, 43-47.

Kennedy, C.R. and Guégan, J.F. (1994) Regional vs. local helminth parasite richness in British freshwater fish: saturated or unsaturated parasite communities? *Parasitology* **109**, 175-185.

Keough, M.J. (1984) Effects of patch size on the abundance of sessile marine invertebrates. *Ecology* **65**, 423-437.

Kimura, S., Hashimoto, Y. and Yamazato, K. (1972) Toxicity of the zoanthid *Palythoa tuberculosa*. *Toxicon* **10**, 611-617.

Klumpp, D.W., McKinnon, D. and Daniel, P. (1987) Damselfish territories: zones of high productivity on coral reefs. *Marine Ecology Progress Series* **40**, 41-51.

Knowlton, N. (1992) Thresholds and multiple stable states in coral reef community dynamics. *American Zoologist* **32**, 674-682.

Knowlton, N. and Jackson, J.B.C. (1994) New taxonomy and niche partitioning on coral reefs: jack of all trades or master of some? *Trends in Ecology and Evolution* **9**, 7-9.

Knowlton, N., Lang, J.C., and Keller, B.D. (1990) Case study of natural population collapse: post-hurricane predation on Jamaican staghorn corals. *Smithsonian Contributions to the Marine Sciences* **31**, 1-25.

Knowlton, N., Lang, J.C., Rooney, M.C. and Clifford, P. (1981) Evidence for delayed mortality in hurricane-damaged Jamaican staghorn corals. *Nature* **294**, 251-252.

Kohn, A.J. (1968) Microhabitat, abundance and food of *Conus* on atoll reefs in the Maldive and Chagos Islands. *Ecology* **49**, 1046-1061.

Kohn, A.J. (1971) Diversity, utilization of resources, and adaptive radiation in shallow-water marine invertebrates of tropical oceanic islands. *Limnology and Oceanography* **16**, 332-348.

Kohn, A.J. and Leviten, P.J. (1976) Effect of habitat complexity on population density and species richness in tropical intertidal predatory gastropod assemblages. *Oecologia* **25**, 199-210.

Krebs, C.J., Boutin, S., Boonstra, R., Sinclair, A.R.E. *et al.* (1995) Impact of food and predation on the snowshoe hare cycle. *Science* **269**, 1112-1115.

Krebs, C.J. (1994) *Ecology: The Experimental Analysis of Distribution and Abundance*, 4th edn, Harper Collins College Publishers, New York.

La Barre, S.C., Coll, J.C. and Sammarco, P.W. (1986) Defensive strategies of soft corals (Coelenterata: Octocorallia) of the Great Barrier Reef. II. The relationship between toxicity and feeding deterrence. *Biological Bulletin* **171**, 565-576.

Laborel, J. (1974) West African reef corals: an hypothesis on their origin. *Proceedings of the Second International Coral Reef Symposium* **2**, 425-443.

Lang, J. (1973) Interspecific aggression by scleractinian corals. 2. Why the race is not only to the swift. *Bulletin of Marine Science* **23**, 260-279.

Lang, J.C. and Chornesky, E.A. (1990) Competition between scleractinian reef corals - a review of mechanisms and effects, in *Ecosystems of the World 25 Coral Reefs*, (ed Z. Dubinsky), Elsevier, Amsterdam, pp. 209-252.

Law, R. and Morton, R.D. (1996) Permanence and the assembly of ecological communities. *Ecology* **77**, 762-775.

Lawton, J.H. (1987) Are there assembly rules for successional communities, in *Colonization, Succession and Stability*, (eds A.J. Cray, M.J. Crawley and P.J. Edwards), Blackwell Scientific Publications, Oxford., pp. 225-244.

Lawton, J.H. (1988) Food webs, in *Ecological Concepts*, (ed. J.M. Cherrett), Blackwell Scientific Publications, Oxford, pp. 43-78.

Lawton, J.H. (1990) Local and regional species-richness of bracken-feeding insects, in *Bracken Biology and Management*, (eds J.A. Thompson and R.T. Smith), Australian Institute of Agricultural Science, Sydney, pp. 197-202.

Lawton, J.H. (1995) The variety of life. *Nature* **373**, 32.

Lessios, H.A. (1984) Spread of *Diadema* mass mortality through the Caribbean. *Science* **226**, 335-337.

Lessios, H.A. (1988) Mass mortality of *Diadema antillarum* in the Caribbean: what have we learned? *Annual Review of Ecology and Systematics* **19**, 371-393.

Levin, S.A. (1992) The problem of pattern and scale in ecology. *Ecology* **73**, 1943-1967.

Levin, S.A. and Paine, R.T. (1974) Disturbance, patch formation, and community structure. *Proceedings of the National Academy of Sciences USA* **71**, 2744-2747.

Levins, R. (1969) Some demographic and genetic consequences of environmental heterogeneity for biological control. *Bulletin of the Entomological Society of America* **15**, 237-240.

Levins, R. (1970) Extinction, in *Some Mathematical Questions in Biology*, (ed M.L. Gerstenhaber), American Mathematical Society, Providence, pp. 77-107.

Levitan, D.R. (1988) Algal-urchin biomass responses following mass mortality of *Diadema antillarum* Philippi at Saint John, U.S. Virgin Islands. *Journal of Experimental Marine Biology and Ecology* **119**, 167-178.

Levitan, D.R. (1989) Density-dependent size regulation in *Diadema antillarum*: effects of fecundity and survivorship. *Ecology* **70**, 1414-1424.

Levitan, D.R. (1991) Influence of body size and population density on fertilization success and reproductive output in a free-spawning invertebrate. *Biological Bulletin* **181**, 261-268.

Lewis, D.H. and Smith, D.C. (1971) The autotrophic nutrition of symbiotic marine coelenterates with special reference to hermatypic corals. I. Movement of photosynthetic products between the symbionts. *Proceedings of the Royal Society of London, Series B, Biological Sciences* **178**, 111-129.

Lewis, J. (1980) Hurricane damage. *Nature* **287**, 480.

Lewontin, R.C. (1969) The meaning of stability, in *Diversity and Stability in Ecological Systems*, Brookhaven Symposia in Biology **22**, 13-24.

Liddell, W.D. and Ohlhorst, S.L. (1987) Patterns of reef community structure, North Jamaica. *Bulletin of Marine Science* **40**, 311-329.

Liddell, W.D. and Ohlhorst, S.L. (1988) Comparison of Western Atlantic coral reef communities. *Proceedings of the Sixth International Coral Reef Symposium* **3**, 281-286.

Liddell, W.D., Ohlhorst, S.L. and Coates, A.G. (1984) Modern and ancient carbonate environments of Jamaica. *Sedimenta* **10**, 1-8.

Lidgard, S. and Jackson, J.B.C. (1989) Growth in encrusting cheilostome bryozoans: I. Evolutionary trends. *Paleobiology* **15**, 255-282.

Lindquist, N. and Hay, M.L. (1996) Palatability and chemical defense of marine invertebrate larvae. *Ecological Monographs* **66**, 431-450.

Littler, M.M., Taylor, P.R. and Littler, D.S. (1986) Plant defense associations in the marine environment. *Coral Reefs* **5**, 63-11.

Lively, C.M. (1986a) Predator-induced shell dimorphism in the acorn barnacle *Chthamalus anisopoma*. *Evolution* **40**, 232-242.

Lively, C.M. (1986b) Competition, comparative life histories, and maintenance of shell dimorphism in a barnacle. *Ecology* **67**, 858-864.

Logan, A. (1984) Interspecific aggression in hermatypic corals from Bermuda. *Coral Reefs* **3**, 131-138.

Lotka, A.J. (1925) *Elements of Physical Biology*, Williams and Wilkins, Baltimore.

Lotka, A.J. (1932) The growth of mixed populations: two species competing for a common food supply. *Journal of the Washington Academy of Sciences* **22**, 461-469.

Lovett Doust, L. (1981) Population dynamics and specialization in a clonal perennial (*Ranunculus repens*) I. The dynamics of ramets in contrasting habitats. *Journal of Ecology* **69**, 743-755.

Low, R.M. (1971) Interspecific territoriality in a pomacentrid reef fish, *Pomacentrus flavicauda* Whitley. *Ecology* **52**, 648-654.

Loya, Y. (1972) Community structure and species diversity of hermatypic corals at Eilat, Red Sea. *Marine Biology* **13**, 100-123.

Loya, Y. (1975) Some possible effects of water pollution on the community of Red Sea corals. *Marine Biology* **29**, 177-185.

Loya, Y. (1976a) Effects of water turbidity and sedimentation on the community structure of Puerto Rican corals. *Bulletin of Marine Science* **26**, 450-466.

Loya, Y. (1976b) The Red Sea coral *Stylophora pistillata* is an *r* strategist. *Nature* **259**, 478-480.

Loya, Y. (1976c) Recolonization of Red Sea corals affected by natural catastrophes and man-made perturbations. *Ecology* **57**, 278-289.

Loya, Y. (1990) Changes in a Red Sea coral community structure: a long-term case history study, in *The Earth in Transition. Patterns and Processes of Biotic Impoverishment*, (ed G.M. Woodwell), Cambridge University Press, New York, pp. 369-384.

Loya, Y. and Slobodkin, L.B. (1971) The coral reefs of Eilat (Gulf of Eilat, Red Sea). *Symposia of the Zoological Society of London* **28**, 117-139.

Lucas, J.S. and Jones, M.M. (1976) Hybrid crown-of-thorns starfish (*Acanthaster planci* x *A. brevispinus*) reared to maturity in the laboratory. *Nature* **263**, 409-412.

MacArthur, R.H. (1955) Fluctuations of animal populations and a measure of community stability. *Ecology* **36**, 533-536.

MacArthur, R.H. (1957) On the relative abundance of bird species. *Proceedings of the National Academy of Sciences USA* **43**, 293-295.

MacArthur, R.H. (1960) On the relative abundance of species. *The American Naturalist* **94**, 25-36.

MacArthur, R.H. (1965) Patterns of species diversity. *Biological Reviews* **40**, 510-533.

MacArthur, R.H. (1972) *Geographical Ecology: Patterns in the Distribution of Species*, Harper & Row, Publishers, Inc., New York.

MacArthur, R.H. and Wilson, E.O. (1963) An equilibrium theory of insular zoogeography. *Evolution* **17**, 373-387.

MacArthur, R.H. and Wilson, E.O. (1967) *The Theory of Island Biogeography*, Princeton University Press, Princeton.

Maragos, J.E. and Jokiel, P.L. (1986) Reef corals of Johnston Atoll: one of the world's most isolated atolls.

Coral Reefs **4**, 141-150.

Martinez, N.D. (1991) Artifacts or attributes? Effects of resolution on the Little Rock Lake food web. *Ecological Monographs* **61**, 367-392.

Martinez, N.D. (1994) Scale-dependent constraints on food-web structure. *The American Naturalist* **144**, 935-953.

Mather, P. and Bennett, I. (eds) (1984) *A Coral Reef Handbook: A Guide to the Fauna, Flora, and Geology of Heron Island and Adjacent Reefs and Cays*, 2nd edn, The Australian Coral Reef Society, Brisbane.

Matsuda, S. (1989) Succession and growth rates of encrusting crustose coralline algae (Rhodophyta, Cryptonemiales) in the upper fore-reef environment off Ishigaki Island, Ryukyu Islands. *Coral Reefs* **7**, 185-195.

May, R.M. (1973) *Stability and Complexity in Model Ecosystems,* Princeton University Press, Princeton.

McIntyre, A. and Molfino, B. (1996) Forcing of Atlantic equatorial and subpolar millennial cycles by precession. *Science* **274**, 1867-1870.

McNaughton, S.J. (1977) Diversity and stability of ecological communities: a comment on the role of empiricism in ecology. *The American Naturalist* **111**, 515-525.

McNaughton, S.J. (1985) Ecology of a grazing ecosystem: the Serengeti. *Ecological Monographs* **55**, 259-294.

Mellinger, M.V. and McNaughton, S.J. (1975) Structure and function of successional vascular plant communities in New York. *Ecological Monographs* **45**, 161-182.

Menge, B.A. (1995) Indirect effects in marine rocky intertidal interaction webs: patterns and importance. *Ecological Monographs* **65**, 21-74.

Menge, B.A., Berlow, E.L., Blanchette, C.A. *et al.* (1994) The keystone species concept: variation in interaction strength in a rocky intertidal habitat. *Ecological Monographs* **64**, 249-286.

Menge, B.A. and Lubchenco, J. (1981) Community organization in temperate and tropical rocky intertidal habitats: prey refuges in relation to consumer pressure gradients. *Ecological Monographs* **51**, 429-450.

Menge, B.A., Lubchenco, J., Gaines, S.D. *et al.* (1986) A test of the Menge-Sutherland model of community organization in a tropical rocky intertidal food web. *Oecologia* **71**, 75-89.

Menge, B.A. and Sutherland, J.P. (1987) Community regulation: variation in disturbance, competition, and predation in relation to environmental stress and recruitment. *The American Naturalist* **130**, 730-757.

Mesolella, K.J. (1967) Zonation of uplifted Pleistocene coral reefs on Barbados, West Indies. *Science* **156**, 638-640.

Mesolella, K.J. (1968) The uplifted reefs of Barbados: physical stratigraphy, facies relationships and absolute chronology. Ph.D. dissertation, Brown University, Providence.

Meyer, J.L., Schultz, E.T. and G.S. Helfman (1983) Fish schools: An asset to corals. *Science* **220**, 1047-1049.

Miller, M.W. and Hay, M.E. (1996) Coral-seaweed-grazer-nutrient interactions on temperate reefs. *Ecological Monographs* **66**, 323-344.

Miller, R.S. (1967) Pattern and process in competition. *Advances in Ecological Research* **4**, 1-74.

Miller, T.E. and Kerfoot, W.C. (1987) Redefining indirect effects, in *Predation: Direct and Indirect Impacts on Aquatic Communities*, (eds W.C. Kerfoot and A. Sih), University Press of New England, Hanover, pp. 33-37.

Mithen, S.J. and Lawton, J.H. (1986) Food-web models that generate constant predator-prey ratios. *Oecologia* **69**, 542-550.

Molfino, B. (1994) Palaeoecology of marine systems, in *Aquatic Ecology: Scale, Pattern and Process*, (eds P.S. Giller, A.G. Hildrew and D.G. Raffaelli), Blackwell Scientific Publications, Oxford, pp. 517-546.

Moll, H. (1986) The coral community structure on the reefs visited during the Snellius II Expedition in eastern Indonesia. *Zoologische Mededelingen Leiden* **60**, 1-25.

Montgomery, W.L. (1990) Zoogeography, behavior and ecology of coral-reef fishes, in *Ecosystems of the World 25 Coral Reefs,* (ed Z. Dubinsky), Elsevier, Amsterdam, pp. 329-364.

Moore, R.C. (ed) (1956) *Treatise on Invertebrate Paleontology,* Part F, *Coelenterata,* University of Kansas Press, Lawrence.

Moore, R.E. and Scheuer, P.J. (1971) Palytoxin: a new marine toxin from a coelenterate. *Science* **171**, 495-498.

Moran, P.J. (1986) The *Acanthaster* phenomenon. *Oceanography and Marine Biology: an Annual Review* **24**, 379-480.

Morrison, D. (1988) Comparing fish and urchin grazing in shallow and deeper coral reef algal communities. *Ecology* **69**, 1367-1382.

Morrissey, J. 1980. Community structure and zonation of macroalgae and hermatypic corals on a fringing reef flat of Magnetic Island (Queensland, Australia). *Aquatic Botany* **8**, 91-139.

Morse, A.N.C. (1992) Role of algae in the recruitment of marine invertebrate larvae, in *Plant-Animal Interactions in the Marine Benthos,* (eds D.M. John, S.J. Hawkins and J.H. Price), The Systematics Association Special Volume No. 46, Clarendon Press, Oxford, pp. 385-403.

Morse, A.N.C. and D.E. Morse (1996) Flypapers for coral and other planktonic larvae. *Bioscience* **46**, 254-262.

Morse, D.E., Hooker, N., Morse, A.N.C. and Jensen, R.A. (1988) Control of larval metamorphosis and recruitment in sympatric agariciid corals. *Journal of Experimental Marine Biology and Ecology* **116**, 193-217.

Morse, D.E. and Morse, A.N.C. (1991) Enzymatic characterization of the morphogen recognized by *Agaricia humilis* (scleractinian coral) larvae. *Biological Bulletin* **181**, 104-122.

Morse, D.E., A.N.C. Morse, P.T. Raimondi and Hooker, N. (1994) Morphogen-based chemical flypaper for *Agaricia humilis* coral larvae. *Biological Bulletin* **186**, 172-181.

Morton, S.R. (1993) Determinants of diversity in animal communities of arid Australia, in *Species Diversity in Ecological Communities: Historical and Geographical Perspectives,* (eds R.E. Ricklefs and D. Schluter), The University of Chicago Press, Chicago, pp. 159-169.

Moyer, J.T., Emerson, W.K. and Ross, M. (1982) Massive destruction of scleractinian corals by the muricid gastropod, *Drupella,* in Japan and the Philippines. *The Nautilus* **96**, 69-82.

Munro, J.L. (1969) The sea fisheries of Jamaica. *Jamaica Journal* **3**, 16-22.

Munro, J.L. (ed) (1983) *Caribbean Coral Reef Fishery Resources.* International Center for Living Aquatic Resources Management Studies and Reviews **7**, Manila.

Munro, J.L. and Williams, D.McB. (1985) Assessment and management of coral reef fisheries: biological, environmental and socio-economic aspects. *Proceedings of the Fifth International Coral Reef Congress, Tahiti* **4**, 545-581.

Muscatine, L. and D'Elia, C.F. (1978) The uptake, retention, and release of ammonium by reef corals. *Limnology and Oceanography* **23**, 725-734.

Muscatine, L., Falkowski, P.G., Porter, J.W. and Dubinsky, Z. (1984) Fate of photosynthetically fixed carbon in light- and shade-adapted colonies of the symbiotic coral *Stylophora pistillata. Proceedings of the Royal Society of London, Series B, Biological Sciences* **222**, 181-202.

Muscatine, L. and Porter, J.W. (1977) Reef corals: mutualistic symbioses adapted to nutrient-poor environments. *Bioscience* **27**, 454-460.

Myers, A.A. and Giller, P.S. (eds) (1988) *Analytical biogeography. An integrated approach to the study of animal and plant distributions.* Chapman and Hall, London.

Myers, A.A. (1994) Biogeographic patterns in shallow-water marine systems and the controlling processes at different scales, in *Aquatic Ecology: Scale, Pattern and Process,* (eds P.S. Giller, A.G. Hildrew and D.G. Raffaelli), Blackwell Scientific Publications, Oxford, pp. 547-574.

Neudecker, S. (1979) Effects of grazing and browsing fishes on the zonation of corals in Guam. *Ecology* **60**, 666-762.

Nicholson, A.J. (1933) The balance of animal populations. *Journal of Animal Ecology* **2**, 132-178.

Odum, H.T. and Odum, E.P. (1955) Trophic structure and productivity of a windward coral reef community on Eniwetok Atoll. *Ecological Monographs* **25**, 291-320.

Ogden, J.C. (1994) Coral reef catastrophe. *Science* **266**, 1931.

Ogden, J.C., Brown, R.A. and Salesky, N. (1973) Grazing by the echinoid *Diadema antillarum* Philippi: formation of halos around West Indian patch reefs. *Science* **182**, 715-717.

Ogden, J.C. and Lobel, P.S. (1978) The role of herbivorous fishes and urchins in coral reef communities. *Environmental Biology of Fishes* **3**, 49-63.

Oliver, Jr., W.A. (1980) The relationship of the scleractinian corals to the rugose corals. *Paleobiology* **6**, 146-160.

Olson, R.R. and McPherson, R. (1987) Potential vs. realized larval dispersal: fish predation on larvae of the ascidian *Lissoclinum patella* (Gottschaldt). *Journal of Experimental Marine Biology and Ecology* **110**, 245-256.

Orians, G.H. and Solbrig, O.T. (eds) (1977) *Convergent Evolution in Warm Deserts. An Examination of Strategies and Patterns in Deserts of Argentina and the United States*, Dowden, Hutchinson & Ross, Inc., Stroudsburg.

Osman, R.W. (1977) The establishment and development of a marine epifaunal community. *Ecological Monographs* **47**, 37-63.

Osman, R.W. and Dean, T.A. (1987) Intra- and interregional comparisons of numbers of species on marine hard substrate islands. *Journal of Biogeography* **14**, 53-67.

Osman, R.W. and Haugsness, J.A. (1981) Mutualism among sessile invertebrates: a mediator of competition and predation. *Science* **211**, 846-848.

Paine, R.T. (1966) Food web complexity and species diversity. *The American Naturalist* **100**, 65-75.

Paine, R.T. (1969) A note on trophic complexity and community stability. *The American Naturalist* **103**, 91-93.

Paine, R.T. (1979) Disaster, catastrophe, and local persistence of the sea palm *Postelsia palmaeformis*. *Science* **205**, 685-688.

Paine, R.T. (1984) Ecological determinism in the competition for space. *Ecology* **65**, 1339-1348.

Paine, R.T. (1988) Food webs: road maps of interactions or grist for theoretical development? *Ecology* **69**, 1648-1654.

Paine, R.T. (1994) *Marine Rocky Shores and Community Ecology: An Experimentalist's Perspective*, Ecology Institute, Oldendorf/Luhe.

Paine, R.T. and Levin, S.A. (1981) Intertidal landscapes: disturbance and the dynamics of pattern. *Ecological Monographs* **51**, 145-178.

Pandolfi, J.M. (1996) Limited membership in Pleistocene reef coral assemblages from the Huon Peninsula, Papua New Guinea: constancy during global change. *Paleobiology* **22**, 152-176.

Park, T. (1954) Experimental studies of interspecies competition II. Temperature, humidity, and competition in two species of *Tribolium*. *Physiological Zoology* **27**, 177-238.

Paul, V.J. (1992a) Seaweed chemical defenses on coral reefs, in *Ecological Roles of Marine Natural Products*, (ed. V.J. Paul), Comstock Publishing Associates, Ithaca, pp. 24-50.

Paul, V.J. (1992b) Chemical defenses in benthic marine invertebrates, in *Ecological Roles of Marine Natural Products*, (ed. V.J. Paul), Comstock Publishing Associates, Ithaca, pp. 164-188.

Paul, V.J. and Van Alstyne, K.L. (1988) Chemical defense and chemical variation in some tropical Pacific species of *Halimeda* (Halimedaceae; Chlorophyta). *Coral Reefs* **6**, 263-270.

Paulay, G. (1997) Diversity and distribution of reef organisms, in *Life and Death of Coral Reefs*, (ed C. Birkeland), Chapman and Hall, New York, pp. 298-353.

Pawlik, J.R. and Fenical, W. (1989) A re-evaluation of the ichthyodeterrent role of prostaglandins in the Caribbean gorgonian coral *Plexaura homomalla. Marine Ecology Progress Series* **52**, 95-98.

Pearson, R.G. (1974) Recolonization by hermatypic corals of reefs damaged by *Acanthaster. Proceedings of the Second International Coral Reef Symposium* **2**, 207-215.

Pearson, R.G. (1981) Recovery and recolonization of coral reefs. *Marine Ecology Progress Series* **4**, 105-122.

Pennings, S.J. (1997) Indirect interactions on coral reefs, in *Life and Death of Coral Reefs*, (ed C. Birkeland), Chapman and Hall, London, pp. 249-272.

Petraitis, P.S., Latham, R.E. and Niesenbaum, R.A. (1989) The maintenance of species diversity by disturbance. *The Quarterly Review of Biology* **64**, 393-418.

Pianka, E.R. (1966) Latitudinal gradients in species diversity: a review of concepts. *The American Naturalist* **100**, 33-46.

Pichon, M. and Morrissey, J. (1981) Benthic zonation and community structure of South Island Reef, Lizard Island (Great Barrier Reef). *Bulletin of Marine Science* **31**, 581-593.

Pimm, S.L. (1982) *Food Webs*, Chapman and Hall, London.

Pimm, S.L. (1984) Food chains and return times, in *Ecological Communities: Conceptual Issues and the Evidence* (eds D.R. Strong, Jr., D. Simberloff, L.G. Abele and A.B. Thistle), Princeton University Press, Princeton, pp. 397-412.

Philander, S.G. (1990) *El Niño, La Niña, and the Southern Oscillation*, Academic Press, San Diego.

Platt, T. and Sathyendranath, S. (1994) Scale, pattern and process in marine ecosystems, in *Aquatic Ecology: Scale, Pattern and Process*, (eds P.S. Giller, A.G. Hildrew and D.G. Raffaelli), Blackwell Scientific Publications, Oxford, pp. 593-600.

Polis, G.A. (1991) Complex trophic interactions in deserts: an empirical critique of food-web theory. *The American Naturalist* **138**, 123-155.

Polis, G.A. and Strong, D.R. (1996) Food web complexity and community dynamics. *The American Naturalist* **147**, 813-846.

Polunin, N.V.C. and Klumpp, D.W. (1992) A trophodynamic model of fish production on a windward reef tract, in *Plant-Animal Interactions in the Marine Benthos*, (eds D.M. John, S.J. Hawkins and J.H. Price), The Systematics Association Special Volume No. 46, Clarendon Press, Oxford, pp. 213-233.

Polunin, N.V.C. and Roberts, C.M. (eds) (1996) *Reef Fisheries*, Chapman and Hall, London.

Porter, J.W. (1972) Ecology and species diversity of coral reefs on opposite sides of the Isthmus of Panama. *Bulletin of the Biological Society of Washington* **2**, 89-116.

Porter, J.W. (1974) Zooplankton feeding by the Caribbean reef-building coral *Montastrea cavernosa. Proceedings of the Second International Coral Reef Symposium* **1**, 111-125.

Porter, J.W. (1976) Autotrophy, heterotrophy, and resource partitioning in Caribbean reef-building corals. *The American Naturalist* **110**, 731-742.

Porter, J.W., Battey, J.F. and Smith, G.J. (1982) Perturbation and change in coral reef communities. *Proceedings of the National Academy of Sciences USA* **79**, 1678-1681.

Porter, J.W. and Targett, N.M. (1988) Allelochemical interactions between sponges and corals. *Biological Bulletin* **175**, 230-239.

Porter, J.W, Woodley, J.D., Smith, G.J. *et al.* (1981) Population trends among Jamaican reef corals. *Nature* **294**, 249-250.

Potts, D.C. (1984) Generation times and the Quaternary evolution of reef-building corals. *Paleobiology* **10**, 48-58.

Power, M.E., Tilman, D., Estes, J.A. *et al.* (1996) Challenges in the quest for keystones. *Bioscience* **46**, 609-620.

Pulliam, H.R. (1988) Sources, sinks, and population regulation. *The American Naturalist* **132**, 652-661.

Raffaelli, D.G., Hildrew, A.G. and Giller, P.S. (1994) Scale, pattern and process: concluding remarks, in *Aquatic Ecology: Scale, Pattern and Process*, (eds P.S. Giller, A.G. Hildrew and D.G. Raffaelli), Blackwell Scientific Publications, Oxford, pp. 601-606.

Rahav, O., Dubinsky, Z., Achituv, Y. and Falkowski, P.G. (1989) Ammonium metabolism in the zooxanthellate coral, *Stylophora pistillata. Proceedings of the Royal Society of London, Series B, Biological Sciences* **236**, 325-337.

Randall, J.E. (1965) Grazing effect on sea grasses by herbivorous reef fishes in the West Indies. *Ecology* **46**, 255-260.

Randall, J.E. (1967) Food habits of reef fishes of the West Indies. *Studies in Tropical Oceanography* **5**, 665-847.

Randall, J.E., Allen, G.R. and Steene, R.C. (1990) *Fishes of the Great Barrier Reef and Coral Sea*, Crawford House Press, Bathurst.

Randall, R.H. (1967) Reef physiography and distribution of corals at Tumon Bay, Guam, before crown-of-thorns starfish *Acanthaster planci* (L.) predation. *Micronesica* **9**, 119-158.

Reaka-Kudla, M.L. (1995) An estimate of known and unknown biodiversity and potential for extinction on coral reefs. *Reef Encounter* (newsletter of the International Society for Reef Studies) **17**, 8-12.

Recher, H.F. (1969) Bird species diversity and habitat diversity in Australia and North America. *The American Naturalist* **103**, 75-80.

Reinthal, P.N., Kensley, B. and Lewis, S.M. (1984) Dietary shifts in the queen triggerfish, *Balistes vetula*, in the absence of its primary food item, *Diadema antillarum. Marine Ecology - Pubblicazioni della Stazione Zoologica di Napoli I* **5**, 191-195.

Reiswig, H. M. (1971) Particle feeding in natural populations of three marine Demosponges. *Biological Bulletin* **141**, 568-591.

Reynolds, C.S. (1994) The role of fluid motion in the dynamics of phytoplankton in lakes and rivers, in *Aquatic Ecology: Scale, Pattern and Process*, (eds P.S. Giller, A.G. Hildrew and D.G. Raffaelli), Blackwell Scientific Publications, Oxford, pp. 141-187.

Reznick, D.N., Shaw, F.H., Rodd, F.H. and Shaw, R.G. (1997) Evaluation of the rate of evolution in natural populations of guppies (*Poecilia reticulata*). *Science* **275**, 1934-1937.

Rhoades, D.F. and Cates, R.G. (1976) Toward a general theory of plant antiherbivore chemistry. *Recent Advances in Phytochemistry* **10**, 168-213.

Rice, B. (1985) No evidence for divergence between Australia and elsewhere in plant species richness at tenth-hectare scale. *Proceedings of the Ecological Society of Australia* **14**, 99-101.

Richardson, C.A., Dustan, P. and Lang, J.C. (1979) Maintenance of living space by sweeper tentacles of *Montastrea cavernosa*, a Caribbean reef coral. *Marine Biology* **55**, 181-186.

Richardson, D.M., Cowling, R.M., Lamont, B.B. and van Hensbergen, H.J. (1995) Coexistence of *Banksia* species in southwestern Australia: the role of local and regional processes. *Journal of Vegetation Science* **6**, 329-342.

Ricklefs, R.E. (1987) Community diversity: relative roles of local and regional processes. *Science* **235**, 167-171.

Ricklefs, R.E. (1990) *Ecology*, 3rd edn, W.H. Freeman and Company, New York.

Ricklefs, R.E. and Cox, G.W. (1977) Morphological similarity and ecological overlap among passerine birds on St. Kitts, British West Indies. *Oikos* **29**, 60-66.

Ricklefs, R.E. and Cox, G.W. (1978) Stage of taxon cycle, habitat distribution, and population density in the avifauna of the West Indies. *The American Naturalist* **112**, 875-895.

Ricklefs, R.E. and Latham, R.E. (1993) Global patterns of diversity in mangrove floras, in *Species Diversity in Ecological Communities: Historical and Geographical Perspectives*, (eds R.E. Ricklefs and D. Schluter), The University of Chicago Press, Chicago, pp. 215-229.

Ricklefs, R.E. and Schluter, D. (eds) (1993) *Species Diversity in Ecological Communities: Historical and Geographical Perspectives*, The University of Chicago Press, Chicago.

Rinkevich, B., Shashar, N. and Liberman, T. (1992) Nontransitive xenogeneic interactions between four common Red Sea sessile invertebrates. *Proceedings of the Seventh International Coral Reef Symposium* **2**, 833-839.

Roberts, C.M. (1997) Connectivity and management of Caribbean coral reefs. *Science* **278**, 1454-1457.

Roberts, C.M. (1998) Fishery and reef management. *Science* **279**, 2022-2023.

Robertson, D.R. (1987) Responses of two coral reef toadfishes (Batrachoididae) to the demise of their primary prey, the sea urchin *Diadema antillarum*. *Copeia* **1987**, 637-642.

Robertson, D.R. (1996) Interspecific competition controls abundance and habitat use of territorial Caribbean damselfishes. *Ecology* **77**, 885-899.

Robertson, D.R. and Gaines, S.D. (1986) Interference competition structures habitat use in a local assemblage of coral reef surgeonfishes. *Ecology* **67**, 1372-1383.

Robertson, D.R., Hoffman, S.G. and Sheldon, J.M. (1981) Availability of space for the territorial Caribbean damselfish *Eupomacentrus planifrons*. *Ecology* **62**, 1162-1169.

Robertson, D.R., Sweatman, H.P.A., Fletcher, E.A. and Cleland, M.G. (1976) Schooling as a mechanism for circumventing the territoriality of competitors. *Ecology* **57**, 1208-1220.

Robles, C. and Robb, J. (1993) Varied consumer effects and the prevalence of intertidal algal turfs. *Journal of Experimental Marine Biology and Ecology* **166**, 65-91.

Rogers, C.S. (1993) Hurricanes and coral reefs the intermediate disturbance hypothesis revisited. *Coral Reefs* **12**, 127-137.

Rogers, C.S., Gilnack, M. and Fitz III, H.C. (1983) Monitoring of coral reefs with linear transects: a study of storm damage. *Journal of Experimental Marine Biology and Ecology* **66**, 285-300.

Rogers, C.S., Fitz III, H.C., Gilnack, M. *et al.* (1984) Scleractinian coral recruitment patterns at Salt River submarine canyon, St. Croix, U.S. Virgin Island. *Coral Reefs* **3**, 69-76.

Romano, S.L. and Palumbi, S.R. (1996) Evolution of scleractinian corals inferred from molecular systematics. *Science* **271**, 640-642.

Root, R.B. (1967) The niche exploitation of the blue-gray gnatcatcher. *Ecological Monographs*, **37**, 317-350.

Rosen, B.R. (1988a) Progress, problems and patterns in the biogeography of reef corals and other tropical marine organisms. *Helgoländer Meeresuntersuchungen* **42**, 269-301.

Rosen, B.R. (1988b) From fossils to earth history: applied historical biogeography, in *Analytical Biogeography. An Integrated Approach to the Study of Animal and Plant Distributions* (eds A.A. Myers and P.S. Giller), Chapman and Hall, London, pp.437-481.

Ross, M.A. and Hodgson, G. (1982) A quantitative study of hermatypic coral diversity and zonation at Apo Reef, Mindoro, Philippines. *Proceedings of the Fourth International Coral Reef Symposium* **2**, 281-291.

Roughgarden, J. (1983) Competition and theory in community ecology. *The American Naturalist* **122**, 583-601.

Roughgarden, J. (1986) A comparison of food-limited and space-limited animal competition communities, in *Community Ecology*, (eds J. Diamond and T.J. Case), Harper & Row, Publishers, Inc., New York, pp. 492-516.

Roughgarden, J. (1989) The structure and assembly of communities, in *Perspectives in Ecological Theory*, (eds J. Roughgarden, R.M. May and S.A. Levin), Princeton University Press, Princeton, pp. 203-226.

Roughgarden, J., Gaines, S.D. and Pacala, S.W. (1987) Supply side ecology: the role of physical transport processes, in *Organization of Communities Past and Present,* (eds J.H.R. Gee and P.S. Giller), Blackwell Scientific Publications, Londen pp. 491-518.

Roughgarden, J., Gaines, S. and Possingham, H. (1988) Recruitment dynamics in complex life cycles. *Science* **241**, 1460-1466.

Rowan, R., and Knowlton, N. (1995) Intraspecific diversity and ecological zonation in coral algal symbiosis. *Proceedings of the National Academy of Sciences* USA **92**, 2850-2853.

Roy, K., Valentine, J.W., Jablonski, D. and Kidwell, S.M. (1996) Scales of climatic variability and time averaging in Pleistocene biotas: implications for ecology and evolution. *Trends in Ecology and Evolution* **11**, 458-463.

Ruddiman, W.F. and McIntyre, A. (1984) Ice-age thermal response and climatic role of the surface Atlantic Ocean, 40° N to 63° N. *Geological Society of America Bulletin* **95**, 381-396.

Russ, G.R. (1982) Overgrowth in a marine epifaunal community: competitive hierarchies and competitive networks. *Oecologia* **53**, 12-19.

Russ, G.R. (1984a) Distribution and abundance of herbivorous grazing fishes in the central Great Barrier Reef. I. Levels of variability across the entire continental shelf. *Marine Ecology Progress Series* **20**, 23-34.

Russ, G.R. (1984b) Distribution and abundance of herbivorous grazing fishes in the central Great Barrier Reef. II. Patterns of zonation of mid-shelf and outershelf reefs. *Marine Ecology Progress Series* **20**, 35-44.

Russ, G.R. (1984c) A review of coral reef fisheries. *UNESCO Reports in Marine Sciences* **27**, 74-92.

Russ, G.R. (1991) Coral reef fisheries: effects and yields, in *The Ecology of Fishes on Coral Reefs,* (ed P.F. Sale), Academic Press, New York, pp. 601-635.

Rützler, K. (1970) Spatial competition among Porifera: solution by epizoism. *Oecologia* **5**, 85-95.

Rützler, K. and Macintyre, I.G. (eds) (1982) The Atlantic barrier reef ecosystem at Carrie Bow Cay, Belize. I. Structure and communities. *Smithsonian Contributions in Marine Science* **12**, 1-539.

Rylaarsdam, K.W. (1983) Life histories and abundance patterns of colonial corals on Jamaican reefs. *Marine Ecology Progress Series* **13**, 249-260.

Ryther, J.H. (1959) Potential productivity of the sea. *Science* **130**, 602-608.

Sale, P.F. (1974) Mechanisms of co-existence in a guild of territorial fishes at Heron Island. *Proceedings of the Second International Coral Reef Symposium* **1**, 193-206.

Sale, P.F. (1977) Maintenance of high diversity in coral reef fish communities. *The American Naturalist* **111**, 337-359.

Sale, P.F. (1980) The ecology of fishes on coral reefs. *Oceanography and Marine Biology: an Annual Review* **18**, 367-421.

Sale, P.F. (1984) The structure of communities of fish on coral reefs and the merit of a hypothesis-testing, manipulative approach to ecology, in *Ecological Communities: Conceptual Issues and the Evidence,* (eds D.R. Strong, Jr., D. Simberloff, L.G. Abele and A.B. Thistle), Princeton University Press, Princeton, pp. 478-490.

Sale, P.F. (ed) (1991a) *The Ecology of Fishes on Coral Reefs,* Academic Press, New York.

Sale, P.F. (1991b) Reef fish communities: open nonequilibrial systems, in *The Ecology of Fishes on Coral Reefs,* (ed P.F. Sale), Academic Press, New York, pp. 564-598.

Sale, P.F. and Cowen, R.K. (1998) Fishery and reef management. *Science* **279**, 2022.

Salisbury, E.J. (1929) The biological equipment of species in relation to competition. *Journal of Ecology* **17**, 197-222.

Salvat, B. (ed)(1987) *Human Impacts on coral reefs: Facts and Recommendations.* Museum National d'Histoirie Naturelle et Ecole Pratique des Hautes Etudes, Antenne de Tahiti, Centre de l'Environnement, Papetoai, Moorea, French Polynesia.

Sammarco, P.W. (1980) *Diadema* and its relationship to coral spat mortality: grazing, competition, and biological disturbance. *Journal of Experimental Marine Biology and Ecology* **45**, 245-272.

Sammarco, P.W. (1982) Echinoid grazing as a structuring force in coral communities: whole reef manipulations. *Journal of Experimental Marine Biology and Ecology* **61**, 31-55.

Sammarco, P.W., Coll, J.C., La Barre, S. and Willis, B. (1983) Competitive strategies of soft corals (Coelenterata: Octocorallia): allelopathic effects on selected scleractinian corals. *Coral Reefs* **1**, 173-178.

Sammarco, P.W., Coll, J.C. and La Barre, S. (1985) Competitive strategies of soft corals (Coelenterata: Octocorallia). II. Variable defensive responses and susceptibility to scleractinian corals. *Journal of Experimental Marine Biology and Ecology* **91**, 199-215.

Sammarco, P.W. and Coll, J.C. (1992) Chemical adaptations in the Octocorallia: evolutionary considerations. *Marine Ecology Progress Series* **88**, 93-104.

Sammarco, P.W., Levinton, J.S. and Ogden, J.C. (1974) Grazing and control of coral reef community structure by *Diadema antillarum* Philippi (Echinodermata: Echinoidea): a preliminary study. *Journal of Marine Research* **32**, 47-53.

Sammarco, P.W. and Williams, A.H. (1982) Damselfish territoriality: influence on *Diadema* distribution and implications for coral community structure. *Marine Ecology Progress Series* **8**, 53-59.

Samuels, C.L. and Drake, J.A. (1997) Divergent perspectives on community convergence. *Trends in Ecology and Evolution* **12**, 427-432.

Sano, M., Shimizu, M. and Nose, Y. (1984) Food habits of teleostean reef fishes in Okinawa Island, southern Japan. *University of Tokyo, University Museum, Bulletin* **25**, 1-128.

Sano, M., Shimizu, M. and Nose, Y. (1987) Long-term effects of destruction of hermatypic corals by *Acanthaster planci* infestation on reef fish communities at Iriomote Island, Japan. *Marine Ecology Progress Series* **37**, 191-199.

Sarà, M. (1970) Competition and cooperation in sponge populations. *Symposia of the Zoological Society of London* **25**, 273-284.

Schluter, D. and Ricklefs, R.E. (1993a) Species diversity. An introduction to the problem, in *Species Diversity in Ecological Communities: Historical and Geographical Perspectives*, (eds R.E. Ricklefs and D. Schluter), The University of Chicago Press, Chicago, pp. 1-10.

Schluter, D. and Ricklefs, R.E. (1993b) Convergence and the regional component of species diversity, in *Species Diversity in Ecological Communities: Historical and Geographical Perspectives*, (eds R.E. Ricklefs and D. Schluter), The University of Chicago Press, Chicago, pp. 230-240.

Schmitz, O.J. (1997) Press perturbations and the predictability of ecological interactions in a food web. *Ecology* **78**, 55-69.

Schoener, T.W. (1983) Field experiments on interspecific competition. *The American Naturalist* **122**, 240-285.

Schoener, T.W. (1989) The ecological niche, in *Ecological Concepts. The Contribution of Ecology to an Understanding of the Natural World*, (ed J.M. Cherrett), Blackwell Scientific Publications, Oxford, pp. 79-110.

Schulze, E.-D. and Mooney, H.A. (eds) (1994) *Biodiversity and Ecosystem Function*, Springer-Verlag, Berlin.

Sebens, K.P. (1982a) Competition for space: growth rate, reproductive output, and escape in size. *The American Naturalist* **120**, 189-197.

Sebens, K.P. (1983b) Intertidal distribution of zoanthids on the Caribbean coast of Panama: effects of predation and desiccation. *Bulletin of Marine Science* **32**, 316-335.

Sebens, K.P. (1986) Spatial relationships among encrusting marine organisms in the New England subtidal zone. *Ecological Monographs* **56**, 73-96.

Sebens, K.P. (1987) Competition for space: effects of disturbance and indeterminate competitive success. *Theoretical Population Biology* **32**, 430-441.

Sebens, K.P. and Miles, J.S. (1988) Sweeper tentacles in a gorgonian octocoral: morphological modifications for interference competition. *Biological Bulletin* **175**, 378-387.

Shapiro, D.Y. (1991) Intraspecific variability in social systems of coral reef fishes, in *The Ecology of Fishes on Coral Reefs*, (ed P.F. Sale), Academic Press, New York, pp. 331-355.

Shelford, V.E. (1913) *Animal Communities in Temperate America*, The University of Chicago Press, Chicago.

Shepherd, S.A. and Thomas, I.M. (eds) (1982) *Marine Invertebrates of South Australia Part I*, D.J. Woolman, Government Printer, South Australia.

Shepherd, S.A. and Veron J.E.N. (1982) Stony Corals (Order Scleractinia or Madreporaria), in *Marine Invertebrates of South Australia Part I*, (eds S.A. Shepherd and I.M. Thomas), D.J. Woolman, Government Printer, South Australia, pp. 169-178.

Sheppard, C.R.C. (1980) Coral cover, zonation and diversity on reef slopes of Chagos Atolls, and population structures of the major species. *Marine Ecology Progress Series* **2**, 193-205.

Sheppard, C.R.C. (1981) Illumination and the community beneath tabular *Acropora* species. *Marine Biology* **64**, 53-58.

Sheppard, C.R.C. (1982) Coral populations on reef slopes and their major controls. *Marine Ecology Progress Series* **7**, 83-115.

Simbotwe, M.P. and G.R. Friend (1985) Comparisons of the herpetofaunas of tropical wetland habitats from Lochinvar National Park, Zambia and Kakadu National Park, Australia. *Proceedings of the Ecological Society of Australia* **14**, 141-151.

Simpson, R.H. and H. Riehl (1981) *The Hurricane and its Impact*, Louisiana State University Press, Baton Rouge.

Sorokin, Y.I. (1993) *Coral Reef Ecology*, Ecological Studies Volume 102, Springer-Verlag, Berlin.

Sousa, W.P. (1979) Disturbance in marine intertidal boulder fields: the nonequilibrium maintenance of species diversity. *Ecology* **60**, 1225-1239.

Sousa, W.P. and Connell, J.H. (1992) Grazing and succession in marine algae, in *Plant-Animal Interactions in the Marine Benthos*, (eds D.M. John, S.J. Hawkins and J.H. Price), Systematics Association Special Volume No. 46, Clarendon Press, pp.425-441.

Spencer, T. (1992) Bioerosion and biogeomorpholgy, in *Plant-Animal Interactions in the Marine Benthos*, (eds D.M. John, S.J. Hawkins and J.H. Price), The Systematics Association Special Volume No. 46, Clarendon Press, Oxford, pp. 492-509.

Stanley, Jr., G.D. and Swart, P.K. (1995) Evolution of the coral-zooxanthellae symbiosis during the Triassic: a geo-chemical approach. *Paleobiology* **21**, 179-199.

Statzner, B. and Borchardt, D. (1994) Longitudinal patterns and processes along streams: modelling ecological responses to physical gradients, in *Aquatic Ecology: Scale, Pattern and Process*, (eds P.S. Giller, A.G. Hildrew and D.G. Raffaelli), Blackwell Scientific Publications, Oxford, pp. 113-140.

Stearns, S.C. (1992) *The Evolution of Life Histories*, Oxford University Press, Oxford.

Stehli, F.G. and Wells, J.W. (1971) Diversity and age patterns in hermatypic corals. *Systematic Zoology* **20**, 115-126.

Steinberg, P.D. (1989) Biogeographical variation in brown algal polyphenolics and other secondary metabolites: comparison between temperate Australasia and North America. *Oecologia* **78**, 373-382.

Steinberg, P.D. (1992) Geographical variation in the interaction between marine herbivores and brown algal secondary metabolites, in *Ecological Roles of Marine Natural Products*, (ed. V.J. Paul), Comstock Publishing Associates, Ithaca, pp. 51-92.

Steinberg, P.D. (1994) Lack of short-term induction of phlorotannins in the Australasian brown algae *Ecklonia radiata* and *Sargassum vestitum. Marine Ecology Progress Series* **112**, 129-133.

Steneck, R.S. (1982) A limpet-coralline alga association: adaptations and defenses between a selective herbivore and its prey. *Ecology* **63**, 507-522.

Steneck, R.S. (1986) The ecology of coralline algal crusts: convergent patterns and adaptive strategies. *Annual Review of Ecology and Systematics* **17**, 273-303.

Stenseth, N.C. (1995) Snowshoe hare populations: squeezed from below and above. *Science* **269**, 1061-1062.

Stimson, J. (1985) The effect of shading by the table coral *Acropora hyacinthus* on understory corals. *Ecology* **66**, 40-53.

Stoddart, D.R. (1969) Ecology and morphology of Recent coral reefs. *Biological Reviews* **44**, 433-498.

Stoddart, D.R. (1972) Catastrophic damage to coral reef communities by earthquake. *Nature* **239**, 51-52.

Stoner, D.S. (1990) Recruitment of a tropical colonial ascidian: relative importance of pre-settlement and post- settlement processes. *Ecology* **71**, 1682-1690.

Stuart, C.T. and Rex, M.A. (1994) The relationship between developmental pattern and species diversity in deep-sea prosobranch snails, in *Reproduction, Larval Biology and Recruitment in the Deep Sea Benthos*, (eds C.M. Young and K.J. Eckelbarger), Columbia University Press, New York, pp. 118-136.

Sugihara, G., Schoenly, K. and Trombla, A. (1989) Scale invariance in food web properties. *Science* **245**, 48-52.

Sutarno, I.N. (1990) Shape and condition of living coral colonies in the waters around Banda Islands, central Maluka, in *Waters of the Maluka and its Environments,* (eds D.P. Praseno and W.S. Atmadja), Indonesian Institute of Sciences (LIPI), Ambon, pp. 135-147.

Sutherland, J.P. (1974) Multiple stable points in natural communities. *The American Naturalist* **108**, 859-873.

Sutherland, J.P. (1981) The fouling community at Beaufort, North Carolina: a study in stability. *The American Naturalist* **118**, 499-519.

Sutherland, J.P. (1987) Recruitment limitation in a tropical intertidal barnacle: *Tetraclita panamensis* (Pilsbry) on the Pacific coast of Costa Rica. *Journal of Experimental Marine Biology and Ecology* **113**, 267-282.

Sutherland, J.P. (1990a) Recruitment regulates demographic variation in a tropical intertidal barnacle. *Ecology* **71**, 955-972.

Sutherland, J.P. (1990b) Perturbations, resistance, and alternative views of the existence of multiple stable points in nature. *The American Naturalist* **136**, 270-275.

Sutherland, J.P. and Karlson, R.H. (1977) Development and stability of the fouling community at Beaufort, North Carolina. *Ecological Monographs* **47**, 425-446.

Sy, J.C., Herrera, F.S. and McManus, J.W. (1982) Coral community structure of a fringing reef at Mactan Island, Cebu, Philippines. *Proceedings of the Fourth International Coral Reef Symposium* **2**, 263-269.

Tanner, J.E., Hughes, T.P. and Connell, J.H. (1994) Species coexistence, keystone species, and succession: a sensitivity analysis. *Ecology* **75**, 2204-2219.

Tanner, J.E., Hughes, T.P. and Connell, J.H. (1996) The role of history in community dynamics: a modelling approach. *Ecology* **77**, 108-117.

Tansley, A.G. (1914) Presidential address. *Journal of Ecology* **2**, 194-202.

Tansley, A.G. (1935) The use and abuse of vegetational concepts and terms. *Ecology* **16**, 284-307.

Tansley, A.G. and Adamson, R.S. (1925) Studies on the vegetation of the English chalk. III. The chalk grasslands of the Hampshire-Sussex border. *Journal of Ecology* **13**, 177-223.

Targett, N.M., Coen, L.D., Boettcher, A.A. and Tanner, C.E. (1992) Biogeographic comparisons of marine algal polyphenolics: evidence against a latitudinal trend. *Oecologia* **89**, 464-470.

Targett, N.M., Boettcher, A.A., Targett, T.E. and Vrolijk, N.H. (1995) Tropical marine herbivore assimilation of phenolic-rich plants. *Oecologia* **103**, 170-179.

Taylor, J.D. (1978) Habitats and diet of predatory gastropods at Addu Atoll, Maldives. *Journal of Experimental Marine Biology and Ecology* **31**, 83-103.

Taylor, J.D. (1984) A partial food web involving predatory gastropods on a Pacific fringing reef. *Journal of Experimental Marine Biology and Ecology* **74**, 273-290.

Terborgh, J.W. and Faaborg, J. (1980) Saturation of bird communities in the West Indies. *The American Naturalist* **116**, 178-195.

Thompson, J.N. (1988) Variation in interspecific interactions. *Annual Review of Ecology and Systematics* **19**, 65-87.

Thompson, J.N. (1994) *The Coevolutionary Process*, The University of Chicago Press, Chicago.

Thresher, R.E. (1980) *Reef Fish: Behavior and Ecology on the Reef and in the Aquarium*. Palmetto Publishing Company, St. Petersburg.

Thresher, R.E. (1991) Geographic variability in the ecology of coral reef fishes: evidence, evolution, and possible implications, in *The Ecology of Fishes on Coral Reefs*, (ed P.F. Sale), Academic Press, New York, pp. 401-436.

Thresher, R.E. and Colin, P.L. (1986) Trophic structure, diversity and abundance of fishes of the deep reef (30-300 m) at Enewetak, Marshall Islands. *Bulletin of Marine Science* **38**, 253-272.

Tilman, D. (1996) Biodiversity: population versus ecosystem stability. *Ecology* **77**, 350-363.

Tilman, D., Wedin, D. and Knops, J. (1996) Productivity and sustainability influenced by biodiversity in grassland ecosystems. *Nature* **379**, 718-720.

Tomascik. T., van Woesik, R. and Mah, A.J. (1996) Rapid coral colonization of a recent lava flow following a volcanic eruption, Banda Islands, Indonesia. *Coral Reefs* **15**, 169-175.

Trench, R.K. (1971) The physiology and biochemistry of zooxanthellae symbiotic with marine coelenterates. II. Liberation of fixed ^{14}C by zooxanthellae *in vitro*. *Proceedings of the Royal Society of London, Series B, Biological Sciences* **177**, 237-250.

Tribble, G.W. and Randall, R.H. (1986) A description of the high-latitude shallow water coral communities of Miyake-jima, Japan. *Coral Reefs* **4**, 151-159.

Turner, S.J. (1994) The biology and population outbreaks of the corallivorous gastropod *Drupella* on Indo-Pacific reefs. *Oceanography and Marine Biology: an Annual Review* **32**, 461-530.

Underwood, A.J. and Denley, E.J. (1984) Paradigms, explanations, and generalizations in models for the structure of intertidal communities on rocky shores, in *Ecological Communities: Conceptual Issues and the Evidence*, (eds D.R. Strong, Jr., D. Simberloff, L.G. Abele and A.B. Thistle), Princeton University Press, Princeton, pp.151-180.

Underwood, A.J., Denley, E.J. and Moran, M.J. (1983) Experimental analyses of the structure and dynamics of mid-shore rocky intertidal communities in New South Wales. *Oecologia* **56**, 202-219.

Valentine, J.W. and Jablonski, D. (1993) Fossil communities: compositional variation at many time scales, in *Species Diversity in Ecological Communities: Historical and Geological Perspectives*, (eds R.E. Ricklefs and D. Schluter), The Chicago University Press, Chicago, pp. 341-349.

Van Alstyne, K.L. and Paul, V.J. (1990) The biogeography of polyphenolic compounds in marine macroalgae: temperate brown algal defenses deter feeding by tropical herbivorous fishes. *Oecologia* **84**, 158-163.

Van Devender, T.R. (1986) Climatic cadences and the composition of Chihuahuan desert communities: the late Pleistocene packrat midden record, in *Community Ecology*, (eds J. Diamond and T.J. Case), Harper and Row, New York, pp. 285-299.

Vanni, M.J. and Layne, C.D. (1997) Nutrient recycling and herbivory as mechanisms in the "top-down" effect of fish on algae in lakes. *Ecology* **78**, 21-40.

Vanni, M.J., Layne, C.D. and Arnott, S.E. (1997) "Top-down" trophic interactions in lakes: effects of fish on nutrient dynamics. *Ecology* **78**, 1-20.

Vermeij, G.J. (1978) *Biogeography and Adaptation: Patterns of Marine Life*, Harvard University Press, Cambridge.

Veron, J.E.N. (1986) *Corals of Australia and the Indo-Pacific*, University of Hawaii Press, Honolulu.

Veron, J.E.N. (1993) A biogeographic database of hermatypic corals; species of the central Indo-Pacific, genera of the world. *Australian Institute of Marine Science Monograph Series* **10**, 1-433.

Veron, J.E.N. (1995) *Corals in Space and Time: The Biogeography and Evolution of the Scleractinia*, Cornell University Press, Ithaca.

Veron, J.E.N. and Done, T.J. (1979) Corals and coral communities of Lord Howe Island. *Australian Journal of Marine and Freshwater Research* **30**, 203-236.

Volterra, V. (1926) Variazioni e fluttuazioni del numero d'individui in specie animali conviventi. *Memorie della R. Accademia dei Lincei* (ser. 6) **2**, 31-113.

Wahle, C.M. (1980) Detection, pursuit, and overgrowth of tropical gorgonians by milleporid hydrocorals: Perseus and Medusa revisited. *Science* **209**, 689-691.

Warren, P.H. (1989) Spatial and temporal variation in the structure of a freshwater food web. *Oikos* **55**, 299-311.

Watt, A.S. (1947) Pattern and process in the plant community. *Journal of Ecology* **35**, 1-22.

Webb III, T. and Bartlein, P.J. (1992) Global changes during the last 3 million years: climatic controls and biotic responses. *Annual Review of Ecology and Systematics* **23**, 141-173.

Wellington, G.M. (1980) Reversal of digestive interactions between Pacific reef corals: mediation by sweeper tentacles. *Oecologia* **47**, 340-343.

Wellington, G.M. (1982) Depth zonation of corals in the Gulf of Panama: control and facilitation by resident reef fishes. *Ecological Monographs* **52**, 223-241.

Wells, J.W. (1954) Recent corals of the Marshall Islands. *United States Geological Survey, Professional Papers* **260**, 385-459.

Wells, J.W. and J.C. Lang (1973) Systematic list of Jamaican shallow-water Scleractinia. *Bulletin of Marine Sciences* **23**, 55-58.

Westoby, M. (1985) Two main relationships among the components of species richness. *Proceedings of the Ecological Society of Australia* **14**, 103-107.

Westoby, M. (1993) Biodiversity in Australia compared with other continents, in *Species Diversity in Ecological Communities: Historical and Geographical Perspectives*, (eds R.E. Ricklefs and D. Schluter), The University of Chicago Press, Chicago, pp. 170-177.

Westneat, M.W. and Resing, J.M. (1988) Predation on coral spawn by planktivorous fish. *Coral Reefs* **7**, 89-92.

Whittaker, R.H. (1960) Vegetation of the Siskiyou Mountains, Oregon and California. *Ecological Monographs* **30**, 279-338.

Whittaker, R.H. (1977) Evolution of species diversity in land communities. *Evolutionary Biology* **10**, 1-67.

Whittaker, R.H. and Likens. G.E. (1973) Primary production: the biosphere and man. *Human Ecology* **1**, 357-369.

Wiens, J.A. (1989) *The Ecology of Bird Communities.* Volume 1. *Foundations and Patterns*, Cambridge University Press, Cambridge.

Wiens, J.A. (1991) Ecological similarity of shrub-desert avifaunas of Australia and North America. *Ecology* **72**, 479-495.

Wilkinson, C.R. (1987) Interocean differences in size and nutrition of coral reef sponge populations. *Science* **236**, 1654-1657.

Williams, A.H. (1978) Ecology of threespot damselfish: social organization, age structure, and population stability. *Journal of Experimental Marine Biology and Ecology* **34**, 197-213.

Williams, A.H. (1979) Interference behavior and ecology of threespot damselfish (*Eupomocentrus planifrons*). *Oecologia* **38**, 223-230.

Williams, A.H. (1980) The threespot damselfish: a noncarnivorous keystone species. *The American Naturalist* **116**, 138-142.

Williams, A.H. (1981) An analysis of competitive interactions in a patchy back-reef environment. *Ecology* **62**, 1107-1120.

Williams, D. McB. (1986) Temporal variation in the structure of reef slope fish communities (central Great Barrier Reef): short-term effects of *Acanthaster planci* infestation. *Marine Ecology Progress Series* **28**, 157-164.

Williams, D. McB. (1991) Patterns and processes in the distribution of coral reef fishes, in *The Ecology of Fishes on Coral Reefs*, (ed P.F. Sale), Academic Press, New York, pp. 437-474.

Williams, D. McB. and Hatcher, A.I. (1983) Structure of fish communities on outer slopes of inshore, mid-shelf and outer shelf reefs of the Great Barrier Reef. *Marine Ecology Progress Series* **10**, 239-250.

Wilson, A.O. (1969) Three coral reefs in Bermuda's North Lagoon: physiography and distribution of corals and calcareous algae. *Bermuda Biological Station for Research, Special Publications* **2**, 51-64.

Winemiller, K.O. (1990) Spatial and temporal variation in tropical fish trophic networks. *Ecological Monographs* **60**, 331-367.

Woodley, J.D. (1980) Hurricane Allen destroys Jamaican coral reefs. *Nature* **287**, 387.

Woodley, J.D. (1992) The incidence of hurricanes on the north coast of Jamaica since 1870: are the classic reef descriptions atypical? *Hydrobiologia* **247**, 133-138.

Woodley, J.D., Chornesky, E.A., Clifford, P.A. *et al.* (1981) Hurricane Allen's impact on Jamaican coral reefs. *Science* **214**, 749-761.

Wootton, J.T. (1992) Indirect effects, prey susceptibility, and habitat selection: impacts of birds on limpets and algae. *Ecology* **73**, 981-991.

Wootton, J.T. (1993) Indirect effects and habitat use in an intertidal community: interaction chains and interaction modifications. *The American Naturalist* **141**, 71-89.

Wootton, J.T. (1994) The nature and consequences of indirect effects in ecological communities. *Annual Review of Ecology and Systematics* **25**, 443-466.

Yamaguchi, M. (1975) Sea level fluctuations and mass mortalities of reef animals in Guam, Mariana Islands. *Micronesica* **11**, 227-243.

Yodzis, P. (1978) Competition for space and the structure of ecological communities. *Lecture Notes in Biomathematics*, Vol. 25, Springer-Verlag, New York.

Yodzis, P. (1989) *Introduction to Theoretical Ecology*, Harper & Row, Publishers, Inc., New York.

Yonge, C.M. (1971) Thomas F. Goreau: a tribute. *Symposia of the Zoological Society of London* **28**, xxi-xxxv.

Zann, L., Brodie, J., Berryman, C. and Nagasima, M. (1987) Recruitment, ecology, growth and behavior of juvenile *Acanthaster planci* (L.) (Echinodermata, Asteroidea). *Bulletin of Marine Science* **41**, 561-575.

SUBJECT INDEX

Population and Community Biology Series

1. R.M. Anderson (ed.): *Population Dynamics of Infectious Diseases: Theory and Applications.* 1982. Hb.
2. S.L. Pimm: *Food Webs.* 1982. Hb/Pb.
3. R.J. Taylor: *Predation.* 1984. Hb/Pb.
4. B.F.J. Manley: *The Statistics of Natural Selection.* 1985. Hb/Pb.
5. P.G.N. Digby and R.A. Kempton: *Multivariate Analysis of Ecological Communities.* 1987. Hb/Pb/Reprint.
6. P.A. Keddy: *Competition.* 1989. Hb/Pb/Reprint.
7. B.F.J. Manley: *Stage-Structured Populations.* 1990. Hb.
8. S.S. Bell, E.D. McCoy and H.R. Mushinsky (eds.): *Habitat Structure: The Physical Arrangement of Objects in Space.* 1991. Hb.
9. D.L. DeAngelis: *Dynamics of Nutrient Cycling and Food Webs.* 1992. Pb.
10. T. Royama: *Analytical Population Dynamics.* 1992. Hb/Pb.
11. D.C. Glenn-Lewin, R.K. Peet and T.T. Veblen (eds.): *Plant Succession: Theory and Prediction.* 1992.
12. M.A. Burgman, S. Ferson and H.R. Akçakaya: *Risk Assessment in Conservation Biology.* 1993. Hb/Reprint.
13. K.J. Gaston: *Rarity.* 1994. Hb/Pb.
14. W.J. Bond and B.W. Van Wilgen: *Fire and Plants.* 1996. Hb.
15. M. Williamson: *Biological Invasions.* 1996. Hb/Pb.
16. P.J. den Boer and J. Reddingius: *Regulation and Stabilization: Paradigms in Population Ecology.* 1996. Hb.
17. W.E. Kunin and K.J. Gaston (eds.): *The Biology of Rarity: Causes and Consequences of Rare-Common Differences.* 1997. Hb.
18. Shripad Tujlapurkar and H. Caswell (eds.): *Structured-Population Models in Marine, Terrestrial, and Freshwater Systems.* 1997. Pb.
19. J.P. Grover: *Resource Competition.* 1997. Hb.
20. J. Fryxell and P. Lundberg: *Behaviour and Trophic Dynamics.* 1998. Pb.
21. T. Czárán: *Spatiotemporal Models of Population and Community Dynamics.* 1998. Pb.
22. M. Scheffer: Dynamics of Shallow Lake Communities. 1998. Hb.
23. R.H. Karlson: Dynamics of Coral Communities. 1999 ISBN 0-7923-5534-2
24. O.T. Sandlund, P.J. Schei and A. Viken (eds.): Invasive Species and Biodiversity Management. 1999 ISBN 0412-84080-4
25. J.J. Worrall (Ed.): Structure and Dynamics of Fungal Populations. 1999 ISBN 0412-80430-1

KLUWER ACADEMIC PUBLISHERS – DORDRECHT / BOSTON / LONDON